Multivariate Bonferroni-Type Inequalities

Theory and Applications

Multivariate Bonferroni-Type Inequalities
Theory and Applications

John Tuhao Chen

Bowling Green State University
Ohio, USA

CRC Press
Taylor & Francis Group
Boca Raton London New York

CRC Press is an imprint of the
Taylor & Francis Group, an **informa** business
A CHAPMAN & HALL BOOK

CRC Press
Taylor & Francis Group
6000 Broken Sound Parkway NW, Suite 300
Boca Raton, FL 33487-2742

First issued in paperback 2019

© 2015 by Taylor & Francis Group, LLC
CRC Press is an imprint of Taylor & Francis Group, an Informa business

No claim to original U.S. Government works

ISBN-13: 978-1-4665-1843-8 (hbk)
ISBN-13: 978-0-367-37852-3 (pbk)

Visit the Taylor & Francis Web site at
http://www.taylorandfrancis.com

and the CRC Press Web site at
http://www.crcpress.com

To the memory of my grandmother, Gua-Bi Lin.

Contents

List of Figures

List of Tables

Preface

Multivariate Bonferroni-type inequalities are endowed with abundant applications in a variety of disciplines. For instance, in clinical trials where two or more symptoms are observed and the joint distribution of the endpoints cannot be legitimately assumed, the assessment of drug efficacy in clinical trials necessitates multivariate Bonferroni-type bounds to control the overall error rate in multiple testing. In engineering, the estimation of the reliability of a composite system requires multivariate inequalities when a series of subsystems are involved and each subsystem contains components in a parallel design. In actuarial science, the evaluation of several risks associated with extreme events demands multivariate inequalities. In the literature, developments of univariate Bonferroni-type inequalities are documented in the book written by Galambos and Simonelli in 1996. It is not the intention of this book to garner all historical pieces. Instead, we focus on and systematically synthesize research discoveries on multivariate Bonferroni-type inequalities published in the past decade. The research developments intrinsically include advancements of bounding techniques with the corresponding innovations in applications.

For bounding techniques, this book addresses the method of linear programming for multivariate bounds, multivariate hybrid bounds, sub-Markovian bounds, as well as bounds using Hamilton circuits. The emergence of new bounding approaches pushes the conventional definitions of optimal inequalities, and demands new insights of the understanding of linear optimality and Fréchet optimality.

Regarding bounding applications, the past decade has witnessed many innovative multiple testing procedures routinely developed based on Bonferroni inequalities, which have profound impacts on clinical trials and drug discoveries. Besides therapeutic windows and minimum effective doses, probability inequalities also command attention in molecular cancer therapy, vascular nursing administration, and the analysis of post-thrombotic syndrome. Furthermore, inequalities also provide a path toward resolving a current challenge facing the statistical society — the analysis of big-data. By partitioning an ocean of data into time segments in which the data is manageable, probability inequalities bridge the unmanageable big-data with existing methodologies for statistically plausible data analyses. In this setting, inequalities are essentially screening instruments used to filter out important information (such as patterns or trends in sales) across time. It is timely to synthesize and update theoretical and applied developments on inequality theory for future investigations.

One of the common concerns is the over-conservativeness of the Bonferroni approach. In fact, with the availability of computing capability today, the involvement of high order inequalities and optimal inequalities significantly sharpens bounding results of the Bonferroni methods.

This book consists of nine chapters covering theory and applications of multivariate Bonferroni-type inequalities. Chapter 1, as a prelude, discusses basic concepts of multivariate Bonferroni-type bounds in various fields. It contains examples pertaining to the analysis of multiple-risks and martingales to showcase the use of probability inequalities in different areas of application. It should be noted that the basic Bonferroni bounds in this chapter provide conservative numerical outcomes. This partly serves as a motivation for the discussion of the improved inequalities in the follow-up chapters.

Chapter 2 introduces notations, basic probability identities, and definitions to facilitate discussions in the sequel. It also introduces the classification of univariate and multivariate bounds with optimality. Chapter 3 focuses on multivariate bounds using the method of indicator functions. This is a conventional bounding method that is refurbished with non-linear Bonferroni-type bounds. Chapter 4 is devoted to the method of linear programming for bivariate upper and lower bounds. It highlights the fact that the product of two univariate optimal lower bounds may not be a bivariate optimal lower bound, signifying that multivariate bounding is not a parallel extension of univariate bounding. Chapters 1 to 4 conclude the description on basic concepts and methods in probability inequalities.

From Chapter 5 to Chapter 9, the book addresses bounding results and applications of multivariate Bonferroni-type inequalities. Chapter 5 describes bivariate upper bounds using bounding techniques described in Chapters 3 and 4. It elucidates bivariate linear optimal upper bounds and concludes with an application in constructing new multiple testing procedures with probability upper bounds. Such a procedure can be applied, for instance, to estimate the therapeutic window of a drug in clinical trials. Chapter 6 goes beyond bivariate upper bounds by considering vectorized upper bounds and hybrid bounds that break the thinking frame of traditional Bonferroni bounds. Following the discussion on probability upper bounds (in Chapters 5 and 6), Chapter 7 presents an optimization algorithm for a bivariate lower bound using the method of linear programming, and extends the idea into multivariate cases. Chapter 8 considers vectorized high dimensional lower bounds with refinements, such as Hamilton-type circuits and Sub-Markovian events. Finally, Chapter 9 discusses case studies involving applications of probability inequalities.

The book is designed to be accessible for graduate students in statistics. The materials in the first chapter and the last four chapters of the book are comprehensible for applied statisticians in various fields. These chapters heuristically put forward statistical thinking on approximating the probability of unions of several sets of events that occur simultaneously. Thus, the prerequisite for these chapters is the basic understanding of probability theory. The rest of the book intertwines bounding techniques with applications, which requires the understanding of matrix algebra, probability theory, and combinatorial arguments. The materials in the book are coherent and self-contained. Readers uninterested in mathematical derivations may skip the proofs.

This monograph stems from the lecture notes of a one-semester graduate level course at Bowling Green State University. It may serve as a graduate textbook or a reference to facilitate researchers, graduate students, and practitioners with updated information on statistical bounding techniques and applications.

I am grateful to a number of people who provided me with help and encouragement. First, I am indebted to Professor Eugene Seneta who brought me into the world of probability inequalities, to Professors Fred M. Hoppe and Charles Dunnett who showed me the beauty and elegance of simultaneous inference, and to Professors Satish Iyengar, Henry Block, and Allan Sampson who taught me skills in statistical consulting, and nurtured me with fruitful and valuable discussions. I am also grateful to my medical collaborators (Dr. David Brent, Dr. Anthony Comerota, Dr. Eileen Walsh, and Dr. William Feeman) for inspiring conversations and valuable projects that stimulated new statistical methodologies related to multivariate Bonferroni inequalities.

I owe a great deal to my wife BingLin for her unselfish support, and to my children: Vincent, for suggestions and opinions on the framework of the book; Belinda for serving as my personal "secretary" (editing and typing the bibliography of this book); and Lincy and Patrick for their understanding and flexibility in rescheduling our playtimes from tomorrow to another tomorrow.

The reviewers' comments and suggestions significantly improved the original manuscript of this book. Their contributions are gratefully acknowledged. Especially, I need to thank Professor Fred M. Hoppe for painstakingly reading almost every page of the book, providing me with extremely helpful comments, and contributing two new proofs in the book; Professor Eugene Seneta for constructive suggestions and corrections on the references; and Professor András Prékopa for suggestions on the research area of discrete moment problems. Of course, I am solely responsible for unavoidable typos and errors in the book. Finally, I would like to thank this book's acquisitions editor, Mr. David Grubbs at Chapman and Hall/CRC, for his kindness and great effort during this project, especially for the initiation of this project during his campus visit to BGSU.

Chapter 1

Introduction

Serving as an essential tool in the development of statistical methodologies, probability inequalities play a critical role in statistics and have a plethora of applications (see, for example, Gumbel (1958), Galambos (1978, 1995), Hoppe (1993)). The elegance of probability inequalities stems from their generality (no special model assumptions), applicability (adaptable to various fields), and reliability (conservative in controlling error rate). Investigation of probability inequalities has always been a primary research area in statistics (Wong and Shen (1995), Block, Costigan, and Sampson (1997)). The discovery of a novel probability inequality correspondingly results in refinements of the associated statistical methodologies, advancing the field of statistical inference. Today, the Bonferroni procedure directly affects many applied fields such as analyzing extreme values in actuarial science (Galambos (1984, 1987, 1994)), establishing bounds for convergence rates of random sequences in limiting theory (Seneta and Weber (1982), Hoppe (2006, 2009)), comparing drug effects for two or more treatments in medical research (Seneta (1993), Chen, Walsh, Comerota, et al (2011), Chen and Comerota (2012)), evaluating business markets in financial analysis (Chen (2010), Gupta, Chen and Troskie (2003)), and assessing system reliabilities in engineering (Castillo (1988), Guerriero, Pozdnyakov, Glaz, et al. (2010), Block, Costigan, and Sampson (1992)), to list just a few.

It is practically impossible to include all inequalities in a book due to the vast number of publications on different types of inequalities and their applications. Therefore this book focuses on multivariate versions of Bonferroni-type inequalities, which are extensions of the univariate Bonferroni method for scenarios involving two or more sets of events (Galambos and Lee, 1992, 1994). Since the Bonferroni procedure is a common approach to adjusting error rates in multiple testing, we include discussions on inequalities bridging simultaneous inference with individual p-values (Glaz and Ravishanker (1991), Guo (1995), Hochberg (1988), among others).

To set the scene for the discussion in later chapters, we select several applications to emphasize the use of Bonferroni inequalities in the first chapter. Materials in this chapter are organized as follows. Section 1.1 focuses on the use of multivariate Bonferroni-type bounds for multiple extreme value analysis, which bridges the multivariate cumulative distribution with the overall risk of multiple natural disasters. Section 1.2 discusses the use of multivariate Bonferroni-type inequality in the inference of the minimum effective dose of a drug. Section 1.3 presents the use of multivariate Bonferroni-type inequalities to evaluate system reliability. Besides measuring one outcome of interest, practical problems (such as the identification of the therapeutic window of a drug or the comparison of pedagogies in education) frequently require the evaluation of probabilities for two correlated outcomes. Therefore, after addressing the need of multivariate Bonferroni-type inequalities when the experimental outcome consists of two correlated measurements (Section 1.4), we describe the use of inequalities on premium reserve processes and the ruin probability in Section 1.5. Finally, Section 1.6 introduces the application of Bonferroni inequality (in conjunction with a martingale inequality) to the analysis of market portfolios.

For convenience, we use the basic multivariate Bonferroni-type (which is usually conservative) inequality in this chapter to illustrate the basic mechanism and underlying principles of applications in various fields. The over-conservativeness of the basic Bonferroni bounds calls for inequality refinements. Refined versions of the corresponding inequalities will be addressed in Chapters 2 to 8. Readers who are familiar with applications of the multivariate Bonferroni inequality may skip this chapter.

1.1 Multiple Extreme Values

A description of probability inequalities analyzing univariate extreme values is well-documented in Galambos (1998). Extreme value problems frequently occur in our daily life. For example, when we consider the probability of flooding in an area, assume that for each day, the highest water level is observable, our primary interest (the chance of being free of flood) is the chance of the lowest reading over a period below the river bank (if any one reading goes above the river bank, flooding occurs). Similar consideration applies to the analysis of other natural disasters, such as heavy rain, extreme temperature, hurricane, etc. In this section, after illustrating the method of risk evaluation for one risk factor (such as flooding) using the univariate Bonferroni bound, we will consider the evaluation of multiple risks (for example, simultaneous occurrence of flooding, a tornado, and extreme temperature) over a period. The latter necessitates the application of multivariate Bonferroni-type bounds.

In the literature, analysis of extreme values sometimes assumes independence among observations for computation convenience. In Section 1.1.1 we show that the convenient assumption of independence overlooks the longitudinal effects of the observations, which may lead to misleading assertions. On the other hand, the use of Bonferroni inequalities avoids making such independence assumption in statistical inference.

1.1.1   Evaluating the Risk of Multiple Disasters

Example 1.1.1   Let X_i, $i = 1, ..., 120$ denote daily measurements of the highest river water level in an area for the next four months. Assume the river bank is 7 feet high. Then, the probability of flooding in the next four months (120 days) in this area is

$$P(\bigcup_{i=1}^{120}\{X_i > 7\}) = P(\max(X_1,...,X_{120}) > 7). \tag{1.1}$$

Flooding occurs if the highest water level exceeds the height of the river bank in any one of the days of interest. If we assume that the water levels X_i, $i = 1,...,120$ identically and independently follow a common distribution (for example, an exponential model with mean $\lambda = 3$ feet), the risk of flooding during the next 120 days can be computed as follows.

$$
\begin{aligned}
P(\bigcup_{i=1}^{120}\{X_i > 7\}) &= 1 - P(\bigcap_{i=1}^{120}\{X_i \le 7\})(\text{probability of complementary events}) \\
&= 1 - (P(X_1 \le 7))^{120}(\text{independence assumption}) \\
&= 1 - (1 - e^{-7/3})^{120}(\text{exponential model assumption}) \\
&= 1 - 0.90303^{120} \\
&= 1 - 4.832 \times 10^{-6} \\
&= 0.999995.
\end{aligned}
$$

Therefore, the chance of flooding is almost 1. However, is this result plausible in practice?   □

Notice that the mean and standard deviation are 3 feet for the exponential model assumed in Example 1.1.1. The height of the river bank (7 feet) is more than twice the average water level (3 feet), and the variability of the water level is 3 feet. On top of the average water level, 6 feet is still below the height of the river bank. However, our computation says that the chance of flooding is close to 1! This is a sharp contrast between the structure of the river bank and the computed probability of flooding in the area. Obviously, there is a flaw in the above computation.

If we examine the model assumptions closely, we can easily see that the independence assumption is problematic. Although the chance of not flooding is 0.90303 each day, the independence assumption makes the probability of not flooding simultaneously over the 120 days as small as 4.832×10^{-6}, which leads to an answer of almost certain flooding sometime over the next 120 days. This conclusion is not in line with the setting of the example. In fact, for this problem, it is implausible to assume independence between the daily readings of the highest water level. Water levels of the river on two consecutive days are highly likely to be correlated with each other. Although the independence assumption eases the computation for the evaluation of flooding risk, it is essentially a wrong assumption.

Without the independence assumption, an alternative way to evaluate the probability of flooding over a period of time is the use of probability inequality, as pointed out in Bailey (1977) or Galambos (1978). Denote A_i the event of $X_i > 7$ for $i = 1,...,n$, $(n = 120)$ in Example 1.1.1, since the union of the events, A_i, $i = 1,...,n$, can be bounded as follows,

$$\max_i P(X_i > 7) \le P(\max(X_1,...,X_n) > 7) = P(\bigcup_{i=1}^{n} A_i) \le \sum_{i=1}^{n} P(A_i),$$

since $\max_i P(X_i > 7) = 0.09697$, and the probability is bounded by 1, the risk of flooding in the next 120 days can be bounded in the range of $(0.09697, 1)$.

Of course, the upper and lower bounds are too conservative to be useful in this example. Instead of being significantly wrong, as with the independence assumption, the inequality method is conservatively correct. The drawback of such bounding methods can be improved in the subsequent chapters by either higher order Bonferroni-type inequalities or sharpened bounds (such as the improved upper bounds delineated in Chapters 5 and 6, or the refined lower bounds discussed in Chapters 7 and 8). Notice that in this example, as shown in (1.1), the underlying principle is to introduce bounds for the probability of the union of a set of events.

When an insurance company evaluates more than one risk factor simultaneously (such as flooding, a tornado, and extreme hot weather in one area), the probability of interest involves the intersection of the union of three sets of events. To see this point, denote X_i the daily highest water level of the river, Y_i the daily strongest wind speed in the area, and Z_i the daily hottest temperature. Let A_i, B_j, C_k be the events $\{X_i > c_1\}$, $\{Y_j > c_2\}$, and $\{Z_k > c_3\}$, respectively, for three constants c_1, c_2, c_3 (such as, $c_1 = 3$ feet, $c_2 = 60$ miles/hour, and $c_3 = 110$ degrees Fahrenheit). The risk that all three natural disasters occur in the coming four months (120 days) is then

$$Risk = P([\bigcup_{i=1}^{120} A_i] \cap [\bigcup_{j=1}^{120} B_j] \cap [\bigcup_{k=1}^{120} C_k]). \tag{1.2}$$

Equation(1.2) measures the risk of the occurrence of flooding, tornado, and extremely hot temperature in the four-month period simultaneously. Similar to the discussion on the implausibility of the independence assumption for the daily measurements of the highest water level of a river, it is not legitimate to assume independence among the three natural disasters in one area due to relationships connecting geographic factors and meteorological factors in the area. Under this scenario, the evaluation of Equation (1.2) relies on the upper and lower bounds of probability inequalities. An upper bound to (1.2) can be constructed as

$$P([\bigcup_{i=1}^{120} A_i] \cap [\bigcup_{j=1}^{120} B_j] \cap [\bigcup_{k=1}^{120} C_k]) \leq \min\{1, S_{111}\}, \tag{1.3}$$

where $S_{111} = \sum_i \sum_j \sum_k P(A_i \cap B_j \cap C_k)$, the basic multivariate Bonferroni summation that will be introduced in Chapter 2. Equation (1.3) means that the risk is bounded by the total chance of the occurrence of the three disasters in any combination across all 120 possible days under consideration.

The multivariate Bonferroni-type inequalities discussed in later chapters of this book (Chapters 5, 6, and 7) can be used to refine upper and lower limits for the term in (1.2). References for the multivariate extreme value issue can be found in Galambos (1985, 1988).

1.1.2 Multivariate Cumulative Distributions

Although the probability of the occurrence of multiple natural disasters in the preceding section is formulated as the joint probability of several sets of events, the same terminology may also be applied to the cumulative distribution of multivariate random variables. We shall set up such a connection in this chapter for the convenience of discussion in succeeding chapters.

Following the notation of Section 1.1.1, if we denote the extreme values

$$X_{(1)}^* = \min(-X_1, ..., -X_{120}),$$

$$Y_{(1)}^* = \min(-Y_1, ..., -Y_{120}),$$

$$Z_{(1)}^* = \min(-Z_1, ..., -Z_{120}),$$

the term in (1.2) becomes

$$P(X_{(1)}^* \leq d_1, Y_{(1)}^* \leq d_2, Z_{(1)}^* \leq d_3), \tag{1.4}$$

where $d_i = -c_i$ for $i = 1, 2, 3$. Notice that the term (1.4) is essentially the multivariate cumulative distribution function of the three random variables $X_{(1)}^*$, $Y_{(1)}^*$ and $Z_{(1)}^*$. In fact, when the independence among the extreme values cannot be plausibly assumed, a multivariate Bonferroni-type inequality is an alternative approach to access the multivariate cumulative distribution for the joint event of multiple extreme values using marginal joint probabilities. To see this point, denote a random vector $\mathbf{X} = (X_1, ..., X_d)$ with associated sample realization $\mathbf{x} = (x_1, ..., x_d)$, and denote the multivariate cumulative distribution function (CDF) as follows.

$$F_d(\mathbf{x}) = P(X_1 \leq x_1, ..., X_d \leq x_d),$$

which leads to the d-variate survivor function

$$S_d(\mathbf{x}) = P(X_1 > x_1, ..., X_d > x_d).$$

For any subset of the index set, $K \subset \{1, ..., d\}$, define the corresponding multivariate CDF and survival function, as

$$F_K(\mathbf{x}) = P(X_i \leq x_i, i \in K),$$

and

$$S_K(\mathbf{x}) = P(X_i > x_i, i \in K).$$

Obviously, when $K = \{1, 2, ..., d\}$, $F_K(\mathbf{x}) = F_d(\mathbf{x})$, and $S_K(\mathbf{x}) = S_d(\mathbf{x})$. On the other hand, $F_K(\mathbf{x})$ and $S_K(\mathbf{x})$ can be deduced from $F_d(\mathbf{x})$ and $S_d(\mathbf{x})$, respectively, by letting x_j tend to infinity for any index j not in K.

As pointed out in Reiss and Thomas (1997) for the analysis of multivariate extreme values, the d-variate CDF takes the following form by the use of the inclusion and exclusion approach:

$$F_d(\mathbf{x}) = 1 - P(\bigcup_{i=1}^{d}\{X_i > x_i\})$$

$$= 1 - \sum_{j \le d}(-1)^{j+1} \sum_{|K|=j} S_K(\mathbf{x}) \tag{1.5}$$

$$S_d(\mathbf{x}) = 1 - P(\bigcup_{i=1}^{d}\{X_i \le x_i\})$$

$$= 1 - \sum_{j \le d}(-1)^{j+1} \sum_{|K|=j} F_K(\mathbf{x}). \tag{1.6}$$

Equations (1.5) and (1.6) express the d-variate CDF as a function of multivariate marginal survival functions with dimensions changing from 1 to d. These two identities essentially come from the univariate Bonferroni-type identity. The multivariate versions of the d-variate CDF and survival functions involve multivariate maxima and multivariate minima (as elucidated in Reiss and Thomas (1997), or Galambos and Xu (1990b)).

Topics related to the above example include vector inequalities, discussed in Chapters 6 and 8. We will also discuss methods to refine upper and lower bounds for the joint multivariate cumulative distribution, without making independence assumptions among different groups of random variables. Publications associated with this discussion include the work on order statistics by Balasubramanian and Bapat (1991), Cochran (1941), David (1956, 1981), and Galambos (1972, 1974).

1.2 Minimum Effective Dose

Another application of the multivariate Bonferroni-type inequality is in the investigation of the minimum effective dose of a drug. The minimum effective dose of a drug refers to the smallest efficacious dosage when all higher dosages are efficacious. For example, if dosages 1.5mg, 3.5mg, and 4.5mg are efficacious but dosage 2.0 mg is not, then the minimum effective dosage is 3.5mg, not 1.5mg.

1.2.1   Minimum Effective Dose of MOTRIN

For illustration, we start this section with the following example from pharmaceutical studies.

Example 1.2.1 Consider a non-steroidal anti-inflammatory drug, MOTRIN (ibuprofen), available in 400 mg, 600 mg, and 800 mg tablets for oral administration. Assume that in an experiment designed to identify the minimum effective dose of the drug, the experimental data follow

$$X_{ij} = \mu_i + \varepsilon_{ij}$$

for $i = 0, 1, ..., k$ and $j = 1, ..., n_i$, where ε_{ij} are i.i.d. random variables, $i = 0$ represents the control group, $i = 1, ..., k$ represents the ith treatment group, and n_i represents the number of observations in the ith group. Assume that the experimenter is interested in identifying the minimum effective dose for a given endpoint such as the amount of morphine required for patients after a surgery when MOTRIN is used as a supplement pain controller after an operation.

Under the normality assumption with equal variance for the random error term ε_{ij}, Dunnett (1955)'s method (Dunnett (1955)) is frequently used to make inference on the minimum effective dose. As usual, denote $\overline{X}_i$ the sample mean of observations in treatment i for $i = 0, 1, ..., k$ and S^2 the pooled sample variance:

$$S^2 = \frac{1}{\sum_{i=0}^{k}(n_i - 1)} \sum_{i=0}^{k} \sum_{j=1}^{n_i} (X_{ij} - \overline{X}_i)^2.$$

The two-sided simultaneous $(1 - \alpha)100\%$ confidence intervals on the treatment effects for all $i = 1, ..., k$ are

$$\mu_i - \mu_0 \in (\overline{X}_i - \overline{X}_0 - d_{\alpha/2}^{(n_i,n_0)} S\sqrt{\frac{1}{n_i} + \frac{1}{n_0}}, \quad \overline{X}_i - \overline{X}_0 + d_{\alpha/2}^{(n_i,n_0)} S\sqrt{\frac{1}{n_i} + \frac{1}{n_0}}), \quad (1.7)$$

where $d_{\alpha/2}^{(n_i,n_0)}$ is the two-sided cutoff value from Dunnett's table (which is commonly available in statistical software such as SAS, Minitab, or R), and the simultaneous one-sided $(1 - \alpha)100\%$ confidence intervals for $i = 1, ..., k$ are

$$\mu_i - \mu_0 \geq \overline{X}_i - \overline{X}_0 - d_{\alpha}^{(n_i,n_0)} S\sqrt{\frac{1}{n_i} + \frac{1}{n_0}} \qquad (1.8)$$

where $d_{\alpha}^{(n_i,n_0)}$ is the one-sided cutoff value from Dunnett's table. $\square$

Although Dunnett's method numerically provides confidence ranges for the inference of the minimum effective dose, the validity of Dunnett's procedure actually relies on the assumption that the error terms identically and independently follow a normal distribution with equal variances. If this assumption is violated, Dunnett's procedure is invalid, which may consequently lead to misleading information on the minimum effective dose. Publications on the identification of the minimum effective dose can be found in Dmitrienko, Tamhane and Bretz (2009), Chen (2008b), Chen and Hoppe (1998), among others.

To further illustrate how the equal variance assumption affects the validity of Dunnett's approach, we consider the following example.

Example 1.2.2 To evaluate the minimum effective dose of IV ibuprofen for the treatment of post-operative pain as measured by reduction in the requirement for the narcotic analgesic and morphine following a surgery, consider a randomized, double-blinded, placebo-controlled clinical trial consisting of three treatment groups: T1: 400 mg IV ibuprofen+ morphine, T2: 600 mg IV ibuprofen+ morphine, T3: 800 mg IV ibuprofen+ morphine, and a control group: placebo+ morphine. The primary endpoint is the requirement of morphine in mg. For illustration, Table 1.2.1 provides

```
MiniTab output for Example 1.2.1

                        Individual 95% CIs For Mean Based on Pooled StDev
Level     N    Mean   StDev  -----+---------+---------+---------+----
control  24  54.652   5.688                                    (---*----)
T1       20  44.349   4.880          (----*---)
T2       18  46.002   5.579              (----*----)
T3       15  40.090   4.136  (----*-----)
                             -----+---------+---------+---------+----
                             40.0      45.0      50.0      55.0


Pooled StDev = 5.188

Dunnett's comparisons with a control

Family error rate = 0.05

Individual error rate = 0.0184

Critical value = 2.41

Control = control


Intervals for treatment mean minus control mean

Level    Lower   Center    Upper   ---+---------+---------+---------+------
T1      -14.092  -10.303   -6.513               (----------*---------)
T2      -12.552   -8.649   -4.747                 (----------*----------)
T3      -18.682  -14.562  -10.442   (----------*-----------)
                                    ---+---------+---------+---------+------
                                    -17.5     -14.0     -10.5      -7.0
```

Figure 1.1 *Minitab output for simultaneous confidence intervals*

a set of simulated data for the four groups, and Figure 1.2.1 dispatches a Minitab output using Dunnett's method as an option in the Minitab – ANOVA command.

As shown in the output, although the individual sample means of the treatment effects (at the range of saving 40 mg to 45 mg requirement for the narcotic analgesic and morphine following the surgery) are significantly different from the control group (at the range around 55 mg), and the sample standard deviations are basically consistent at the range of around 4 mg to 6 mg, the simultaneous confidence intervals of the treatment differences are at the range from 8.65 mg to 18.68 mg. If the reduc-

Table 1.1 *Effect of IV ibuprofen on controlling post-operative pain*

Treatment	Sample mean	S.D.	Simultaneous bounds
Placebo+ morphine	54.65	5.69	
400 mg IV ibuprofen+ morphine	44.35	4.88	(-14.09, -10.30)
600 mg IV ibuprofen+ morphine	46.00	5.58	(-12.55, -8.65)
800 mg IV ibuprofen+ morphine	40.09	4.14	(-18.68, -14.56)

tion of 10 mg morphine requirement is clinically pre-specified as being efficacious, the minimum effective dose of IV ibuprofen is 800 mg. Although the treatment of 400 mg IV ibuprofen has a reduction range from 10.30 mg to 14.09 mg, in terms of minimum effective dose, it is desirable to have all higher doses efficacious (Hsu and Berger 1999). However, the inefficacious response to a 600 mg IV ibuprofen treatment rules out the 400 mg treatment as the minimum effective dose.   □

In this particular example, we use Dunnett's method (1955) which relies on the two-sided confidence intervals to identify the minimum effective dose. However, current practices usually use one-sided hypothesis testing to detect the minimum effective dose (for example, Hsu and Berger (1999), Hsu (1996), Tamhane and Dunnett (1999)). Although the two-sided test has the advantage of controlling the two-sided margin errors while the one-sided test has the advantage of being more powerful in detecting the difference, the underpinning principle of simultaneous adjustment for the overall conclusion remains the same.

The following are two critical assumptions in Example 1.2.2 for the validity of the analysis for the minimum effective dose for IV Ibuprofen:

• The normality assumption for the random error associated with each observation; and

• The assumption of equal variances— the homoscedasticity assumption.

Although the above assumptions seem common in practice, ignoring such critical model assumptions may lead to invalid or misleading inference results. For example, the normality assumption for binary data is obviously inappropriate; the independence assumption is also violated for data with longitudinal effects. Even if the p-value is less than 0.05 in the output of Dunnett's procedure, when the model assumption is wrong, the significant p-value means nothing. It is actually answering a question different from the original one. A scenario that John Tukey refers to as *an exact solution to a wrong problem*.

An alternative approach that avoids risky model assumptions is the method of probability inequalities. Although the inequality method sometimes is too conservative under certain conditions, at the very least, it provides a correct analysis (As Tukey would say, *An approximate solution to the right question*). Toward this end, we address the relation between the use of a Bonferroni-type inequality and model assumptions for the detection of the minimum effective dose in the following two subsections. Specifically, we discuss inequality methods dealing with the issue of the

Data for Example 1.2.1

control	T1	T2	T3
57.21	46.52	51.69	43.71
53.00	37.58	52.64	34.28
59.50	43.78	49.67	35.21
54.78	38.32	45.06	40.65
48.07	47.90	35.97	45.77
50.84	36.63	41.87	37.64
53.18	57.52	40.16	37.13
44.01	40.56	45.87	37.77
48.69	48.89	53.16	40.28
54.32	44.79	41.51	37.81
51.93	43.58	50.45	38.32
41.62	41.47	42.57	43.89
52.75	45.36	55.52	45.80
52.99	48.17	52.25	36.23
59.35	43.50	42.34	46.86
54.87	40.20	41.49	
57.15	43.13	43.58	
62.12	50.27	42.22	
57.43	42.77		
51.93	46.04		
59.48			
60.42			
59.57			
66.43			

Figure 1.2 *Simulated data for MED of IV ibuprofen*

lack of normality assumption in Section 1.2.2; inequality methods handling violations of the homoscedasticity assumption in Section 1.2.3; and multivariate inequality methods for inference with multiple endpoints in Section 1.2.4.

1.2.2   Minimum Effective Dose without Normality

Notice that Dunnett's procedure is inappropriate when the normality assumption cannot be plausibly assumed for the underlying population, such as the population with binary data (Tamhane and Dunnett 1999); skew data (for example, Gupta and Chen, 2001, 2003, 2004); or data with exponential models (Margolin and Maurer, 1976). Other related discussions on discrete and continuous skew models can be found in Olkin and Sobel (1965), or positive skewness models in Nguyen et al. (2003), to list just a few. Recent discussions on the minimum effective dose for binary data can be found in Chen (2008b), and Chen (2008a) discussed the use of the partitioning method for both efficacy and toxicity associated with the minimum effective dose. In this case, for the comparison of any two populations, when the inference on two population medians is of interest, one may use the Mann-Whitney-Wilcoxon statistic to construct a distribution-free pairwise comparison between the placebo and a treatment group, and then use the Bonferroni method to adjust for multiplicity. References in this regard also include the general method of Bonferroni adjustment as described in Sison and Glaz (1995), Worsley (1982), Pozdnyakov and Glaz (2007). When additional model assumptions are satisfied, the method for distribution-free multiple comparisons between the treatments and the control is applicable to detect the minimum effective dose. We will discuss this topic in detail in Chapter 9.

1.2.3   Inequality Methods for Behren-Fisher Problem

For some dose-response studies, the assumption of equal variances in Section 1.2.1 is questionable, even for double-blinded experiments. As mentioned in Scheffé (1953), for patients treated with different dosages of a medicine, it is theoretically convenient but practically incorrect to assume identical response variability for all patients recruited in the study.

Dunnett's method is appropriate for analyses of data from agriculture experiments, in which the assumption of equal variances among experimental plots is plausible. However, analogically extending Dunnett's procedure to analyze dose responses at different dose levels for human bodies demands careful examination. In fact, body responses at different dose levels usually vary at discernible ranges.

The consequences and costs of misapplying Dunnett's procedure are self-evident. Unfortunately, in statistical literature, many published analyses give only a cursory nod to such model assumptions. A common excuse for the mis-application of the procedure is *"this is the only method currently available."* Such careless analyses may eventually produce misleading conclusions on the inference of the minimum effective dose.

In clinical trials where endpoints do not satisfy the homescedasticity assumption, even with double-blinded experiments, it is risky to use Dunnett's test with an unverified assumption on the homoscedasticity. In this case, any significant test result is misleading. Due to implausibility of the homoscedasticity assumption, the inference conclusion is actually *an exact solution to a significantly wrong problem.*

When the difference between two treatments is of interest and the homoscedasticity assumption is invalid, for the comparison of the mean effect between a treatment and a control, the uniformly most powerful invariant t-statistic is no longer suitable when the variance of the treatment group is different from that of the control group. In this case, when the sample size is large enough, the Welch's asymptotic t- statistic can be used. However, when the sample size is moderate, the testing problem actually involves the Behrens-Fisher statistic, an open problem in statistics (Casella and Berger, 2002) — although a partial solution such as the Bayesian approach is available using the Markov Chain Monte Carlo. To control the accuracy and power of the test, current approaches include the two-stage sampling method of Chapman's procedure for the mean difference between the two groups.

When more than two treatments are involved, the inference of minimum effective dose necessitates a Bonferroni adjustment to assess the overall confidence level. In the case where two or more endpoints in multiple treatments are of primary interest, multivariate Bonferroni-type inequalities are needed. Further discussion on this topic can be found in Section 9.3. References in this regard include Dudewicz and Ahmed (1998, 2007), Chapman (1950), Tao, Guo and Gao (2002).

1.2.4 Adjusting Multiplicity for Two or More Substitutable Endpoints

The preceding sections focus on multiple comparisons of two or more populations with one clinical endpoint. When two or more clinical endpoints with multiple treatments are of interest in a study, current methods are unable to provide a simultaneous confidence set for both endpoints. For example, consider the substitutive nature of nicotine and morphine in analgesia. When either the reduction in the requirement of narcotic analgesic or the reduction in the requirement of morphine is measured separately for each patient, the simultaneous confidence set takes the form of

$$P([\bigcap_{i=1}^{n} A_i] \bigcup [\bigcap_{j=1}^{n} B_j]) \geq 1 - \alpha, \tag{1.9}$$

where A_1, ..., A_n represent the events where the nicotine efficacies are covered by the corresponding confidence intervals, and B_1, ..., B_n represent the events where the morphine efficacies are bounded by the corresponding confidence intervals, respectively. In general, when more than two endpoints are involved in a study, the evaluation of the probability of the joint event of multiple unions necessitates the use of improved multivariate Bonferroni-type inequalities. More refined outcomes will be discussed in later chapters.

To control the simultaneous confidence level when two or more treatment groups are involved in a clinical trial, it is necessary to use the Bonferroni correction to control the simultaneous confidence level.

For a set of observation Y, where y_{ij} is the observation of the jth patient in the ith comparison group, $i = 0, 1, ..., k$ and $j = 1, ..., n_i$ with $i = 0$ representing the control/placebo group. Let A_i be the event $\{Y : \mu_i - \mu_0 \in (L_i, U_i)\}$ for the treatment i, where L_i and U_i are the lower and upper confidence limits, respectively. The construction of the confidence limits, for pairwise comparisons, is based on the model assumption (such as the t-statistic, Z-statistic for parametric models, or the Mann-Whitney-Wilcoxon statistic for nonparametric analysis). If each confidence interval has a coverage of $1 - \alpha$, for $i = 1, ..., k$,

$$P(A_i) \geq 1 - \alpha$$

then, without adjustment, the simultaneous confidence intervals have coverage probability

$$
\begin{aligned}
P(\bigcap_{i=1}^{k}\{\omega \in A_i\}) &= 1 - P(\bigcup_{i=1}^{k}\{\omega \notin A_i\}) \\
&\geq 1 - \sum_{i=1}^{k} P(\omega \notin A_i) \\
&\geq 1 - k\alpha.
\end{aligned}
$$

Thus, without an appropriate adjustment, the simultaneous confidence set may dramatically reduce the overall confidence level. If $k = 10$, and $\alpha = 0.05$, then in the worst scenario, such as the situation where all the events for summation are highly correlated, the overall significance level may drop to $1 - 10 \times 0.05 = 50\%$.

When two or more substitutive endpoints are of interest (say, t endpoints), the identification of the minimum effective dose of a drug requires the evaluation of

$$P(\text{one of the endpoints is efficacious})$$

$$= P(\bigcap_{i=1}^{k}\{\mathbf{y} \in A_{i1}\} \bigcup \cdots \bigcup \bigcap_{i=1}^{k}\{\mathbf{y} \in A_{it}\}).$$

In Chapters 5 and 7, we discuss the evaluation of this type of problem in a more general setting, and we present refined upper and lower bounds for improving the accuracy of the evaluation of the joint probability in Chapters 6 and 8.

1.3   System Reliability

In the previous two sections, we described the use of multivariate Bonferroni-type bounds for the evaluation of risk for multiple disasters in actuarial science, and for the identification of minimum effective doses when multiple endpoints are of interest in medical research. In this section, we shall introduce the use of multiple Bonferroni-type inequalities for assessing system reliability in engineering. System reliability is the probability that a multi-component system functions (see for example, Ross (1993), Kwerel (1975c), or Chan, Chan and Lin (1988)) over a period of time. For instance, as shown in Figure 1.3, a parallel system works if and only if at least one of its components works, and a series system works if and only if all its components work. The reliability of a system is usually determined by the reliability of its components, along with the design of the circuit.

1.3.1   Basic Systems

Consider a parallel system and a series system, which are the *building blocks* of composite systems. In this section, we shall use examples to describe how to apply the Bonferroni-type inequalities to evaluate the reliability of a basic system.

Example 1.3.1 In a parallel system, let A_i be the event that component i, $i = 1, ..., k$

Series system

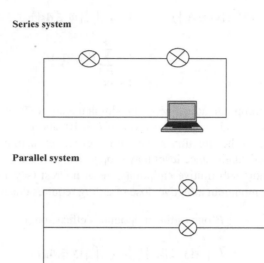

Parallel system

Figure 1.3 *Series system and parallel system*

works over a fixed period of time, say 12 months. Because any functional component will make the system work in a parallel system, the system reliability is $P(\bigcup_{i=1}^{n} A_i)$, which can be bounded by

$$\max\{P(A_1),...,P(A_k)\} \leq P(\bigcup_{i=1}^{k} A_i) \leq \sum_{i=1}^{k} P(A_i). \qquad (1.10)$$

The right-hand side of Inequality (1.10) is the simplest form of the Boole-Bonferroni inequality (the case of multivariate Bonferroni-type inequality when the number of groups of events is more than one). It can be improved by higher order inequalities discussed in Chapters 2, 3, and 8.   □

Example 1.3.2 In a series system, let A_i be the event that component i, $i = 1,...,k$

works over a fixed period of time, say 12 months. In this case, a simple way to access system reliability, $P(\bigcap_{i=1}^{n} A_i)$, is the following,

$$P(\bigcap_{i=1}^{n} A_i) = 1 - P(\bigcup_{i=1}^{n} A_i^c)$$

$$\geq 1 - \sum_{i=1}^{n} P(A_i^c)$$

$$= 1 - \sum_{i=1}^{n} (1 - P(A_i)),$$

thus

$$1 - \sum_{i=1}^{n} (1 - P(A_i)) \leq P(\bigcap_{i=1}^{n} A_i) \leq \min\{P(A_1), ..., P(A_n)\}. \qquad (1.11)$$

The left-hand side of the Inequality (1.11) is another form of the Bonferroni-type inequality; it can be improved by higher order inequalities discussed in Chapter 2, 3, and 8.   □

Notice that we do not use any independence assumption in examples 1.3.1 and 1.3.2. The generality of such settings avoids the independence assumption on reliability theory (the assumption that failure times are independent among components in a system). The plausibility of the independence assumption is questionable, especially for a repaired system. More discussion on the independence assumption can be found in Example 1.1.1.

The bounding technique in Examples 1.3.1 and 1.3.2 pertains to the domain of univariate Bonferroni-type bounds. They can be conveniently extended to the following setting for the reliability of relatively more complicated systems, the composite systems, which necessitate the study of multivariate Bonferroni-type inequalities.

1.3.2   Composite Systems

The preceding section introduces the evaluation of system reliabilities for the parallel and series systems, respectively. However, in practice, most of the systems consist of two or more subsystems. For instance, the system described in Figure 1.4 is a composition of more than one subsystems that control the functioning status of a computing system. The evaluation of such systems requires the multivariate Bonferroni-type inequalities, as illustrated in the following example.

Example 1.3.3 As shown in Figure 1.4, the functioning of the whole system requires the functioning of the three subsystems at the same time, and the functioning of any component in a subsystem suffices for the functioning of the subsystem. The reliability of the system shown in Figure 1.4 can be specified as follows.

Let A_i and B_j, $i, j = 1, 2, 3$, be the events that the three components in the first and second subsystem work, respectively. Let C_k, $k = 1, ..., 4$, be the event that the

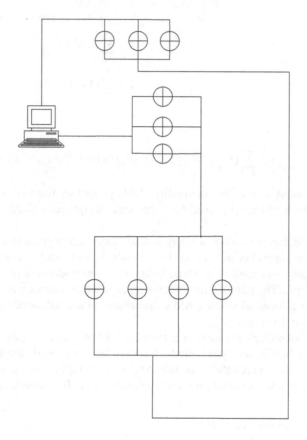

Figure 1.4 *Composite systems*

ith component in the last subsystem works. The system reliability reads

$$P([\bigcup_{i=1}^{3} A_i] \bigcap [\bigcup_{j=1}^{3} B_j] \bigcap [\bigcup_{k=1}^{4} C_k]). \tag{1.12}$$

With the assumption of independence, the evaluation of the system reliability in

(1.12) can be found in, for example, Ross (1993),

$$P([\bigcup_{i=1}^{3} A_i] \cap [\bigcup_{j=1}^{3} B_j] \cap [\bigcup_{k=1}^{4} C_k]) = P(\bigcup_{i=1}^{3} A_i)P(\bigcup_{j=1}^{3} B_j)P(\bigcup_{k=1}^{4} C_k).$$

However, when the interaction of components in the same circuit is taken into consideration, the independence assumption is questionable. Without the independence assumption, the evaluation of the reliability (1.12) can be obtained as follows.

$$P([\bigcup_{i=1}^{3} A_i] \cap [\bigcup_{j=1}^{3} B_j] \cap [\bigcup_{k=1}^{4} C_k])$$

$$= P(\bigcup_{i=1}^{3} \bigcup_{j=1}^{3} \bigcup_{k=1}^{4} [A_i \cap B_j \cap C_k])$$

$$\leq \sum_{i=1}^{3} \sum_{j=1}^{3} \sum_{k=1}^{4} P(A_i \cap B_j \cap C_k). \qquad (1.13)$$

The bound in (1.13) is the simplest form of the multivariate Bonferroni bound. It can be refined or sharpened by multivariate upper and lower bounds discussed in later chapters of this book. □

The following example computes the numerical value of the reliability when the distribution of the failure of each component is specified.

Example 1.3.4 Following the setting in Example 1.3.3, we assume that the lifetime of each component follows an exponential model with parameters $\theta_1 = 0.624$, $\theta_2 = 1.5$, and $\theta_3 = 0.687$. Also assume that if subsystem C is functioning for more than one month, there is a 40% chance that both the subsystems A and B will function for more than one month. Denote $A_i = \{X_i > 1\}$, $B_j = \{Y_j > 1\}$ and $C_k = \{Z_k > 1\}$, then according to (1.12), the system reliability is

$$P([\bigcup_{i=1}^{3} A_i] \cap [\bigcup_{j=1}^{3} B_j] \cap [\bigcup_{k=1}^{4} C_k])$$

$$= P([\bigcup_{i=1}^{3} A_i] \cap [\bigcup_{j=1}^{3} B_j] | \bigcup_{k=1}^{4} C_k)P(\bigcup_{k=1}^{4} C_k)$$

$$= 0.4P(\max(Z_1,...,Z_4) > 1)$$

$$= 0.4(1 - P(\max(Z_1,...,Z_4) \leq 1)$$

$$= 0.4(1 - (1 - e^{-0.687})^4)$$

$$= 0.3756.$$

Now, when we have no information on the conditional probability that describes the relationship of the subsystems, we use the inequality method to estimate the system reliability.

$$P([\bigcup_{i=1}^{3} A_i] \cap [\bigcup_{j=1}^{3} B_j] \cap [\bigcup_{k=1}^{4} C_k])$$

$$= \quad 1 - P([\bigcap_{i=1}^{3} A_i^c] \bigcup [\bigcap_{j=1}^{3} B_j^c] \bigcup [\bigcap_{k=1}^{4} C_k^c])$$

$$\geq \quad 1 - P(\bigcap_{i=1}^{3} A_i^c]) - P([\bigcap_{j=1}^{3} B_j^c]) - P([\bigcap_{k=1}^{4} C_k^c])$$

$$= \quad 1 - P(X_1 < 1, X_2 < 1, X_3 < 1) - P(Y_1 < 1, Y_2 < 1, Y_3 < 1) - P(Z_1 < 1, ..., Z_4 < 1)$$

$$= \quad 1 - (1 - e^{-0.624})^3 - (1 - e^{-1.5})^3 - (1 - e^{-0.687})^4$$

$$= \quad 0.3701,$$

which is consistent with the computed result of the system reliability with conditional information on the subsystems. However, under the independence assumption, we get

$$P([\bigcup_{i=1}^{3} A_i] \bigcap [\bigcup_{j=1}^{3} B_j] \bigcap [\bigcup_{k=1}^{4} C_k])$$

$$= \quad P([\bigcup_{i=1}^{3} A_i]) P([\bigcup_{j=1}^{3} B_j]) P([\bigcup_{k=1}^{4} C_k])$$

$$= \quad P(\max(X_1, X_2, X_3) > 1) P(\max(Y_1, Y_2, Y_3) > 1) P(\max(Z_1, ..., Z_4) > 1)$$

$$= \quad (1 - (1 - e^{-0.624})^3)(1 - (1 - e^{-1.5})^3)(1 - (1 - e^{-0.687})^4)$$

$$= \quad 0.5037.$$

Notice that in this example, for illustration purposes, we assume a conditional probability on the relation of the three subsystems. Strictly speaking, the complete assignment of probabilities should follow the $2^3 = 8$ elementary conjunctions of the three subsystems, which will be further discussed in Chapter 2. Note that in the above estimation of the system reliability, the only term that does not require any information on the correlation among the three subsystems is the Bonferroni adjustment. Although it may be conservative in some settings, in this particular example, when the 40% conditional probability among the three subsystems is legitimate, the performance of the Bonferroni bound is much closer than the convenient assumption of independence.   □

References for the discussion in this section include Boros and Prékopa (1989), Chung (1941), Glaz (1990), Glaz and Pozdnyakov (2005).

1.4   Education Reform and Theoretical Windows

In the preceding sections, the discussions of the risk evaluation of extreme values, minimum effective dose, and system reliability, focus on the inference problem involving one measurement of interest. However, in practice, we frequently confront inference problems involving two correlated measurements. For example, how carrying a gun and having a substance abuse problem relates to adolescent suicide (see, for example, Brent et al. (1999), Chen (2001)). Similar studies can be found in Jensen and Jones (1969) for the simultaneous estimation of variances, Chen, Hoppe, Iyengar, et al (2003) for the impact of past suicide attempts on adolescent suicide among male and female, and Chen, Iyengar and Brent (2007) for the evaluation of component population attributable risks, among others. To specify the situation, consider the efficacy measurement and toxicity measurement of a drug at a particular dose. For certain drugs, the efficacy has no meaning unless the drug is safe for the patient.

The two-measurement problem is not limited to medical researches. In this section, we concentrate on two applications of multivariate Bonferroni-type inequalities for inference involving two correlated measurements. Section 1.4.1 compares learning outcomes between two types of teaching pedagogy in educational reform, and Section 1.4.2 discusses the inference for the therapeutic window of a drug.

1.4.1 Learning Outcomes of Different Pedagogies

In educational reforms, consider the comparison between lecture-based pedagogy, web-based teaching-learning delivery, and the activity-based pedagogy in teaching statistics. Some educators prefer more lecture-based teaching while others advocate web-based or activity-based teaching. Obviously, either completely lecture-based or completely activity-based will lack the advantages of the other pedagogy. To this end, the issue of interest is to see the effect of these pedagogies in teaching. To seek the effects of different weights that balance lecturing and activities in a class, assume that five classes are studied in terms of students' understanding of basic statistical theory and their computing ability for data analysis. One extreme is to train students with theoretical material without hands-on ability; using software for data analysis, in which teaching statistics is treated like a mathematics class. Another extreme is to train students with computing skills without touching the underlying statistical theory, in which teaching statistics is treated like a cooking-class and students essentially serve as extensions of a machine in the process of data analysis. Either method ends up with distorted statistical education.

Notice that the outcome measurement mainly consists of two different components of each student: the student's understanding of statistical theory and the student's hands-on ability to analyze data, such as using software. The two components are highly correlated. Consider k different teaching methods to be compared with a standard teaching method (traditional teaching method, the control group). Since the outputs are not continuous (they are ordinary or categorical data), the assumption of normality is inappropriate. Similar to the discussion in Section 1.2.3, the nonparametric method Mann-Whitney-Wilcoxon statistic is applied to compare the median differences in learning scores of the two measurements regarding theoretical background and hands-on computing skills. Let A_i and B_j be the event that the nonparametric confidence intervals cover the true median difference of learning scores of the two measurements.

$$A_i : \quad mt_i - mt_0 \in (LT_i, UT_i)$$

$$B_j : \quad ms_j - ms_0 \in (LS_j, US_j),$$

where $mt_i - mt_0$ $i = 1,...,k$ and $ms_j - ms_0$, $j = 1,...,k$ are the differences of median learning scores, reflecting students' understanding of the theoretical background and practical skills, respectively. In this case, the overall probability of interest can be estimated as follows,

$$P(\bigcap_{i=1}^{k}\bigcap_{j=1}^{k}(A_i\cap B_j)) = 1 - P(\bigcup_{i=1}^{k}\bigcup_{j=1}^{k}(A_i\cap B_j)^c)$$

$$\geq 1 - \sum_{i=1}^{k}\sum_{j=1}^{k}P((A_i\cap B_j)^c)$$

$$= 1 - \sum_{i=1}^{k}\sum_{j=1}^{k}(1 - P(A_i\cap B_j))$$

$$= S_{11} - (k^2 - 1), \qquad (1.14)$$

where $S_{11} = \sum_{i=1}^{k}\sum_{j=1}^{k}P(A_i\cap B_j)$ is the first degree bivariate Bonferroni-type summation. S_{11} will be formally defined and delineated in Chapter 2. Methods (for example, Nakamura and Douke 2003) improving the bound in (1.14) will be elucidated in later chapters.

Notice that the formulation here is different from that of Section 1.2.4, where the two endpoints are substitutive for the treatment of post-operation pains, however, occurrences of the two measurements in educational reform are simultaneous for the discussion in this section.

1.4.2   Therapeutic Windows of a Drug

In Section 1.2, we discussed the issue of using an inequality method to identify the minimum effective dose of a drug. The method essentially bounds the upper limit and lower limit of the probability of the union of a set of events. A related but discernible topic pertaining to the domain of simultaneous inference is the estimation of the therapeutic window of a drug, in which the measurement of each patient consists of two components, the efficacy measurement and toxicity measurement.

The drug therapeutic window is determined by the range between the minimum effective dose and the maximum tolerated dose. The maximum tolerated dose is the highest safe dose (a dose that is safe and all lower doses are safe). For example, when dosages 1, 2, 4 are safe but dosage 3 is not, the maximum tolerated dose is 2 due to the toxicity of dosage 3. An effective dose is useful only when it is safe for patients. We use the following example to illustrate this concept.

Example 1.4.1 *Letrozole* is a drug used to treat advanced breast cancer in post-menopausal women. The use of *Letrozole* to treat cancer is grounded on the theory that receptors for specific hormones needed for cell growth are located on the surface of tumor cells. *Letrozole* works by blocking an active hormone (enzyme aromatase in the breast) which is used to convert androgens into estrogen, so that tumors depending on estrogen for growth will shrink. However, the functioning of *Letrozole* also affects normal activities of healthy cells. Thus when the efficacy of *Letrozole* is of interest, the side-effects of the drug (which include bone pain, back pain, and fatigue) should also be considered. Overdose of a drug may damage the functioning mechanisms of the cell system. □

For certain chemical compounds, the side effect (or toxicity) hinders the application of higher dosages of the drug. As with *Letrozole* (Example 1.4.1), when the chemical compound shrinks the tumor cells, it also hurts the growth of healthy cells. Too much chemical may create new health issues for the patient. With this in mind, we need to identify the therapeutic window.

In a clinical trial where each patient is evaluated with two measurements regarding the efficacy and toxicity, denote A_i the event that the ith confidence interval of efficacy covers the unknown treatment difference. Namely, $A_i : \mu_i - \mu_0 \in C_i$ where C_i is the confidence interval for the treatment difference. Denote B_i the event that the ith confidence interval of toxicity covers the unknown difference of the mean measurements of toxicity. The therapeutic window of a drug refers to the dosage range in which the drug is both safe and efficacious.

Example 1.4.2 Assume that for *Letrozole* in hormone therapy for breast cancer, on the basis of a set of data the following 95% simultaneous confidence intervals are obtained for the efficacy and toxicity of the four dosages under investigation: $\mu_1 - \mu_0 \in (1.2, 3.4)$, $\mu_2 - \mu_0 \in (1.4, 3.7)$, $\mu_3 - \mu_0 \in (1.8, 4.1)$, $\mu_4 - \mu_0 \in (1.7, 4.4)$, $\eta_1 - \eta_0 \in (0.4, 1.2)$, $\eta_2 - \eta_0 \in (0.9, 1.5)$, $\eta_3 - \eta_0 \in (1.1, 1.9)$, $\eta_4 - \eta_0 \in (1.3, 2.2)$, where μ_i and η_i are the mean measurements of efficacy and toxicity at dose i, respectively. If the increase of 1.3 unit of efficacy is considered to be efficacious (so the minimum effective dose is 2) and the increase of 1.2 unit in toxicity measurement is considered to be safe (so the maximum tolerated dose is 3), the therapeutic window is from dosage 2 to dosage 3 with confidence level 95%. □

In the construction of the simultaneous confidence intervals for the detection of the therapeutic window, the critical point is the evaluation of the joint probability

$$P(\bigcap_{i=1}^{4} \bigcap_{j=1}^{4} (A_i \bigcap B_j)), \tag{1.15}$$

where A_i and B_j are events that the mean measurements of the efficacy and toxicity fall in the corresponding confidence intervals, respectively. The special feature in this scenario is the correlation between the two measurements for each patient. Similar to the evaluation of (1.14), the evaluation of the joint probability (1.15) using the marginal probabilities of the joint events $P(A_i \bigcap B_j)$ for $i, j = 1, ..., n$ will be discussed in Chapters 4, 5, and 7.

It should be mentioned that the inference problem in Section 1.2 for the minimum effective dose of two substitutive medicines is different from the two measurement problem discussed in this section. For the therapeutic window, it is mandatory for the dose to satisfy both efficacious and safe. However, for the minimum effective dose using two substitutive medicines, it suffices to have one of the drugs work for the patient. Related references for probability inequalities include Chung (1943a, 1943b), Galambos (1975b), Erdos, Neveu, and Rényi (1963), Glaz, Pozdnyakov, and Wallenstein (2009).

1.5   Ruin Probability and Multiple Premiums

In this section, we discuss the use of inequalities in stochastic processes, a relatively more advanced field in statistical inference. Readers unfamiliar with the concept of the martingale may skip this section.

Considering an insurer package that consists of k portfolios with ruin times τ_s^1, ..., τ_s^k, the T-year ruin probability reads

$$RP(s) = P(\bigcup_{i=1}^{k}\{\tau_s^i \leq T\}),$$

assuming the ruin of any one of the portfolios ruins the package. Thus, the evaluation of the ruin probability via the inequality method concentrates on the performance of the individual portfolio in the package.

Reiss and Thomas (1997) describe a model in which an insurer estimates an adequate initial reserve for a portfolio over a time frame (say 10 years) with a pre-determined ruin probability (say, 1% or 5%). Denote $S(t) = \sum_{i \leq N(t)} X_i(t)$, the total amount of $N(t)$ claims of a portfolio over a period t, where $X_i(t)$ is the random claim size. Assume that the claim size process $\{X(t)\}$ is independent, with claim arrival time process $\{T_1, ..., T_n, ...\}$. Further, assume that the claim sizes follow the same distribution $E(X_i) = E(X)$, so that the net premium becomes $E(S(t)) = E(X)E(N(t))$.

If the claim number process follows a time homogeneous Poisson process with intensity λ, then $E(N(t)) = \lambda t$ and the net premium reads

$$E(S(t)) = E(X)\lambda t.$$

For a portfolio where the policy holder pays a total premium to the insurer in order to gain compensations for future losses, the benefit of the insurer comes from the difference between the total premium and the net premium. Denote

$$\frac{total\,premium}{net\,premium} = \gamma,$$

where $\gamma - 1$ is the safety loading of the insurer.

If the insurer reserves s amount for a given portfolio, the reserve process of the insurer reads

$$U(t) = s + \gamma E(S(t)) + I(t) - C(t) - S(t),$$

where $I(t)$ is the interest income for the accumulated reserve and $C(t)$ is the cost including expenses, taxes, dividends, etc. up to time t. If $U(t) < 0$, the portfolio is ruined. Thus, the T-year ruin event occurs when

$$\tau_s \leq T, \qquad \tau_s = inf\{t : U(t) < 0\}$$

and the T-year's ruin probability is $P(\tau_s \leq T)$.

Example 1.5.1 Consider the model in which an insurer invests an initial reserve ($2,000,000) for a portfolio over a long period of time, and updates the portfolio once the investment return reaches certain volume, say $20,000,000. Assuming that his *monthly* net profit (from interest, premium, etc., minus costs) is $5,000 and the average amount of insurance claim from this portfolio is $10,000 with probability 0.2, the chance that the insurer goes broke with this portfolio before reaching his goal of $20,000,000 can be obtained using De Moivre's martingale as follows.

With the initial reserve of $2,000,000, denote by X_t the insurer's reserve at time t (from the beginning of the portfolio). For each period, the insurer either obtains a net profit of $5,000 with probability $1 - 0.2 = 0.8$, or bears with a net loss of $5,000 ($10,000 - $5,000 = $5,000$) with a probability 0.2. Assume that the arrivals of the insurance claims are independent. Then taking $5,000 as a scale parameter, the net revenue of the insurer in each month Y_t is actually a random walk with a winning probability of 0.8.

Since the initial reserve is $2,000,000, which is 400 units of $5,000, denote $X_0 = 400$, and $X_t = \sum_{i=0}^{t} Y_t$, the reserve of the insurer at time t. Now $E(|X_t|) < \infty$ and $E(X_{t+1}|X_t, ..., X_0) = E(Y_{t+1} + X_t|X_t, ..., X_0) = 0.6 + X_t$, it follows that $\{X_t, t \geq 0\}$ is not a martingale. However, notice that

$$E(4^{-X_{t+1}}|4^{-X_t}, 4^{-X_{t-1}}, ...4^{-X_0}) = 4^{(-1-X_t)} \times 0.8 + 4^{-X_t+1} \times 0.2$$
$$= 4^{(-X_t)}. \tag{1.16}$$

Thus, $\{Z_t, t \geq 0\}$ with $Z_t = 4^{-X_t}$ is a martingale. Taking expectations in both sides of the martingale equation gets

$$E(Z_t) = E(Z_{t-1}) = ... = E(Z_0) = 4^{-400}.$$

If the insurer goes broke at time T, it follows from the above equation that $E(Z_T) = 4^{-400}$. Let p_0 be the probability that the insurer goes broke before reaching his goal of $20,000,000 (which is 4,000 units in the scale of $5,000), we have

$$E(Z_T) = 4^{-0} \times p_0 + 4^{-4000} \times (1 - p_0). \tag{1.17}$$

Combining the equation (1.17) with $E(Z_T) = 4^{-400}$ gets $p_0 = \frac{4^{-400} - 4^{-4000}}{1 - 4^{-4000}}$, which is almost zero. So, in this case, it is almost certain that the insurer will not go broke since the initial reserve is relatively large, compared with the loss.

Example 1.5.2 In Example 1.5.1, if the arrival rate of insurance claim is 0.5, $Z_t = 1$ for all $t \geq 0$, the martingale approach does not work since the associated series does not converge. In this case, we consider the following method to estimate the ruin probability and the ruin time T.

By the martingale property of $\{X_t, t \geq 0\}$, we have $E(X_t) = E(X_0)$ for any time t. Thus for the ruin time T,

$$E(X_T) = 0 \times p_0 + 4000 \times (1 - p_0) = 400 (= E(X_0)),$$

which results in $p_0 = 90\%$. Thus, the chance of ruin becomes very substantial (90%) when the arrival rate of insurance claims is assumed to be 50%.

Now, let $S_t = X_t^2 - t$ where $X_t = \sum_{i=0}^{t} Y_i$, and $\{Y_i, i \geq 0\}$ is a random walk with winning probability 0.5. Since

$$
\begin{aligned}
E(S_{t+1}|S_t, S_{t-1}, ..., S_0) &= E((X_{t+1}^2 - t - 1|S_t, ..., S_0) \\
&= E(Y_{t+1}^2 + 2Y_{t+1}X_t + X_t^2 - t - 1|S_t, ..., S_0) \\
&= X_t^2 - t \\
&= S_t,
\end{aligned}
\tag{1.18}
$$

the process $\{S_t, t \geq 0\}$ is a martingale. Taking expected values in both sides of equation (1.18) at time T gets

$$
\begin{aligned}
0^2 \times p_0 + 4000^2 \times (1 - p_0) - E(T) &= E(X_T^2 - T) \\
&= E(X_0^2) \\
&= 400^2,
\end{aligned}
$$

thus, $E(T) = 1440,000$; this means that in the long run, on average, the insurer stays for 1,440,000 months (=12000 years) before going broke. □

Now, back to the consideration of the ruin probability of k portfolios in the package, the insurer's initial reserve is determined by setting $RP(s) = q$ for a pre-specified ruin probability q and the ruin occurs in a particular year (say the ith year) is

$$
P(\bigcup_{i=1}^{k} \{\tau_s^i \leq T + 1, \tau_s^i \geq T\}).
\tag{1.19}
$$

Denote $A_i = \{\tau_s^i \leq T + 1\}$, $B_i = \{\tau_s^i \geq T\}$, the above ruin probability of the package in (1.19) can be evaluated by the ruin probability of each portfolio by

$$
P(\text{at least one portfolio ruins at time T}) = P(\bigcup_{i=1}^{k}(A_i \cap B_i)) \leq S_{11},
\tag{1.20}
$$

where S_{11} is the first degree bivariate Bonferroni summation. Certainly, bound (1.20) can be improved by refined bivariate upper and lower Bonferroni-type inequalities. Studies related to this discussion include Chung and Erdos (1952), Rényi (1967), Serfling (1974), and Galambos (1974, 1975a).

1.6   Martingale Inequality and Asset Portfolio

In this section, we show the connection of Bonferroni inequality to martingale inequality in the evaluation of asset portfolio.

Consider the situation where k asset prices are involved in an investment portfolio, in which the Bonferroni inequality can be used in evaluating the chance of large deviation in the portfolio. For any $x \geq 0$,

$$P([max_{i=1,\ldots,k}|X_{i,n} - X_{i,0}|] \geq x) = P(\bigcup_i A_i)$$
$$\leq \sum_i P(A_i)$$

where A_i denotes the set $\{|X_{i,n} - X_{i,0}| \geq x\}$.

Let P_t be an asset price and $X_t = \frac{P_t - P_{t-1}}{P_t}$ be the rate of the return of the asset from period $t-1$ to period t. Denote I_{t-1} an information set available at period $t-1$. The efficient market hypothesis $E(X_t|I_{t-1}) = E(X_t)$ necessitates the martingale property of the return rates $\{X_t, t = 0, 1, \ldots\}$. This means that no systematic trading strategy that exploits the conditional mean dynamics can be more profitable in the long run than holding the market portfolio.

For the martingale return rate $\{X_t, t = 0, 1, \ldots\}$, it is known (Hoeffding's martingale inequality) that if at each period t, the differences of return rates do not surpass a certain threshold, $P(|Y_n - Y_{n-1}| \leq K_n) = 1$ for all n with a sequence of real numbers $\{K_1, \ldots, K_n, \ldots\}$, then the chance of large return rate after n period of time is bounded by:

$$P(|X_n - X_0| \geq x) \leq 2\exp(\frac{-\frac{1}{2}x^2}{\sum_{i=1}^n K_i^2}),$$

for any $x \geq 0$.

Now consider the situation in (1.21), where for any $x \geq 0$, we have

$$P([max_{i=1,\ldots,k}|X_{i,n} - X_{i,0}|] \geq x) \leq 2\sum_j \exp(\frac{-\frac{1}{2}x^2}{\sum_{i=1}^n K_{j,i}^2}). \qquad (1.21)$$

Here, the univariate first degree Bonferroni inequality Bonferroni inequality is utilized in deriving the inequality (1.21). Further discussions on the improvement of accuracy will be given in upcoming chapters. Published results related to this topic include Feller (1957), Galambos (1966), Móri and Székely (1983), Pozdnyakov, Glaz et al. (2005), Kaehler and Maller (2010), among others.

Chapter 2

Fundamentals

This chapter provides basic terminology and methods for the discussion of Bonferroni-type inequalities. As illustrated in Chapter 1, Bonferroni-type inequalities play a critical role in various applied fields including medicine, finance, education, economics, and engineering. The main principle underpinning the applications of probability inequalities in Chapter 1 is the evaluation of the joint probability by means of marginal distributions (or joint probabilities with reduced dimensions). For example, in a dose-response study, when the experimenter is interested in constructing a simultaneous confidence set that estimates differences of drug effects among several treatments with a control, the joint probability of confidence intervals for the comparison between the ith dose effect and the placebo, A_i, for $i = 1, ..., k$, reads

$$P(\bigcap_{i=1}^{k} A_i) = 1 - P(\bigcup_{i=1}^{k} A_i^c).$$

Evaluation of the simultaneous confidence level on the left-hand side is converted into the evaluation of the probability of the union of a set of events on the right-hand side.

Note that the probability on the right-hand side in the above equation can be bounded by

$$P(\bigcup_{i=1}^{k} A_i^c) \le \sum_{i=1}^{k} P(A_i^c). \tag{2.1}$$

Thus the probability of simultaneous coverage can be correspondingly estimated by

$$P(\bigcap_{i=1}^{k} A_i) \ge 1 - \sum_{i=1}^{k} P(A_i^c),$$

which is the usual Bonferroni correction (Bonferroni (1936) or Boole (1854)) by taking $1 - \frac{\alpha}{k}$ as the confidence level for each individual comparison.

Notice that the above bound could be too conservative when the number of comparisons, k, is large. In that situation, any inequality improving (2.1) will lead to a set of simultaneous confidence intervals that refines the conventional Bonferroni method with an improved accuracy level of the simultaneous confidence ranges.

The above discussion of simultaneous confidence intervals (for example, Chen and Hoppe (2004), Csorgo et al. (1998)) is applicable to each of the scenarios discussed in Chapter 1. For example, a sharpened upper bound may lead to narrow confidence intervals, which may subsequently enhance the power or reduce the sample size on the evaluation of the system reliability in Section 1.3, when additional information on the sub-population of joint events becomes available. Such demands in applications partly motivate investigations in the refinement of probability bounds.

To give a glance at the refinement of the inequality (2.1), consider the following bounds that use higher degree Bonferroni summations when it is possible to obtain information on the joint probabilities on $P(A_{i_i} \cap ... \cap A_{i_t})$ for an integer $1 \le t \le \frac{k}{2}$. In this case, the evaluation of the quantity $P(\bigcup_{i=1}^{k} A_i^c)$ in bound (2.1) can be improved by the following Bonferroni-type upper and lower bounds on $A_1^c, ..., A_k^c$:

$$\sum_{j=1}^{2t} (-1)^{j+1} \sum_{1 \le i_1 \le ... \le i_j \le k} P(A_{i_1}^c \cap ... \cap A_{i_j}^c)$$

$$\le P(\bigcup_{i=1}^{k} A_i^c)$$

$$\le \sum_{j=1}^{2t-1} (-1)^{j+1} \sum_{1 \le i_1 \le ... \le i_j \le k} P(A_{i_1}^c \cap ... \cap A_{i_j}^c). \qquad (2.2)$$

Bound (2.2) obviously extends Bound (2.1) by using joint probabilities with reduced dimensions. Before we go any further in the discussion of the bounding theory related with (2.2), there are several technical terms that we need to define. These technical terms will facilitate discussions in the rest of the book.

- *Bonferroni summation* : Consider an arbitrary number of events $A_1, ..., A_n$ in a probability space $(\Omega, \mathcal{F}, P)$. In general, the Bonferroni summation refers to the quantity

$$S_t = \begin{cases} \sum_{1 \le i_1 < ... < i_t \le n} P(A_{i_1}...A_{i_t}), & 1 \le t \le n \\ 1, & t = 0 \\ 0, & t > n. \end{cases}$$

When necessary, we also use $S_t(A)$ for S_t to address the associated set of events $\{A_i, i = 1, ..., n\}$. Bonferroni summations are building blocks for the construction of the classical Bonferroni inequality. For example, with Bonferroni summations $S_1, ..., S_{2t}$, the bound in (2.2) can be expressed as

$$\sum_{j=1}^{2t} (-1)^{j+1} S_j \le P(\bigcup_{i=1}^{k} A_i^c) \le \sum_{j=1}^{2t-1} (-1)^{j+1} S_j, \qquad (2.3)$$

where the Bonferroni summation $S_j(A^c) = \sum_{1 \le i_1 \le ... \le i_j \le k} P(A_{i_1}^c \cap ... \cap A_{i_j}^c)$ (Bonferroni (1936)). Here $S_j(A^c)$ is the Bonferroni summation for the event set $\{A_i^c, i = 1, ..., k\}$.

- *Bonferroni-type Bounds*: Consider an arbitrary number of events $A_1, ..., A_n$ in a probability space $(\Omega, \mathcal{F}, P)$, in which v is the number of the events that occur at a given sample point. The Bonferroni-type inequalities refer to upper or lower bounds for $P(v \geq r)$ with $r = 1, ..., n$, or for $P(v = r)$ with $r = 0, 1, ..., n$, in terms of Bonferroni summations. A Bonferroni-type bound typically consists of components of the linear combination of Bonferroni summations such as the bound in (2.2) (See, for example, Galambos and Mucci (1980), Jogdeo (1977), or Hoppe and Seneta (2012)).

- *Degree of an inequality*: For a Bonferroni-type bound, the maximal number of events involved in the intersection for the Bonferroni summations of a bound is called the degree of the bound. For example, the Boole's upper bound

$$P(\bigcup_{i=1}^{n} A_i) \leq \sum_{i=1}^{n} P(A_i)$$

is a degree-one upper bound because it only involves the first degree Bonferroni inequality Bonferroni summation (see for example, Kounias (1969), Móri and Székely (1985))

$$S_1 = \sum_{i=1}^{n} P(A_i).$$

When the joint probabilities of any two events are available, for instance, the upper bound (2.1) can be extended to

$$\begin{aligned} P(\bigcup_{i=1}^{k} A_i^c) &\leq \sum_{i=1}^{k} P(A_i^c) - \sum_{1 \leq i < j \leq k} P(A_i^c \bigcap A_j^c) + \sum_{1 \leq i < j < t \leq k} P(A_i^c \bigcap A_j^c \bigcap A_t^c) \\ &= S_1 - S_2 + S_3, \end{aligned} \tag{2.4}$$

which is a degree-three Bonferroni-type upper bound. (For example, Rényi (1958), Recsei and Seneta (1987)).

By using more information (in this case, the the joint probabilities of three events), the degree-three bound (2.4) extends the degree-one bound (2.1). Normally, higher degree is associated with a sharper bound because of the utilization of more information.

With the basic terminologies defined above, this chapter is organized as follows. Section 2.1 focuses on concepts and inequalities on univariate Bonferroni-type bounds, which includes basic bounding techniques such as binomial moments. It introduces basic relationships among univariate Bonferroni-type bounds, non-linear combination bounds, and classical bounds (such as the Dawson-Sankoff bound and the Sobel-Uppuluri-Galambos upper bounds).

After establishing fundamental concepts of univariate bounds, comparisons of various types of univariate bounds naturally necessitate comparison standards (criteria), which lead to the definition of optimality for probability inequalities. Toward this end, Section 2.2 addresses three basic versions of optimality: Fréchet optimality, linear optimality, and linear programming optimality. Optimality is critical in measuring whether one might gain any bounding improvement, because once an optimal bound is found, there is no space for improvement under the same conditions.

The introduction of multivariate bounding and optimization theory is arranged in Sections 2.3 and 2.4. After the foundation of multivariate bounds in Section 2.3, Section 2.4 sheds new light on the multi-strata correlation between sets of events for multivariate optimality. This section also contains an interesting phenomenon that multivariate bounds are not always analogous to corresponding univariate bounds.

2.1   Univariate Bonferroni-type Bounds

In this section, we discuss basic issues on univariate bounding theory, addressing bounds with linear combinations of univariate Bonferroni summations and bounds with non-linear functions of the Bonferroni summations.

2.1.1   Linear Combination Bounds

Two Bonferroni-type bounds will be introduced in this section. One is the classical Bonferroni bound and another is an improved version of the Bonferroni bound, the Sobel-Uppuluri-Galambos bound. Since this book focuses on *Multivariate* Bonferroni-type bounds, we only mention the univariate bounds as an introduction to smooth the discussion in the sequel. More illustrations on univariate bounds can be found in Galambos (1978), Galambos and Simonelli (1996), or the papers of Tan and Xu (1989), and Wegner (2005), among others. Alternatively, the univariate bounds and their derivations can be obtained as special cases of the multivariate bounds explained in the later chapters of this book.

We start with the classical Bonferroni inequalities. Consider an arbitrary number of events $A_1, ..., A_n$ in a probability space $(\Omega, \mathcal{F}, P)$. Denote v the number of events that occur at a given sample point. As mentioned in (2.3), the classical Bonferroni bounds read

$$\sum_{j=1}^{2t}(-1)^{j+1}S_j \leq P(\bigcup_{i=1}^{n}A_i) \leq \sum_{j=1}^{2t-1}(-1)^{j+1}S_j,$$

for a set of Bonferroni summations S_j, $j \in \{1, 2, ..., 2t\}$ with $1 \leq t \leq \frac{n}{2}$. Note that the event $\bigcup_{i=1}^{n}A_i = \{v \geq 1\}$, the above inequality becomes the following proposition.

Proposition 2.1.1 Classical Bonferroni Inequalities.

$$\sum_{j=1}^{2t}(-1)^{j+1}S_j \leq P(v \geq 1) \leq \sum_{j=1}^{2t-1}(-1)^{j+1}S_j. \qquad (2.5)$$

Proof: Consider the decomposition of the following quantity

$$\sum_{k=0}^{2t}(-1)^k S_k$$

$$= \sum_{k=0}^{2t}(-1)^k \sum_{r=k}^{n}\binom{n}{k}P(v=r)$$

$$= \sum_{r=0}^{n} P(v = r) \sum_{k=0}^{\min(2t,r)} (-1)^k \binom{r}{k}$$

since when $k > r$, the quantity $\binom{r}{k} = 0$, we have

$$\sum_{k=0}^{2t} (-1)^k S_k$$

$$= \sum_{r=0}^{n} P(v = r) \sum_{k=0}^{2t} (-1)^k \binom{r}{k}$$

$$= P(v = 0) + \sum_{r=1}^{n} P(v = r) \sum_{k=0}^{2t} (-1)^k \binom{r}{k}.$$

Now, notice that

$$\binom{r}{0} - \binom{r}{1} + \binom{r}{2} - \binom{r}{3} + \ldots + \binom{r}{2t}$$

$$= \binom{r-1}{0} - \binom{r}{1} + \binom{r}{2} - \binom{r}{3} + \ldots + \binom{r}{2t}$$

$$= -\binom{r-1}{1} + \binom{r}{2} - \binom{r}{3} + \ldots + \binom{r}{2t}$$

$$= \binom{r-1}{2} - \binom{r}{3} + \ldots + \binom{r}{2t}$$

$$= -\binom{r-1}{3} + \binom{r}{4} - \binom{r}{5} + \ldots + \binom{r}{2t}$$

$$= \binom{r-1}{4} - \binom{r}{5} + \ldots + \binom{r}{2t}$$

$$= \ldots$$

$$= \binom{r-1}{2t}$$

$$> 0,$$

we have

$$\sum_{k=0}^{2t} (-1)^k S_k > P(v = 0),$$

which implies that

$$\sum_{k=1}^{2t} (-1)^{k+1} S_k \leq 1 - P(v = 0) = P(\bigcup_{i=1}^{n} A_i).$$

Similarly when the summation is for odd terms, the quantity in (2.5) can be decomposed as follows.

$$\sum_{k=0}^{2t-1} (-1)^k S_k$$

$$= P(v=0) + \sum_{r=1}^{n} P(v=r) \sum_{k=0}^{2t-1} (-1)^k \binom{r}{k}.$$

Now, notice that for the summation of odd terms,

$$\binom{r}{0} - \binom{r}{1} + \binom{r}{2} - \binom{r}{3} + \dots + \binom{r}{2t-1}$$

$$= \binom{r-1}{0} - \binom{r}{1} + \binom{r}{2} - \binom{r}{3} + \dots + \binom{r}{2t-1}$$

$$= -\binom{r-1}{1} + \binom{r}{2} - \binom{r}{3} + \dots + \binom{r}{2t-1}$$

$$= \binom{r-1}{2} - \binom{r}{3} + \dots + \binom{r}{2t-1}$$

$$= -\binom{r-1}{3} + \binom{r}{4} - \binom{r}{5} + \dots + \binom{r}{2t-1}$$

$$= \binom{r-1}{4} - \binom{r}{5} + \dots + \binom{r}{2t-1}$$

$$= \dots$$

$$= -\binom{r-1}{2t-1}$$

$$< 0.$$

Thus,

$$\sum_{k=0}^{2t-1} (-1)^k S_k$$

$$= P(v=0) + \sum_{r=1}^{n} P(v=r) \sum_{k=0}^{2t-1} (-1)^k \binom{r}{k}$$

$$< P(v=0).$$

We have

$$\sum_{k=1}^{2t-1} (-1)^{k+1} S_k \geq 1 - P(v=0) = P(\bigcup_{i=1}^{n} A_i).$$

This completes the proof of Proposition 2.1.1 (which illustrates the classical Bonferroni bounds). □

Actually, bounds in Proposition 2.1.1 have an extended version that controls the probability of the event $\{v \geq r\}$ for any positive integer r satisfying $1 \leq r \leq n$.

For the above mentioned integers n and r, and any integer j satisfying $j \leq n$, for notation convenience, denote

$$A_r^n(j) = \sum_{k=0}^{j} (-1)^k \binom{k+r-1}{r-1} S_{k+r},$$

and

$$B_r^n(j) = \sum_{k=0}^{j} (-1)^k \binom{k+r}{r} S_{k+r},$$

Galambos (1978) describes two types of classical inequalities as follows.

$$A_r^n(2u-1) \leq P(v \geq r) \leq A_r^n(2u), \tag{2.6}$$

and

$$B_r^n(2u-1) \leq P(v=r) \leq B_r^n(2u), \tag{2.7}$$

where u is any integer satisfying $2u \leq n$, when $u = 0$, $A_r^n(2u-1) = B_r^n(2u-1) = 0$.

The probability bound in the inequality form (2.6) is for $P(v \geq r)$, which is the probability of the occurrence of **at least** r out of the total of n events. And the bound in form (2.7) is for the upper and lower bounds of $P(v = r)$, which is the probability of the occurrence of **exactly** r out of the total of n events. For instance, financial analysis regarding asset prices (or medical analysis regarding the efficacy of a new drug) requires simultaneous adjustment for multiplicity, which consequently demands inequalities of the form of (2.6). On the other hand, in the analysis on system reliability for exact number of malfunctioning subsystems (or inference in genetics regarding the exact number of gene mutations), estimating the exact number of malfunctioning systems (or gene mutations) necessitates inequalities of the form of (2.7).

Note that when $r = 1$, the bound (2.6) becomes

$$\sum_{k=0}^{2u-1} (-1)^k S_{k+1} \leq P(v \geq 1) \leq \sum_{k=0}^{2u} (-1)^k S_{k+1},$$

which is exactly the classical Bonferroni inequality (2.5) because the event of at least one occurrence of A_i's is the occurrence of the union of the set of events $\{A_i, i = 1, ..., n\}$,

$$P(v \geq 1) = P(\bigcup_{i=1}^{n} A_i).$$

Moreover, letting $r = 1$ and $u = 0$ in the inequality (2.6) leads to the well-known Boole's inequality

$$P(\bigcup_{i=1}^{n} A_i) \leq \sum_{i=1}^{n} P(A_i) = S_1.$$

It should be noted that although the two bounds (2.6) and (2.7) serve different purposes (one for the probability of at least r occurrences and the other for exactly r occurrences), the generation of the two types of inequalities may intertwine well with each other. For instance, the inequalities of the form (2.7) can be used to generate upper and lower bounds for the union of a set of events, $P(\bigcup_{i=1}^{n} A_i)$ as shown below.

Example 2.1.1. Letting $u = 1$ and $r = 1$ in the bound (2.7) yields

$$B_0^n(1) \leq P(v=0),$$

where

$$B_0^n(1) = 1 - S_1.$$

Thus

$$P(v \geq 1) = 1 - P(v = 0) \leq 1 - (1 - S_1) = S_1.$$

This is another way to obtain the Boole's inequality. □

Besides reverting into the classical Bonferroni inequality and the Boole's inequality, the general forms (2.6) and (2.7) have the following refined versions, which are known as the Sobel-Uppuluri-Galambos bounds (Sobel and Uppuluri (1972), Galambos (1978), Recsei and Seneta (1987)).

Proposition 2.1.2 Sobel-Uppuluri-Galambos Inequalities

$$A_r^n(2u-1) + \frac{2u}{n-r}\binom{2u+r-1}{r-1}S_{2u+r}$$

$$\leq \quad P(v \geq r) \tag{2.8}$$

$$\leq \quad A_r^n(2u) - \frac{2u+1}{n-r}\binom{2u+r}{r-1}S_{2u+r+1},$$

and

$$B_r^n(2u-1) + \frac{2u}{n-r}\binom{2u+r}{r}S_{2u+r}$$

$$\leq \quad P(v = r) \tag{2.9}$$

$$\leq \quad B_r^n(2u) - \frac{2u+1}{n-r}\binom{2u+r+1}{r}S_{2u+r+1}.$$

Bounds (2.8) and (2.9) improve bounds (2.6) and (2.7), respectively, when additional information on the $2u+r+1$ degree of Bonferroni summation S_{2u+r+1} is available. For the relationship between these two types of bounds, as illustrated in Example 2.1.1, upper and lower bounds for $P(v \geq 1)$ can be obtained via lower and upper bounds for $P(v = 0)$ in the way of $P(v \geq 1) = 1 - P(v = 0)$. The following example shows that in some cases, the bound obtained from (2.9) may be sharper than the corresponding bound in (2.8).

Example 2.1.2. Consider the upper bound for $P(v = 0)$ in the case where $r = 0$ and $u = 1$ in bound (2.9). We have

$$P(v = 0) \quad \leq \quad B_0^n(2) - \frac{3}{n}S_3$$

$$= \quad 1 - S_1 + S_2 - \frac{3}{n}S_3. \tag{2.10}$$

Thus the degree-three upper bound of $P(v \geq 1)$ reads

$$P(v \geq 1) \quad = \quad 1 - P(v = 0)$$

$$\geq \quad 1-(1-S_1+S_2-\frac{3}{n}S_3)$$

$$= \quad S_1-S_2+\frac{3}{n}S_3. \tag{2.11}$$

Now, the corresponding degree-three lower bound of (2.8) at $r=1$ and $u=1$ reads

$$P(v \geq 1) \quad \geq \quad A_1^n(1)+\frac{2}{n-1}S_3$$

$$= \quad S_1-S_2+\frac{2}{n-1}S_3, \tag{2.12}$$

which is weaker than the lower bound (2.11) when $n \geq 3$. Notice that both upper and lower bounds use the information on S_1, S_2, and S_3 only. □

So far, we have discussed bounds of the linear combinations of the Bonferroni summations, which are within the thinking framework of Boole-Bonferroni following the logical stream of inclusion-and-exclusion. In what follows in this section, we introduce univariate bounds that go beyond the framework of Boole-Bonferroni by using non-linear combinations of Bonferroni summations.

2.1.2 Non-linear Combination Bounds

The result in this section is in the spirit of the early work of Renyi (1958), Galambos and Renyi (1968), Galambos (1969), and Hunter (1976) on general bounds using Borel functions of Bonferroni summations. Instead of using a linear combination of Bonferroni summations, this idea goes beyond the framework of conventional linear combination bounds to create upper or lower bounds for the probability of the union of a set of events (Dawson-Sankoff, 1967).

We start the discussion with a degree-three bound that improves the Sobel-Uppuluri-Galambos lower bound, and conclude the section with bounds related to Hunter's upper bound.

When S_1, S_2 and S_3 are used, the Sobel-Uppuluri-Galambos lower bound (2.11) uses the linear combination of the three Bonferroni summations:

$$P(\bigcup_{i=1}^{n} A_i) \geq S_1 - S_2 + \frac{3}{n}S_3. \tag{2.13}$$

In the case where $S_2 \geq S_1$ and S_3 is relatively small, the bound (2.13) becomes very weak (its value is close to zero or even negative). Seneta and Chen (2000) refine (2.13) by using a framework of non-linear structure. The multivariate version of the bound (2.13) and its improvements are addressed in detail in Seneta and Chen (2000).

For any set of events $A_1, ..., A_n$, and any positive integer k, let $K = \{t_1, ..., t_k\}$ with $1 \leq t_1 \leq ... \leq t_k \leq n$. Denote $F(\mathbf{x})$ a Borel function for $\mathbf{x} \in R^k$ such that $F(\mathbf{0}) = 0$. Let c be a constant satisfying

$$c = [\max_{t_1-1 \leq J \leq n} (F(\binom{J}{t}, t \in K))]^{-1}$$

if the maximum is not zero; and zero, otherwise, where

$$F(\binom{J}{t}, t \in K) = F(\binom{J}{t_1}, \binom{J}{t_2}, ..., \binom{J}{t_k}).$$

Proposition 2.1.3 Let t_1 be the smallest integer in the set K above. If the expected value $E[F(\binom{v}{t}), t \in K)]$ can be written as a function of the Bonferroni summation, $G(.)$:

$$G(S_t, t \in K) = E[F(\binom{v}{t}, t \in K)],$$

and $G(0) = 0$, we have a general bound

$$P(v \geq t_1) \geq cG(S_t, t \in K).$$

Proof: For any constant c, in order for $cG(S_t(A), t \in K)$ to be a lower bound of $P(v \geq t_1)$, it suffices to show that

$$cF(\binom{v}{t}, t \in K) \leq I(v \geq t_1). \tag{2.14}$$

To show (2.14), notice that for any integer v satisfying $v < t_1$, $I(v \geq t_1) = 0$ and $\binom{v}{t} = 0$ for any $t \in K$, where $K = \{t_1, ..., t_k\}$. Now, since $t_i \geq t_1 > v$, we have

$$F(\binom{v}{t}, t \in K) = F(0) = 0,$$

which means that (2.14) holds for $v < t_1$.

For the value of v that satisfies $v \geq t_1$, $I(v \geq t_1) = 1$ and

$$
\begin{aligned}
F(\binom{v}{t}, t \in K) &\leq \max_{t_1 \leq J \leq n} F(\binom{J}{t}, t \in K) \\
&\leq \max_{t_1 - 1 \leq J \leq n} F(\binom{J}{t}, t \in K) \\
&= \delta,
\end{aligned}
$$

where we denote

$$\delta = \max_{t_1 - 1 \leq J \leq n} F(\binom{J}{t}, t \in K),$$

so that

$$\delta \geq F(\binom{t_1 - 1}{t}, t \in K) = F(0) = 0. \tag{2.15}$$

When $\delta > 0$, letting $c = \delta^{-1}$ yields

$$cF(\binom{v}{t}, t \in K) \leq 1 = I(v \geq t_1);$$

when $\delta = 0$, letting $c = \delta = 0$ also yields this result. This completes the proof of Proposition 2.1.3.   □

Notice that the condition of $G(0) = 0$ in Proposition 2.1.2.1 can be removed by the following result.

Proposition 2.1.4: For any set of events A_1, ..., A_n, and any positive integer k, let $K = \{t_1, ..., t_k\}$ with $1 \le t_1 \le ... \le t_k \le n$. For any $\mathbf{x} \in R^k$, if a Borel function $F(\mathbf{x})$ satisfies $F(0) = 0$, $E[F(\binom{v}{t}), t \in K)] > 0$, and $E[F(\binom{v}{t}), t \in K)]$ can be expressed as a function of the Bonferroni summations:

$$G(S_t, t \in K) = E[F(\binom{v}{t}, t \in K)],$$

then the lower bound

$$P(v \ge t_1) \ge dG(S_t, t \in K),$$

is valid for the constant d and function $G(.)$, where

$$d = \begin{cases} [\max_{t_1 \le J \le n} F(\binom{J}{t}), t \in K)]^{-1}, & \text{if the maximum is not zero} \\ 0, & \text{otherwise.} \end{cases}$$

Proof: Proposition 2.1.4 follows by noticing that $G(S_t, t \in K) \ge 0$ implies the condition that $\max_{t_1 \le J \le n} F(\binom{J}{t}), t \in K) \ge 0$ in (2.15). If the maximum is negative, we have $F(.) < 0$ for any variable. Now, taking expectation of $E(F(\binom{v}{t}), t \in K))$ yields $G(S_t, t \in K) \le 0$, which is in contradiction with the assumption of $G(S_t, t \in K) = E[F(\binom{v}{t}), t \in K)] > 0$ as stated in the condition for Proposition 2.1.4.

Remark: Due to different ranges for maximization, the constant of adjustment d in Proposition 2.1.4 may be larger than the constant c in Proposition 2.1.3. Thus bounds obtained from Proposition 2.1.4 may be sharper than bounds obtained from Proposition 2.1.3. However, the condition that $G(S_t, t \in K) \ge 0$ in Proposition 2.1.4 is more restrictive.

The following example shows how to use Proposition 2.1.3 to obtain an explicit degree-three bound when $S_2 > S_1$, and S_3 is relatively small. This is a situation where the Sobel-Uppuluri-Galambos bound (2.13) becomes insensible.

Example 2.1.3. For any set of events A_1, ..., A_n,

$$P(\bigcup_{i=1}^{n} A_i) \ge \max_{k \ge 5}(\frac{6}{(k+1)(k-4)}(-S_1 + S_2 - \frac{2}{k}S_3)). \tag{2.16}$$

Proof: Consider a function $G(.)$ of the Bonferroni summations S_1, S_2 and S_3 of form

$$G(S_1, S_2, S_3) = -S_1 + S_2 - \frac{2}{k}S_3,$$

for any positive integer k. Denote

$$F(v) = -v + \frac{v(v-1)}{2} - \frac{2}{k}\frac{v(v-1)(v-2)}{6}.$$

Notice that for $3 \le v \le n$

$$
\begin{aligned}
F(v) - F(v-1) &= -1 + (v-1) - \frac{2}{k}\frac{(v-1)(v-2)}{2} \\
&= (v-2)(1 - \frac{v-1}{k}) \\
&> 0 \quad \text{if} \quad v < k+1; \\
&= 0 \quad \text{if} \quad v = k+1; \\
&< 0 \quad \text{if} \quad v > k+1.
\end{aligned}
$$

So, for $1 \le k \le n$,

$$
\begin{aligned}
\max_{3 \le v \le n} F(v) &= F(k) \\
&= (k+1)(k-4)/6;
\end{aligned}
$$

and if $k > n$

$$\max_{3 \le v \le n} F(v) = F(n) < (k+1)(k-4)/6.$$

Thus, by Proposition 2.1.3, the new lower bound for $5 \le k$ using S_1, S_2 and S_3 is

$$P(v \ge 1) \ge \max_{k \ge 5} \frac{6}{(k+1)(k-4)}(-S_1 + S_2 - \frac{2}{k}S_3),$$

which is the lower bound (2.16). □

The following numerical example shows that there exists a probability space where the non-linear combination bound (2.16) is better than the linear combination bound (2.13). The example contains a technique specifying a set of events in a probability space by specifying the probabilities of the elementary conjunctions of the events.

Example 2.1.4 Consider a set of seven events A_1, ..., A_7 in a probability space where all non-zero probabilities of elementary conjunctions of intersections involving A_1, ..., A_7 and their complements are given as follows.

$$P(A_1 A_2 A_3 A_4 A_5 A_6 A_7^c) = 0.05 \quad P(A_1 A_2 A_3^c A_4 A_5^c A_6^c A_7) = 0.1$$
$$P(A_1^c A_2 A_3 A_4 A_5 A_6 A_4) = 0.05 \quad P(A_1 A_2 A_3 A_4 A_5^c A_6^c A_7) = 0.05$$
$$P(A_1 A_2 A_3 A_4 A_5^c A_6^c A_7^c) = 0.1 \quad P(A_1 A_2 A_3 A_4 A_5 A_6^c A_7^c) = 0.15$$
$$P(A_1 A_2 A_3 A_4^c A_5 A_6^c A_7^c) = 0.1 \quad P(A_1 A_2 A_3 A_4^c A_5 A_6^c A_7) = 0.05$$

$$P(A_1A_2A_3^cA_4A_5A_6^cA_7^c) = 0.1 \quad P(A_1^cA_2A_3A_4A_5A_6^cA_7) = 0.01$$

$$P(A_1A_2A_3^cA_4^cA_5A_6A_7^c) = 0.1 \quad P(A_1^cA_2^cA_3A_4A_5A_6A_7) = 0.04$$

$$P(A_1^cA_2^cA_3^cA_4^cA_5^c) = 0.1.$$

Thus the first three Bonferroni summations for the events specified above are

$$S_1 = 4.1 \quad S_2 = 7.5 \quad S_3 = 7$$

and

$$P(\bigcup_{i=1}^{7} A_i) = 0.9.$$

Under this scenario, the Sobel-Uppuluri-Galambos bound (2.13) reads (Sobel and Uppuluri, 1972),

$$P(\bigcup_{i=1}^{7} A_i) \geq S_1 - S_2 + \frac{3}{7}S_3 = -0.4.$$

which is not a meaningful lower bound for the probability of the union of A_1, ..., A_7 because any probability is non-negative. However, the bound (2.16) offers a lower bound as

$$P(\bigcup_{i=1}^{7} A_i) \geq \max_{k \geq 5} (\frac{6}{(k+1)(k-4)}(-S_1 + S_2 - \frac{2}{k}S_3)) = -S_1 + S_2 - \frac{2}{5}S_3 = 0.6,$$

which is, at the very least, a meaningful lower bound.   □

In Example 2.2.7, we will discuss the optimality of the lower bound obtained in Example 2.1.3 for $P(v \geq t_1)$.

The above discussion focuses on the *lower* bound with forms of non-linear combinations of Bonferroni summations. To further delineate the issue of inequality refinements and boundary improvements, we now switch to the discussion on *upper* bounds with non-linear combination structures.

The most well-known improvement of the Bonferroni-type upper bound is the Hunter (1976)'s bound.

$$P(\bigcup_{i=1}^{n} A_i) \leq S_1 - \sum_{(i,j) \in \tau^*} P(A_i \cap A_j), \qquad (2.17)$$

where τ^* is the spanning tree of a graph with each A_i as a vertex and $P(A_i \cap A_j)$ as the weight of the edge connecting A_i and A_j, which maximizes the right-hand side of the inequality.

When Hunter's upper bound is translated (Margaritescu 1986, 1987; Seneta 1988) from its original graph-theoretic setting, it states

$$P(v \geq 1) \leq S_1 - \max_{\pi} \sum_{i=2}^{n} \max_{1 \leq s \leq i-1} P(A_i \cap A_s), \qquad (2.18)$$

where π is the set of all permutations of $A_1, ..., A_n$.

The structure of inequality (2.18) resembles the structure of a degree-two Bonferroni lower inequality

$$P(v \geq 1) \geq S_1 - S_2,$$

since it involves the probabilities of intersections for at most two events. However, it does not involve the complete summation of S_2. For this reason, the inequality (2.18) is named a hybrid bound to reflect the fact that it involves the S_i's in conjunction with other information (non-Bonferroni summation) in providing a degree-two upper bound (see, for example, Hoppe and Seneta (1990), Hoppe (1985)).

The upper bounds (2.17) and (2.18) can be extended to $P(v = r)$ and $P(v \geq r)$ for any integer r by noticing the following two identities.

$$P(v = r) = S_r - \sum_{1 \leq i_1 < ... < i_r \leq n} P(\bigcup_k A_k A_{i_1} \ldots A_{i_r}), \qquad (2.19)$$

where the union is over all $k \notin \{i_1 \ldots i_r\}$ and $S_r = \sum_{1 \leq i_1 < ... < i_r \leq n} P(A_{i_1} \ldots A_{i_r})$.

To see the identity (2.19), notice that

$$
\begin{aligned}
P(v = r) &= P(\text{Exactly r of the events occur}) \\
&= \sum_{1 \leq i_1 \leq ... \leq i_r \leq n} P(A_{i_1} \cap \ldots \cap A_{i_r} \cap (\bigcap_{k \notin \{i_1,...,i_r\}} A_k^c)) \\
&= \sum_{1 \leq i_1 \leq ... \leq i_r \leq n} [P(A_{i_1} \cap \ldots \cap A_{i_r}) - P(\bigcup_{k \notin \{i_1,...,i_r\}} A_k A_{i_1} \ldots A_{i_r})] \\
&= S_1 - \sum_{1 \leq i_1 \leq ... \leq i_r \leq n} P(\bigcup_{k \notin \{i_1,...,i_r\}} A_k A_{i_1} \ldots A_{i_r}).
\end{aligned}
$$

For the term $P(v \geq r)$, the corresponding identity becomes

$$P(v \geq r) = S_r - \sum_{1 \leq i_1 < ... < i_r \leq n} P(\bigcup_k A_k A_{i_1} \ldots A_{i_r}), \qquad (2.20)$$

where the union is over all integers $k \notin \{i_1 \ldots i_r\}$ and $k \leq i_r - 1$.

By bounding the extreme right-hand terms in the identities (2.19) and (2.20), respectively, Hoppe and Seneta (1990) produce a "hybrid" degree $r + 1$ upper bound from each of these identities for the probabilities $P(v = r)$ and $P(v \geq r)$, respectively.

Now we can compare the above two hybrid upper bounds with the Bonferroni-type bounds discussed in Section 2.1. In the case where $r = 1$, the identity (2.20) leads to (2.18). For general r, the upper hybrid bound derived from (2.20) is sharper than the following Bonferroni-type bound

$$P(v \geq r) \leq S_r - (\frac{r+1}{n-r} - \binom{n}{r+1}^{-1}) S_{r+1}, \qquad (2.21)$$

see, for example, Margaritescu(1988, 1989) or Kwerel (1975a).

The multivariate version of Margaritescu's bound will be discussed in Chapter 6, along with the multivariate version of the hybrid-type upper bound given in Hoppe and Seneta (1990). Recent research results of non-linear combination bounds include refined lower bounds of Chen (2003) and Hoppe (2009).

As showed in Margaritescu(1988), the bound (2.21) is optimal among bounds of form $aS_r + cS_{r+1}$ for any constants a and b which depend on the total number of events n. The fact that this optimal bound (2.21) can be improved by an upper bound derived from (2.20) partly stimulates a question on the definition of optimality.

While it is true that the non-linear bounds reach a more accurate evaluation for the probability of the union of a set of events, at the same time, bounds with linear combinations of the Bonferroni summations actually pertain to a subspace of bounds taking non-linear combinations of the Bonferroni summations. Under this scenario, we need to clearly define the optimality of probability bounds. A discernible feature of an optimal bound is that it can not be improved once the optimality bounding condition is satisfied.

2.2   Univariate Optimality

As discussed in Section 2.1, bounds with linear combinations of the Bonferroni summations S_i's are endowed with symmetric structure and mathematical elegance. However, these bounds are generally too conservative, especially for scenarios where the joint probabilities among some events are relatively higher/lower than those of other events. As illustrated in Example 2.1.4, the improvement of non-linear combination bounds over linear combination bounds exemplifies the issue on the extent of bound improvement, and conditions of optimal bounds. The identification of an optimal bound terminates the process of refinement because being optimal implies that there is no space for uniform improvement under the optimality conditions (see, for example, Fréchet (1940, 1943), Galambos and Xu (1993), Hoppe and Nediak (2008)).

For a Bonferroni-type bound, the bound is essentially a function of Bonferroni summations. The starting point in the definition of optimality is the specification of the domain where optimality is claimed. As discussed in Section 2.1, we can always improve a bound with more information on the Bonferroni summations. Thus the specification for the domain of optimization is critical in inequality optimizing theory.

To legitimately formulate a probability bound as a function of Bonferroni summations, we introduce the concept of consistency for any given set of Bonferroni summations.

- *consistent Bonferroni summations*: For a set of t values $s_1, ..., s_t$, if there exists a probability space and a set of events such that the corresponding Bonferroni summations equal the given values. $S_i = s_i$ for $i = 1, ..., t$, such a set of given s_i values, $\{s_1, ..., s_t\}$, is called a set of *consistent* values of Bonferroni summations.

To understand the concept of consistency of Bonferroni summations, consider that in the case where five events $A_1, ..., A_5$ are of interest, there is no event satisfying $S_1 = 4$ and $S_2 = 10$ because when $n = 5$, the first and second Bonferroni summations need to satisfy the following relationship:

$$S_2 = E\left(\binom{v}{2}\right) \le E\left(\frac{v(v-1)}{2}\right) \le E\left(\frac{v(5-1)}{2}\right) = 2S_1.$$

Recall that a Bonferroni summation S_i is also a binomial moment of the occurrence of a set of events, so we start the definition of optimality with the bounding domain restricted to a set of consistent Bonferroni summations.

- *Functional formulation of probability bounds*: Let n and r be two fixed integers, $1 \le r \le n$. Denote T a subset of the set $\{0, 1, ..., n\}$. Let L be a class of functions, $G(.)$, defined on $\{S_i, i \in T\}$ such that for any collection of events $A_1, ..., A_n$ in any probability space

$$P(v \ge r) \ge G(S_i, i \in T)$$

where v is the number of the events $A_1, ..., A_n$ which occur at a sample point. The domain of G consists of the set of consistent values of S_i ($i \in T$) that correspond to a set of events in a probability space.

Definition 2.2.1 (Fréchet Optimality): For a class of upper bounds, L, suppose $G_1 \in L$ is an upper bound for $P(v \ge r)$ such that for any particular set of consistent Bonferroni summations S_i's where $i \in T$, there exists a set of events $A_1^*, A_2^*, ...,$ A_n^* (in a probability space) such that

$$S_i(A_1^*, ..., A_n^*) = S_i \quad i \in T$$

and

$$P(v(A^*) \ge r) = G_1(S_i, i \in T),$$

where $v(A^*)$ is the number of occurrences of the set of events $\{A_i, i = 1, ..., n\}$, then G_1 is said to be a Fréchet optimal upper bound for $P(v \ge r)$. □

A parallel definition of the Fréchet optimal lower bound is self-evident.

Since Definition 2.2.1 starts with and depends on the existence of a set of events that make the inequality achieve the boundary, it is necessary to classify the group of bounds achievable for a set of consistent Bonferroni summations.

- *Achievable Bounding Class*: For any given set of consistent Bonferroni summations $\{S_i, i \in T\}$, if there exists a bound $G^*(.)$ in a class of functions L so that $G^*(S_i, i \in T) = P(v \ge r)$ for a set of events in a probability space, then the class of bounds L is termed an *achievable bounding class* for the given set of consistent Bonferroni summations.

The following example illustrates the concept of an achievable bounding class.
Example 2.2.1: Consider the case where $T = \{1, 2\}$ and $\{S_t, t \in T\} = \{S_1, S_2\}$. Denote the class of upper bounds for $P(v \ge 1)$: $L = \{G : G(S_1, S_2) = aS_1 + bS_2\}$ for any constant a and b. In Section 2.1, we know

$$P(v = 0) \ge B_0^n(1) + \frac{2}{n}S_2 = 1 - S_1 + \frac{2}{n}S_2,$$

by taking $r = 0$ and $u = 1$ in (2.9). Thus

$$P(v \geq 1) = 1 - P(v = 0) \geq S_1 - \frac{2}{n}S_2.$$

Now consider the upper bound $G_1 = S_1 - \frac{2}{n}S_2$. When $n = 5$, $S_1 = 0.8$ and $S_2 = 1$, we have

$$S_1 - \frac{2}{5}S_2 = 0.4,$$

which is achievable for $\{S_1, S_2\} = \{0.8, 1\}$. To see this point, consider the set of events $A_1^*, \ldots, A_5^*$ with all non-zero probabilities of elementary conjunctions specified as follows:

$$P(A_1^* \cap \ldots \cap A_5^*) = 0.1$$

$$P(A_1^* \cap (A_2^*)^c \ldots \cap (A_5^*)^c) = 0.3$$

$$P((A_1^*)^c \cap \ldots \cap (A_5^*)^c) = 0.6$$

In this way, $P(A_1^*) = 0.4$ and $P(A_i^*) = 0.1$ for $i = 2, 3, 4, 5$. Thus we have

$$S_1(A^*) = 0.4 + 0.1 + 0.1 + 0.1 + 0.1 = 0.8$$

$$S_2(A^*) = \sum_{1 \leq i < j \leq 5} P(A_i A_j) = 10 \times 0.1 = 1,$$

and

$$P(v(A^*) \geq 1) = 1 - P(v(A^*) = 0) = 1 - P((A_1^*)^c \cap \ldots \cap (A_5^*)^c) = 0.4,$$

which is the same value as the upper bound $S_1 - \frac{2}{5}S_2$. Thus the upper bound $G_1(S_1, S_2) = S_1 - \frac{2}{5}S_2$ is achievable for the consistent Bonferroni summations $\{S_1, S_2\} = \{0.8, 1\}$.

On the other hand, consider a set of five events $A_1 = A_2 = \ldots = A_5$ with $A_1 = A_2 = A_3 = \Omega$ (the whole sample space), and $A_4 = A_5 = \emptyset$. Here, $S_1 = 3$ and $S_2 = 3$, and we have a set of consistent Bonferroni summations $\{S_1, S_2\} = \{3, 3\}$, and the upper bound $G_1(S_1, S_2) = S_1 - \frac{2}{5}S_2 = 1.8 > 1$, which is unachievable for any events, thus the upper bound $G_1(S_1, S_2) = S_1 - \frac{2}{5}S_2$ is not achievable for the consistent Bonferroni summations $\{3, 3\}$. □

Obviously, for any set of consistent Bonferroni summations $\{S_i, i \in T\}$, when the optimal upper bound G^* is achievable for this set of Bonferroni summations, we have

$$G^*(S_i, i \in T) = P(v(A^*) \geq r) \leq G(S_i, i \in T)$$

for any upper bound $G \in L$. In this sense, the optimality of G^* in the set of bounds L is apparent (the value of an optimal upper bound is lower than that of any upper bound).

In Example 2.2.1, the upper bound

$$G_1(S_1, S_2) = S_1 - \frac{2}{5}S_2$$

is achievable for some set of consistent Bonferroni summations and not achievable for others. As pointed out by Kwerel (1975a) (also Prékopa (1988), Kounias and Marin (1976)), a Fréchet optimal *upper* bound for $P(v_1 \geq 1)$ in terms of S_1 and S_2 is the adjusted degree-two Sobel-Uppuluri upper bound:

$$P(v \geq 1) \leq \min(S_1 - \frac{2}{n}S_2, 1), \tag{2.22}$$

which is achievable for any S_1 and S_2 uniformly after being truncated (adjusted) for the concept of Fréchet optimality.

More discussion on the Fréchet optimality for multivariate bounds can be found in Section 2.4 and Chapter 4. The definition of Fréchet optimality focuses on any set of achievable bounds. We now use some examples to elucidate the concept of a Fréchet optimality.

Example 2.2.2   Deriving Dawson-Sankoff bound

Before verifying the Fréchet optimality, we specify the set of bounds L using two univariate Bonferroni summations S_1 and S_2. According to Proposition 2.1.3, we have the following.

For any set of events $A_1, ..., A_n$,

$$P(\bigcup_{i=1}^{n} A_i) \geq \frac{2}{k+1}(S_1 - \frac{1}{k}S_2), \tag{2.23}$$

for any integer k, $1 \leq k$. This bound (2.23) is then maximized over k at $k = k^* = [\frac{2S_2}{S_1}] + 1$, where $1 \leq k^* \leq n$.

Proof: Besides the original proof of Dawson-Sankoff and a different proof given in Galambos (1977), we present two different methods of derivation for this bound.
Method 1: Using Proposition 2.1.3.

Consider $F(v) = k\binom{v}{1} - \binom{v}{2}$, $0 \leq v \leq n$, for positive integer $k \geq 1$. This gives the function of S_1, S_2 in Proposition 2.1.3:

$$G(S_t, t \in K) = kS_1 - S_2.$$

Now, considering the maximum value of the function $F(v)$ over $1 \leq v \leq n$, we have

$$\begin{aligned} F(v) - F(v-1) &= kv - v(v-1)/2 - [k(v-1) - (v-1)(v-2)/2] \\ &= k - (v-1). \end{aligned}$$

So,

$$\begin{aligned} F(v) - F(v-1) \quad &> \quad 0 \quad \text{if} \quad v < k+1 \\ &= \quad 0 \quad \text{if} \quad v = k+1 \\ &< \quad 0 \quad \text{if} \quad v > k+1. \end{aligned}$$

Therefore, if $k \leq n$, and

$$\max_{0 \leq v \leq n} F(v) \quad = \quad F(k)$$

$$= \quad k \binom{k}{1} - \binom{k}{2}$$

$$= \quad k(k+1)/2.$$

If $k > n$,

$$\max_{0 \leq v \leq n} F(v) = F(n) < k(k+1)/2.$$

Now by Proposition 2.1.3, $cG(S_t, t \in K)$ is a lower bound, with $c = (k(k+1)/2)^{-1}$. The lower bound thus reads

$$cG(S_t, t \in K) \quad = \quad (k(k+1)/2)^{-1}(kS_1 - S_2)$$

$$= \quad \frac{2}{k+1}(S_1 - S_2/k),$$

which is the lower bound (2.23).

Method 2: Use the binomial moments of the Bonferroni summations.

Alternatively, the lower bound (2.23) can be derived as follows (this derivation is courtesy of a student, Mr. Yang Liu, in a graduate course of probability inequalities at Bowling Green State University).

Consider the fact that

$$S_1 = E(v) \quad S_2 = E\left(\binom{v}{2}\right),$$

denote $p_1 = P(v = i)$, we have

$$\frac{2}{k+1}S_1 - \frac{2}{k(k+1)}S_2$$

$$= \quad \sum_{i=1}^{n} \left[\frac{2}{k+1}\left(v - \frac{1}{k}\binom{v}{2}\right)\right]p_i$$

$$= \quad \sum_{i=1}^{n} \left[\frac{v(2k - v + 1)}{k(k+1)}\right]p_i$$

$$\leq \quad \sum_{i=1}^{n} p_i$$

$$= \quad P(v \geq 1).$$

The inequality above is due to the fact that the maximum value of the function $f(i) = i(2k - i + 1)$ reaches its maximum value at $k(k+1)$ when i takes the integer in the set $\{1, 2, ..., n\}$, in conjunction with the following relation,

$$\left|i - \frac{2k+1}{2}\right| = \left|i - k - \frac{1}{2}\right| \geq \frac{1}{2},$$

for any integers i and k.   □

Since $2S_2 \leq (n-1)S_1$, it follows that $k^* \leq n$. In the next example, we will show that the Dawson-Sankoff lower bound is achievable at $k^* = [\frac{2S_2}{S_1}] + 1$, which resembles the Fréchet optimality (see, for example, Galambos and Simonelli (1996)).

Example 2.2.3   Optimal point of the Dawson-Sankoff lower bound

With the same setting of Example 2.2.2, the optimal point of the Dawson-Sankoff bound locates at $k^0 = [\frac{2S_2}{S_1}] + 1$ and $k^0 \leq n$.

To see the optimal point is $k^0 = [\frac{2S_2}{S_1}] + 1$, denote

$$f(k) = \frac{2}{k+1}S_1 - \frac{2}{k(k+1)}S_2.$$

We have

$$
\begin{aligned}
&f(k) - f(k+1) \\
&= \frac{2}{k+1}S_1 - \frac{2}{k(k+1)}S_2 - [\frac{2}{k+2}S_1 - \frac{2}{(k+2)(k+1)}S_2] \\
&= \frac{2}{(k+1)(k+2)}S_1 - \frac{4}{k(k+1)(k+2)}S_2 \\
&= \frac{2}{(k+1)(k+2)}[S_1 - \frac{2}{k}S_2].
\end{aligned}
$$

Thus, when

$$k > k^0 = [\frac{2S_2}{S_1}] + 1,$$

$$k > \frac{2S_2}{S_1},$$

and

$$S_1 - \frac{2}{k}S_2 > 0,$$

which implies that $f(k) \geq f(k+1)$. Thus $f(k^0) \geq f(j)$ for all $j > k^0$.

When

$$k < k^0 = [\frac{2S_2}{S_1}] + 1,$$

$$k \leq \frac{2S_2}{S_1},$$

and

$$S_1 - \frac{2}{k}S_2 \leq 0,$$

which implies that $f(k) \leq f(k+1)$. Thus $f(k^0) \geq f(j)$ for all $j < k^0$.

To see $k^0 \leq n$, recall the property of binomial moment of the Bonferroni summation ,

$$S_i = E[\binom{v}{i}] = \sum_{k=1}^{n} \binom{k}{i} P(v = k).$$

Thus

$$2S_2 = 2 \sum_{k=1}^{n} \frac{k(k-1)}{2} P(v=k) \leq (n-1) \sum_{k=1}^{n} k P(v=k) = (n-1)S_1.$$

Therefore, noting $[\frac{2S_2}{S_1}] \leq \frac{2S_2}{S_1}$ yields

$$k^0 = [\frac{2S_2}{S_1}] + 1 \leq n.$$

This completes Example 2.2.3. □

With Examples 2.2.2 and 2.2.3, we can further discuss the Fréchet optimality property of the Dawson-Sankoff bound as follows.

Example 2.2.4   Fréchet optimal property of the Dawson-Sankoff lower bound

For a given probability space $(\Omega, \mathcal{F}, P)$, and a sequence of events $\{A_1, \ldots A_n\}$ in this space,

$$P(v \geq 1) \geq \frac{2}{k+1} S_1 - \frac{2}{k(k+1)} S_2$$

where $k \geq 1$ is an integer, S_1 and S_2 are the Bonferroni summations.

Consider

$$S_1 = \sum_{i=1}^{n} i p_i \qquad S_2 = \sum_{i=2}^{n} \binom{i}{2} p_i,$$

where $p_i = P(V = i)$ is the probability that exactly v occurrences of the event set $\{A_1, \ldots, A_n\}$.

For any integer k satisfying $1 \leq k \leq n-1$, setting $p_1 = \ldots = p_{k-1} = 0$ and $p_{k+1} = \ldots = p_n = 0$ gets

$$S_1 = kP_k + (k+1)p_{k+1}$$

and

$$S_2 = \binom{k}{2} p_k + \binom{k+1}{2} p_{k+1}.$$

This results in the solution of

$$p_k = S_1 - \frac{2}{k} S_2$$

$$p_{k+1} = -\frac{k-1}{k+1} S_1 + \frac{2}{k+1} S_2,$$

for any $k = 1, \ldots, n-1$.

Now, taking $k = k^* = [\frac{2S_2}{S_1}] + 1$ yields

$$0 \leq p_k \leq 1 \qquad 0 \leq p_{k+1} \leq 1,$$

and by Example 2.2.2,

$$p_k + p_{k+1} \leq P(\bigcup_{i=1}^{n} A_i) \leq 1.$$

The above conditions ensure that there exist two events C and D in the sample space such that $P(C) = p_{k^*}$, and $P(D) = p_{k^*+1}$ and $P(C \cap D) = 0$. Now let

$$A_1^* = C \bigcup D \quad \dots \quad A_{k^*}^* = C \bigcup D \quad A_{k^*+1}^* = D$$

$$A_{k^*+2}^* = \emptyset \quad \dots \quad A_n^* = \emptyset,$$

for the set of events $\{A_i^*, i = 1, \dots, n\}$, all non-zero probabilities in $p_i(A^*) = P(\nu(A^*) = i)$ for $i = 1, \dots, n$, are

$$p_{k^*}(A^*) = P(C) = p_{k^*} \quad p_{k^*+1}(A^*) = P(D) = p_{k^*+1}.$$

Thus

$$S_1(A^*) = S_1 \quad S_2(A^*) = S_2$$

and

$$P(\bigcup_{i=1}^{n} A_i^*) = \frac{2}{k^*+1} S_1 - \frac{2}{k^*(k^*+1)} S_2.$$

Summarizing the above discussions, we have that for any given consistent Bonferroni summations S_1 and S_2, we can choose a probability space and a sequence of events $A_1', \dots, A_n'$, and $k^0 = [\frac{2S_2}{S_1}] + 1$, such that

$$P(\cup_i A_i') = \frac{2}{k^0+1} S_1 - \frac{2}{k^0(k^0+1)} S_2.$$

According to Definition 2.2.1, the Dawson-Sankoff lower bound

$$\frac{2}{k^0+1} S_1 + \frac{2}{k^0(k^0+1)} S_2$$

is a Fréchet optimal lower bound, taking $r = 1$, and $T = \{1,2\}$ in the definition.
□

It should be noted that the Fréchet optimality of the Dawson-Sankoff lower bound depends on the availability of the marginal information S_1 (summation of individual probabilities) and S_2 (summation of joint probabilities of any two events). If further information is available such as S_3 (as shown in the next example), the Dawson-Sankoff bound can be improved even though it is optimal for the bounding domain on S_1 and S_2.

Example 2.2.5   Optimizing a degree-three lower bound
For $n \geq 3$ and any set of events $A_1, \dots, A_n$, and for any positive integer $1 \leq k \leq n-1$,

$$P(\bigcup_{i=1}^{n} A_i) \geq (S_1 - S_2/k + \frac{2S_3}{k(n-1)(n-2)}) \frac{1}{k+1} (\frac{1}{2} + \frac{k-1}{3(n-1)(n-2)})^{-1}, \quad (2.24)$$

and for $k = n$ the lower bound is the same as for $k = n - 1$. The optimal value of the bound (2.24) (for $n \geq 3$, S_1, S_2, and S_3) is achieved at an integer k located in the set

$$\{[\frac{S_2 - S_3/\binom{n-1}{2}}{S_1}] + 1, \ [\frac{S_2 - S_3/\binom{n-1}{2}}{S_1}] + 2, \ ..., \ [\frac{2(S_2 - S_3/\binom{n-1}{2})}{S_1}] + 1\}.$$

Proof: Recall from the definition of Bonferroni summation that $2S_2 \leq (n - 1)S_1$. By Example 2.2.3, the Dawson-Sankoff degree-two optimizing value $k^* = [\frac{2S_2}{S_1}] + 1$ satisfies $1 \leq k^* \leq n$. Now, when the additional information for the probabilities of the intersections of any three events is available, for $n \geq 3$, notice that

$$\frac{S_2 - S_3/\binom{n-1}{2}}{S_1} > 0,$$

because

$$S_3 = E(\binom{v}{3}) < \frac{(n-2)}{3} E(\binom{v}{2}) \leq \binom{n-1}{2} E(\binom{v}{2}) = \binom{n-1}{2} S_2.$$

Now, consider the bound of the following type

$$G(S_1, S_2, S_3) = S_1 - S_2/k + \frac{2S_3}{k(n-1)(n-2)}.$$

The corresponding $F(.)$ function in Proposition 2.1.3 becomes

$$F(v) = v - \frac{v(v-1)}{2k} + \frac{v(v-1)(v-2)}{3k(n-1)(n-2)}.$$

Therefore

$$F(v) - F(v-1) = 1 - \frac{v-1}{k} + \frac{(v-1)(v-2)}{k(n-1)(n-2)},$$

and

$$F(v) - F(v-1) > 0 \quad \text{if} \quad 2 \leq v \leq k+1;$$

$$F(v) - F(v-1) \leq -\frac{1}{k}(1 - \frac{(v-1)(v-2)}{(n-1)(n-2)})$$

$$\leq 0 \quad \text{if} \quad n \geq v \geq k+2.$$

Summarizing the above discussion yields

$$\max_{3 \leq v \leq n} F(v) = F(k+1) \quad \text{when} \quad k+1 \leq n.$$

Since

$$F(k+1) = (k+1)(\frac{1}{2} + \frac{k-1}{3(n-1)(n-2)})$$

when $k+1 \le n$, by Proposition 2.1.3, we get the lower bound (2.24). If $k+1 > n$,

$$\max_{3 \le v \le n} F(v) = F(n).$$

Now, for the optimization of the bound (2.24), consider

$$g(k) = (S_1 - S_2/k + \frac{2S_3}{k(n-1)(n-2)}) \frac{1}{k+1} h(k)$$

where

$$h(k) = (\frac{1}{2} + \frac{k-1}{3(n-1)(n-2)})^{-1}. \qquad (2.25)$$

Consider the difference

$$
\begin{aligned}
& g(k) - g(k+1) \\
= \ & S_1 \{ \frac{h(k)}{k+1} - \frac{h(k+1)}{k+2} \} - \\
& -(S_2 - \frac{S_3}{\binom{n-1}{2}}) \{ \frac{h(k)}{k(k+1)} - \frac{h(k+1)}{(k+1)(k+2)} \}
\end{aligned}
\qquad (2.26)
$$

Now, from (2.25):

$$\{ \frac{h(k)}{k+1} - \frac{h(k+1)}{k+2} \} > 0, \qquad (2.27)$$

since $S_2 - \frac{S_3}{\binom{n-1}{2}} > 0$, we have

$$
\begin{aligned}
& g(k) - g(k+1) \\
\le \ & S_1 \{ \frac{h(k)}{k+1} - \frac{h(k+1)}{k+2} \} - (S_2 - S_3 / \binom{n-1}{2}) \{ \frac{h(k)}{k(k+1)} - \frac{h(k+1)}{k(k+2)} \} \\
= \ & \{ S_1 - \frac{1}{k}(S_2 - S_3 / \binom{n-1}{2}) \} \{ \frac{h(k)}{k+1} - \frac{h(k+1)}{k+2} \} \},
\end{aligned}
$$

thus if

$$\{ S_1 - \frac{1}{k}(S_2 - S_3 / \binom{n-1}{2}) \} \le 0,$$

we have $g(k) - g(k+1) \le 0$. This argument provides the stated lower bound for an optimizing value of k.

Next, when

$$k \ge [\frac{2(S_2 - S_3 / \binom{n-1}{2})}{S_1}] + 1 \ge \frac{2(S_2 - S_3 / \binom{n-1}{2})}{S_1},$$

it follows that

$$-(S_2 - S_3 / \binom{n-1}{2}) \ge -kS_1/2.$$

From (2.26), we get

$$
\begin{aligned}
g(k) - g(k+1) \;\geq\; & S_1\{\frac{h(k)}{k+1} - \frac{h(k+1)}{k+2}\} - \frac{kS_1}{2}\{\frac{h(k)}{k(k+1)} - \frac{h(k+1)}{(k+1)(k+2)}\} \\
=\; & S_1\{\frac{h(k)}{2(k+1)} - h(k+1)\{\frac{1}{k+2} - \frac{k}{2(k+1)(k+2)}\}\} \\
=\; & \frac{S_1}{2(k+1)}\{h(k) - h(k+1)\} \\
>\; & 0.
\end{aligned}
$$

So the function $g(k)$ decreases for

$$
k \geq \frac{2[S_2 - ((\binom{n-1}{2})^{-1})S_3]}{S_1} + 1,
$$

Thus the optimal point is located in the set

$$
\{[\frac{S_2 - S_3/\binom{n-1}{2}}{S_1}] + 1, \; [\frac{S_2 - S_3/\binom{n-1}{2}}{S_1}] + 2, \; ..., \; [\frac{2(S_2 - S_3/\binom{n-1}{2}))}{S_1}] + 1\}.
$$

This completes the proof of Example 2.2.5.   □

From the range of optimality domain in the lower bound (2.24) and the expression for the optimizing value ($k^* = [\frac{2S_2}{S_1}] + 1$) for the degree-two Dawson-Sankoff bound, it follows that the degree-two optimal point k^* is within the range of the degree-three optimal range. The upper limit of the degree-three optimal range is no greater than n. We now present a numerical example to examine the performance of the new bound (2.24) and the Dawson-Sankoff bound at their optimal value k^*. The example shows that the degree-two optimal value k^* for Dawson-Sankoff lower bound, can also be an optimizing value for the refined degree 3 bound (2.24).

Example 2.2.6   Numerical comparisons of optimal bounds.

Consider a set of events A_1, ..., A_6 in a probability space where all non-zero probabilities of elementary conjunctions of intersections involving A_1, ..., A_6 and their complements are specified as follows.

$$P(A_1^c A_2 A_3 A_4 A_5 A_6) = 0.05 \quad P(A_1^c A_2^c A_3^c A_4^c A_5 A_6^c) = 0.1$$

$$P(A_1^c A_2 A_3^c A_4^c A_5^c A_6^c) = 0.2 \quad P(A_1^c A_2^c A_3^c A_4 A_5^c A_6^c) = 0.1$$

$$P(A_1^c A_2^c A_3 A_4^c A_5^c A_6^c) = 0.05 \quad P(A_1 A_2 A_3 A_4 A_5 A_6^c) = 0.05$$

$$P(A_1^c A_2^c A_3^c A_4^c A_5^c A_6) = 0.05 \quad P(A_1 A_2^c A_3^c A_4^c A_5^c A_6^c) = 0.1$$

So, $S_1 = 1.1$, $S_2 = 1$ and $S_3 = 1$. In this case, the Dawson-Sankoff bound reads

$$P(\bigcup_{i=1}^{6} A_i) \geq \frac{2}{3}(S_1 - S_2/2) = 0.4,$$

and

$$P(\bigcup_{i=1}^{6} A_i) = 0.7,$$

but the new bound (2.24) reads, at the Dawson-Sankoff optimizing value $k^* = 2$,

$$0.7 = P(\bigcup_{i=1}^{6} A_i) \geq (S_1 - S_2/2 + S_3/20) \times (\frac{1}{2} + 1/60)^{-1}/3 = 0.4194,$$

which is higher than the value reached by the Dawson-Sankoff lower bound. The domain containing an optimizing value specified by Example 2.2.5 is $\{1,2\}$ in this case, and 2 is the optimizing-value. However here $S_1 > S_2$, and applying (2.24) we obtain the numerical lower bound 0.6, which is closer still to the true value of 0.7. Clearly the bound (2.24) makes better use, in this instance, of the value of S_3.   □

It should be noted that the utilization of additional term S_3 does not guarantee the sharpness of the degree-three lower bound. As illustrated in the following example, the optimal value of a degree-three lower bound can be weaker than the Dawson-Sankoff lower bound for a set of consistent Bonferroni summations.

Example 2.2.7   Degree-two and degree-three lower bounds

Now consider the optimization of the bound (2.16) over k (in Example 2.1.3). Write

$$f(k) = \frac{6}{(k+1)(k-4)}(-S_1 + S_2 - \frac{2}{k}S_3),$$

we have

$$f(k) - f(k-1)$$

$$= \frac{6}{(k+1)(k-4)}(-S_1 + S_2 - \frac{2}{k}S_3) - \frac{6}{k(k-5)}(-S_1 + S_2 - \frac{2}{k-1}S_3)$$

$$= 6(S_2 - S_1)(\frac{1}{(k+1)(k-4)} - \frac{1}{k(k-5)}) - 12S_3[\frac{1}{k(k+1)(k-4)}$$

$$- \frac{1}{k(k-1)(k-5)}]$$

$$= \frac{12}{k(k+1)(k-4)(k-5)}[-(k-2)(S_2 - S_1) + \frac{3(k-3)S_3}{k-1}].   \qquad (2.28)$$

Now let

$$k^* = [\frac{3S_3}{S_2 - S_1}] - 1.$$

For $k \geq 5$, $\frac{(k-2)(k-1)}{k-3} \leq k+1$. If $k \leq k^*$,

$$\frac{(k-2)(k-1)}{k-3} \leq k+1 \leq k^* + 1 = [\frac{3S_3}{S_2 - S_1}],$$

so

$$\frac{(k-2)(k-1)}{k-3} \leq \frac{3S_3}{S_2 - S_1},$$

which implies that

$$(k-2)(k-1) \leq (k-3)\frac{3S_3}{S_2 - S_1}. \qquad (2.29)$$

The equation (2.29) combined with (2.28) yields

$$f(k) - f(k-1) \geq 0 \quad \text{for} \quad 5 < k \leq k^*.$$

Thus

$$f(k^*) \geq f(k) \quad \text{for any} \quad k \leq k^*.$$

Therefore if $k^* \geq n$ the maximizing value is n. If $k \geq k^* + 1$, since $\frac{k(k-1)}{k-2} \geq k+1$, we have

$$\frac{(k-1)k}{k-2} \geq k+1 \geq k^* + 2 = [\frac{3S_3}{S_2 - S_1}] + 1 \geq \frac{3S_3}{S_2 - S_1},$$

which means

$$(k-1)k \geq 3(k-2)S_3/(S_2 - S_1). \qquad (2.30)$$

Substituting $k+1$ for k in (2.28), in conjunction with (2.30), yields

$$f(k+1) - f(k) \leq 0 \quad \text{for any} \quad k \geq k^* + 1.$$

Therefore the optimal lower bound can be reached at k^* or $k^* + 1$, if $k^* + 1 \leq n$.

Now consider the events specified in Example 2.1.4 where a set of events A_1, ..., A_7 in a probability space with all non-zero probabilities of elementary conjunctions of intersections is given as follows.

$$P(A_1 A_2 A_3 A_4 A_5 A_6 A_7^c) = 0.05 \quad P(A_1 A_2 A_3^c A_4 A_5^c A_6^c A_7) = 0.1$$

$$P(A_1^c A_2 A_3 A_4 A_5 A_6 A_4) = 0.05 \quad P(A_1 A_2 A_3 A_4 A_5^c A_6^c A_7) = 0.05$$

$$P(A_1 A_2 A_3 A_4 A_5^c A_6^c A_7^c) = 0.1 \quad P(A_1 A_2 A_3 A_4 A_5 A_6^c A_7^c) = 0.15$$

$$P(A_1 A_2 A_3 A_4^c A_5 A_6^c A_7^c) = 0.1 \quad P(A_1 A_2 A_3 A_4^c A_5 A_6^c A_7) = 0.05$$

$$P(A_1 A_2 A_3^c A_4 A_5 A_6^c A_7^c) = 0.1 \quad P(A_1^c A_2 A_3 A_4 A_5 A_6^c A_7) = 0.01$$

$$P(A_1 A_2 A_3^c A_4^c A_5 A_6 A_7^c) = 0.1 \quad P(A_1^c A_2^c A_3 A_4 A_5 A_6 A_7) = 0.04$$

$$P(A_1^c A_2^c A_3^c A_4^c A_5^c) = 0.1.$$

Thus

$$S_1 = 4.1 \quad S_2 = 7.5 \quad S_3 = 7$$

and

$$k^* = [\frac{3 \times 7}{7.5 - 4.1}] - 1 = 5.$$

In this example, the optimal value of bound (2.16) offers

$$P(\bigcup_{i=1}^{7} A_i) \geq -S_1 + S_2 - \frac{2}{5}S_3 = 0.6.$$

When $S_2 - S_1 > 0$ the optimal value of the bound in (2.16) is achieved at

$$k = k^* = \left[\frac{3S_3}{S_2 - S_1}\right] - 1$$

or at $k = k^* + 1$ when $k^* + 1 \leq n$; and at $k = n$ if $k^* \geq n$. However, the Dawson-Sankoff bound has $k^* = 4$ and yields a lower bound of 0.89 which is much closer to the true value $(P(\bigcup_{i=1}^{7} A_i) = 0.9)$. On the other hand, the refined version (2.24) at $k = 4$ gives the numerical lower bound of 0.879, which is lower than 0.89. In this case, the high value of S_3 is offset by the relatively high value of $n = 7$ in this example. The domain containing the optimizing value according to Example 2.2.5 is $\{2,3,4\}$, and the optimizing value among these is 4. □

Now, we discuss another definition of optimality using linear combinations of Bonferroni summations, which is uniformly optimal for all consistent Bonferroni summations.

Definition 2.2.2 (Linear Optimality): Let L be a set of upper bounds under consideration and assume all bounds $G \in L$ are of the form

$$G = \sum_{i \in T \setminus \{0\}} d_i S_i$$

where T is the index set of Bonferroni summations, d_i's are constants possibly depending on n but not on the events A_1, ..., A_n under consideration. The bound G depends on all S_i, $i \in T$ only. If there exists a bound $G_2 \in L$, which satisfies $G_2 \leq G$ for any $G \in L$. Then G_2 is a linear optimal lower bound for $P(v \geq r)$. □

Correspondingly, a parallel definition of a linearly optimal lower bound is self-evident.

In the literature, the question of an optimal linear upper bound for $P(v \geq r)$ is implicitly addressed in Sathe, Pradhan and Shah (1980). Margaritescu (1988) showed that the upper bound (2.21) is linear optimal for fixed r, n and $T = \{1, 2\}$. Specializing in the case $r = 1$, we find that

$$P(v \geq 1) \leq S_1 - \frac{2}{n} S_2, \tag{2.31}$$

with the right-hand side being a linear optimal upper bound. Comparing with (2.22), it follows that the adjusted optimal linear bound is Fréchet optimal.

We now use an upper bound to illustrate the concept of linear optimality.

Example 2.2.8   Linear optimal upper bound

The upper bound $S_1 - \frac{2}{n} S_2$ is a linear optimal upper bound for the probability of the union of a set of events A_1, ..., A_n.

Proof: First, to consider the validity of the quantity $S_1 - \frac{2}{n} S_2$ as an upper bound, we need to show that

$$P(\bigcup_{i=1}^{n} A_i) \leq S_1 - \frac{2}{n} S_2.$$

To this end, notice that the lower bound of $P(v=0)$ reads, by (2.9) and taking $u=1$,

$$
\begin{aligned}
P(v=0) \quad &\geq \quad B_0^n(2u-1)+\frac{2u}{n}S_{2u} \\
&= \quad B_0^1(1)+\frac{2}{n}S_2 \\
&= \quad \sum_{k=0}^{1}(-1)^k S_k + \frac{2}{n}S_2 \\
&= \quad 1-S_1+\frac{2}{n}S_2.
\end{aligned}
$$

Thus,

$$
P(v\geq 1)\leq S_1 -\frac{2}{n}S_2.
$$

Alternatively, notice that

$$
\begin{aligned}
S_1 -\frac{2}{n}S_2 \quad &= \quad E(v-\frac{2}{n}\frac{v(v-1)}{2}) \\
&= \quad E(v-\frac{v}{n}(v-1)) \\
&\leq \quad E(v-(v-1)) \\
&= \quad E(I_{\{v\geq 1\}}) \\
&= \quad P(v\geq 1).
\end{aligned}
$$

This shows that the quantity $S_1 - \frac{2}{n}S_2$ is a valid upper bound for $P(v\geq 1)$. Now to show that this bound is also a linear optimal bound, we consider any upper bound of the form

$$
P(v\geq 1)\leq d_1 S_2 + d_2 S_2
$$

for any set of events in any probability space, where d_1 and d_2 are constants depending on the number of events n only. The following shows that any such constants d_1 and d_2 in an upper bound will have to satisfy the condition that the associated upper bound value is higher than the value of the upper bound $S_1 - \frac{2}{n}S_2$. Therefore, $S_1 - \frac{2}{n}S_2$ is a linear optimal upper bound.

Consider a set of n events: $A_1 = \Omega$, $A_2 = ... = A_n = \emptyset$. In this case, $S_1 = 1$ and $S_2 = 0$, thus, by the condition of upper bound, we have $d_1 \geq 1$.

Consider another set of n events: $A_1 = ... = A_n = \Omega$, we have $S_1 = n$ and $S_2 = \frac{n(n-1)}{2}$, and

$$
d_1 S_1 + d_2 \frac{n(n-1)}{2} \geq 1.
$$

Thus

$$
\begin{aligned}
d_1 S_1 + d_2 S_2 \quad &\geq \quad d_2 S_1 + \frac{1-d_1 n}{\frac{n(n-1)}{2}}S_2 \\
&= \quad d_1(S_1 - \frac{2}{n-1}S_2) + \frac{2}{n(n-1)}S_2
\end{aligned}
$$

$$\geq S_1 - \frac{2}{n-1}S_2 + \frac{2}{n(n-1)}S_2$$

$$= S_1 - \frac{2}{n}S_2.$$

This shows that the upper bound $S_1 - \frac{2}{n}S_2$ is lower than or equal to any upper bounds taking the form of $d_1 S_1 + d_2 S_2$ for constants d_1 and d_2. Thus it is a linear optimal upper bound. □

Example 2.2.8 discusses linear optimal upper bounds for $P(v \geq 1) = P(\bigcup_{i=1}^n A_i)$. However, for the corresponding linear optimal lower bound, as shown in Example 2.2.4, the optimal value of k^* in the Dawson-Sankoff lower bound depends on the values of S_1 and S_2, thus it is not linear optimal in the sense defined in Definition 2.2.2. Nevertheless, the follow example shows that the Dawson-Sankoff lower bound is linear optimal when the bound G is extended to a group of bounds.

Example 2.2.9   Linear optimal lower bound

For any constants a and b, let v be the number of occurrences of a set of events. If $P(v \geq 1) \geq aS_1 + bS_2$ for any events, then there exists one integer k, $k = 1, 2, ..., n-1$ so that

$$aS_1 + bS_2 \geq \frac{2}{k+1}S_1 - \frac{2}{k(k+1)}S_2.$$

In this sense, the group of lower bounds

$$G = \{\frac{2}{k+1}S_1 - \frac{2}{k(k+1)}S_2, k = 1, 2, ..., n-1\},$$

or

$$G = \max_{k=1,2,...,n-1}[\frac{2}{k+1}S_1 - \frac{2}{k(k+1)}S_2]$$

is the linear optimal lower bound.

Proof: Since $P(v \geq 1) \geq aS_1 + bS_2$ is valid for any events, for any integer $k = 1, ..., n-1$, consider $A_1 = ... = A_k = \Omega$, and $A_{k+1} = ... = A_n = \emptyset$, we have

$$1 \geq ak + bk(k-1)/2.$$

When k is very large, keeping the above inequality leads to $b < 0$. Thus, $a > 0$ (otherwise the lower bound becomes trivial because any negative number is of course a lower bound of any probability).

Taking $k = 1$ gets $a \leq 1$, thus $0 < a < 1$.

The condition $1 \geq ak + bk(k-1)/2$ leads to $b \leq \frac{2(1-ak)}{k(k-1)}$ for $k = 2, ..., n$. Thus

$$aS_1 + bS_2 \leq a(S_1 - \frac{2}{k-1}S_2) + \frac{2}{k(k-1)}S_2, \qquad (2.32)$$

for any $2 \leq k \leq n$. We now consider the following three situations for the possible value of a.

Case I: The constant a is in the range $0 \leq a \leq \frac{2}{n}$. In this case, taking $k = n$ in (2.32) gets

$$
\begin{aligned}
aS_1 + bS_2 &\leq a(S_1 - \frac{2}{k-1}S_2) + \frac{2}{k(k-1)}S_2 \\
&= a(S_1 - \frac{2}{n-1}S_2) + \frac{2}{n(n-1)}S_2 \\
&\leq \frac{2}{n}(S_1 - \frac{2}{n-1}S_2) + \frac{2}{n(n-1)}S_2 \\
&= \frac{2}{n}S_1 - \frac{2}{n(n-1)}S_2,
\end{aligned}
$$

which is a Dawson-Sankoff lower bound when $k = n-1$. Note that the second inequality in the above derivation comes from the fact that

$$
S_1 - \frac{2}{n-1}S_2 = E(v) - \frac{2}{n-1}E(\frac{v(v-1)}{2}) \geq 0.
$$

Case II: The constant a is in the range $\frac{2}{n} \leq a \leq 1$. In this case, the interval $[\frac{2}{n}, 1]$ can be divided by $n-1$ points $\frac{1}{t}$ with $t = 2, ..., n-1$. Thus there exists an integer $p \in \{2, ..., n-1\}$ so that

$$
\frac{2}{p+1} \leq a \leq \frac{2}{p}. \tag{2.33}
$$

For this quantity p, if $S_1 - \frac{2}{k-1}S_2 \geq 0$, from (2.33), we have $a \leq \frac{2}{p}$, thus

$$
\begin{aligned}
aS_1 + bS_2 &\leq a(S_1 - \frac{2}{p-1}S_2) + \frac{2}{p(p-1)}S_2 \\
&\leq \frac{2}{p}(S_1 - \frac{2}{p-1}S_2) + \frac{2}{p(p-1)}S_2 \\
&= \frac{2}{p}S_1 - \frac{2}{p(p-1)}S_2.
\end{aligned}
$$

This is a Dawson-Sankoff lower bound when the integer k takes the value $p-1$.

Now, if $S_1 - \frac{2}{p-1}S_2 \leq 0$ for the quantity p, from (2.33), we have $a \geq \frac{2}{p+1}$. Thus

$$
\begin{aligned}
aS_1 + bS_2 &\leq a(S_1 - \frac{2}{p-1}S_2) + \frac{2}{p(p-1)}S_2 \\
&\leq \frac{2}{p+1}(S_1 - \frac{2}{p-1}S_2) + \frac{2}{p(p-1)}S_2 \\
&= \frac{2}{p+1}S_1 - \frac{2}{p(p+1)}S_2.
\end{aligned}
$$

This is a Dawson-Sankoff lower bound when the integer k takes the value p.

Summarizing the above discussion concludes Example 2.2.9. $\qquad$ □

The property of linear optimality appears weaker than that of Fréchet optimality.

For example, the Fréchet optimal Dawson-Sankoff lower bound (Example 2.2.2) is not linear, and the statement of Theorem 1 of Galambos (1977) makes clear that the Dawson-Sankoff bound is better than any linear lower bound.

Besides linear optimality and Fréchet optimality, in the literature there is another way to formulate the optimality of probability bounds, which is defined in the language of linear programming (Kwerel (1975a), Prékopa (1990b)) as follows.

Definition 2.2.3 (Linear Programming Optimality): For a set of consistent Bonferroni summations S_i, $i = 1,...,t$ where t is a prefixed integer $1 \leq t \leq n$, denote $p_i = P(v = i)$, $i = 0, 1, ..., n$. The optimal upper bound of $P(v \geq 1)$ can then be formulated as

$$\max(p_1 + ... + p_n)$$

subject to $\sum_{i=0}^{n} p_i = 1$ and

$$p_1 + 2p_2 + ... + tp_t + ... + np_n = S_1$$
$$p_2 + ... + \binom{t}{2}p_t + ... + \binom{n}{2}p_n = S_2$$
$$..... \quad ...$$
$$p_t + ... + \binom{n}{t}p_n = S_t.$$

A more general version of optimality grounded on this approach in the bivariate case and its connection with Fréchet optimality is given in Chapter 5.

The definition of optimality in Definition 2.2.1 appears to be too complicated to verify due to the construction (existence) of the probability space. Because of this, Fréchet optimality has not been used explicitly for years. However, this version of optimality can be treated in general in the language of linear programming (Definition 2.2.3), as implied in the accounts of Hailperin (1965), Kwerel (1975b), Kounias and Sotirakoglou (1989). The approach of defining optimality in Definition 2.2.3 shows the numerical achievement of probability upper/lower bounds. We return to this idea and elucidate it in Chapter 4 when we define bivariate Fréchet optimality.

So far, for a set of events $\{A_i\}$, $i = 1,...,n$, we have discussed three kinds of optimality. The first one is Fréchet (or absolute) optimality, which was introduced by Fréchet in 1935 for bounds involving only the probability of each individual event A_i, $i = 1,...,n$. Along with Fréchet optimality are linear optimality (Definition 2.2.2) and optimality defined through linear programming (Definition 2.2.3). All these definitions have their distinguishable rationales under difference circumstances to characterize the property of optimality.

The above introduction focuses on the three definitions of optimality. Examples regarding optimality defined in linear programming are deferred to Chapter 4, which is devoted to the technique of linear programming in probability bounding theory.

With the background of univariate Bonferroni-type bounds and optimality, we can now introduce notations and basic results for multivariate bounds.

2.3 Multivariate Bounds

In this section, we shall introduce multivariate Bonferroni-type bounds that involve several sets of events. As shown in Chapter 1, the extension of Bonferroni-type bounds from univariate to multivariate has broad applications in various fields (Galambos and Xu (1990a,1995), Lee (1997)).

2.3.1 Complete Bonferroni Summations

Similar to the univariate bounding theory, we shall define the concept of *multivariate Bonferroni summation*, a building block that is frequently applied in the multivariate bounding theory (Erdos (1946), Erdos and Kac (1940)).

Definition 2.3.1 (Multivariate Bonferroni Summation): Consider J sets of events A_{ij} for $i = 1, ..., M_j$ and $j = 1, ..., J$ in an arbitrary probability space $(\Omega, \mathcal{F}, P)$. Denote m_j an integer satisfying $0 \le m_j \le M_j$ for pre-fixed integers M_j with $j = 1, ..., J$. Also denote $v_1, ..., v_J$ the number of events $\{A_{i1}\}, ..., \{A_{iJ}\}$, respectively, which occur. Define the J-variate Bonferroni summation as follows.

$$S_{m_1...m_J} = \sum P(\bigcap_{j=1}^{J} \bigcap_{s=1}^{m_j} A_{i_s j}), \qquad (2.34)$$

where the summation is over the range :

$$1 \le i_1 < ... < i_{m_j} \le M_j \quad \text{for} \quad j = 1, ..., J.$$

In the cases where it is necessary to address the effect of j in the sequence $i_1, ..., i_{m_j}$ for the jth set of events, we also use $t(1, j), ..., t(m_j, j)$ to replace $i_1, ..., i_{m_j}$ when necessary.

The following example illustrates the concept of multivariate Bonferroni summation defined in (2.34).

Example 2.3.1 Let $A_1, ..., A_n$ and $B_1, ..., B_m$ be two sets of events in an arbitrary probability space $(\Omega, \mathcal{F}, P)$. Denote v_1 and v_2 the numbers of $\{A_i\}$, $\{B_j\}$ which occur, respectively. For any two integers r and u, consider bounds for

$$P(v_1 \ge r, v_2 \ge u) \quad \text{and} \quad P(v_1 = r, v_2 = u). \qquad (2.35)$$

Paralleling the quantities S_t in Section 2.1 and according to (2.34), the bivariate Bonferroni summation refers to

$$S_{r,u} = \begin{cases} \sum_{\substack{1 \le i_1 < ... < i_r \le n \\ 1 \le j_1 < ... < j_u \le m}} P(A_{i_1}...A_{i_r} B_{j_1}...B_{j_u}) & 1 \le r \le n, 1 \le u \le m \\ S_u(B) & r = 0, \quad 0 \le u \le m \\ S_r(A) & u = 0, \quad 0 \le r \le n \\ 1 & r = u = 0 \\ 0 & r > n \quad \text{or} \quad u > m. \end{cases} \qquad (2.36)$$

$S_{r,u}(A, B)$ may also be used to indicate the associated event sets $\{A_i\}$, and $\{B_j\}$. Comparing the terms (2.36) with (2.34), it is obvious that (2.36) is a special case of (2.34) when $J = 2$, $M_1 = n$ and $M_2 = m$. □

The above example shows the analogy between the multivariate Bonferroni summation and the univariate Bonferroni summation defined in Section 2.1. Furthermore, similar to the univariate Bonferroni summation, the multivariate Bonferroni summation, as shown in the following lemmas, is endowed with the interpretation of the joint binomial moments in terms of J sets of events $\{A_{ij}, i = 1, ..., M_j, j = 1, ..., J\}$ in a product space. (See, for example, Meyer (1969)).

Lemma 2.3.1 Consider J sets of events A_{ij} for $i = 1, ..., M_j$ and $j = 1, ..., J$ in an arbitrary probability space $(\Omega, \mathcal{F}, P)$. Denote by m_j an integer satisfying $0 \le m_j \le M_j$ for pre-fixed integers M_j with $j = 1, ..., J$. Also denote $v_1, ..., v_J$ the number of events $\{A_{i1}\}, ..., \{A_{iJ}\}$, respectively, which occur at a sample point. For the multivariate Bonferroni summation defined in (2.34), we have

$$S_{m_1, ..., m_J} = E(\prod_{j=1}^{J} \binom{v_j}{m_j}))$$

$$= \sum_{j=1}^{J} \sum_{t_j=1}^{M_j} (\prod_{j=1}^{J} \binom{t_j}{m_j}) P(v_j = t_j, \quad 1 \le j \le J).$$

Proof: Consider any given integer j and a sequence of m_j index $1 \le t(1, j) < ... < t(m_j, j) \le M_j$. Since the indicator function of the intersection of all events satisfies the following condition

$$I(\bigcap_{j=1}^{J} \bigcap_{s=1}^{m_j} A_{t(s,j),j}) = \begin{cases} 1, & \text{if } \bigcap_{j=1}^{J} \bigcap_{s=1}^{m_j} A_{t(s,j),j} \text{ occurs;} \\ 0, & \text{otherwise,} \end{cases}$$

we can define a J-variate function as follows

$$F(m_1, ..., m_J) = \sum I(\bigcap_{j=1}^{J} \bigcap_{s=1}^{m_j} A_{t(s,j),j}) \tag{2.37}$$

where the summation $\sum$ is for $1 \le t(1, j) < ... < t(m_j, j) \le M_j$.

Now, notice that for an indicator function of any event C in the σ-field, $E(I_C) = P(C)$. From the definition of multivariate Bonferroni summation (2.34) in conjunction with the relationship between the probability and expectation of the associated indicating function, taking the expected value in (2.37) yields

$$S_{m_1, ..., m_J} = E(F(m_1 ... m_J)), \tag{2.38}$$

by replacing index $i_1, ..., i_{m_j}$ with the new index $t(1, j), ..., t(m_j, j)$.

On the other hand, by the multiplication law, we have

$$F(m_i ... m_J) = \sum I(\bigcap_{j=1}^{J} \bigcap_{s=1}^{m_j} A_{t(s,j),j})$$

$$= \prod_{j=1}^{J} \binom{v_j}{m_j}, \tag{2.39}$$

because for the jth class, all possible numbers of spontaneous occurrence of distinct m_j events out of the total of v_j events, is $\binom{v_j}{m_j}$. Now, taking expectations on both sides of the equation (2.39) gets

$$E(F(m_1 \ldots m_J)) = \sum_{j=1}^{J} \sum_{t_j=1}^{M_j} (\prod_{j=1}^{J} \binom{t_j}{m_j}) P(v_j = t_j), \quad 1 \leq j \leq J),$$

completing the proof of Lemma 2.3.1.   □

2.3.2 Partial Bonferroni Summations

Consider, $J = 2$, the bivariate binomial moment $S_{r,u}$ defined in the previous section. When both classes of events are in a symmetric or equivalent status, it is natural to utilize the Bonferroni summation, $S_{r,u}$. However, in some applications, it makes more sense to concentrate on one of the event-classes before summing for the information regarding the second class. Consider the use of probability inequality in detecting minimum effective dose of a drug. If the clinical outcome involves two endpoints: primary endpoint (such as the while blood cell counts in cancer research) and secondary endpoint (such as the systolic blood pressure), medical investigations normally consider the efficacy of the primary endpoint before taking the secondary endpoint into consideration. Another such scenario is the detection of the therapeutic window of a drug, where the efficacy is meaningful only when the chemical component is within a pre-specified toxicity level for the experimental subjects. An efficacious dose of a cancer treatment is meaningless if it also kills key functioning cells of a human body. For such applications, it makes more sense to sum the information over one of the event-classes at a fixed level of another index. Toward this end, we introduce the concept of *partial Bonferroni summation* as follows.

Definition 2.3.2 (Partial Bonferroni Summation): Considering two sets of events $\{A_i, i = 1, \ldots, n\}$ and $\{B_j, j = 1, \ldots, m\}$, the quantity of partial Bonferroni summation $S_{r,(u)}$ is defined as

$$S_{r,(u)} = E(\binom{v_1}{r} \delta_{v_2,u}) = \sum_{k=0}^{n} \binom{k}{r} P(v_1 = k, v_2 = u),$$

where the number of occurring events is fixed at u for the event class $\{B_j\}$. And

$$S_{(r),u} = E(\binom{v_2}{u} \delta_{v_1,r}) = \sum_{t=0}^{m} \binom{t}{u} P(v_1 = r, v_2 = t),$$

where the number of occurring events is fixed at r for the event class $\{A_i\}$.

In the definition, $\delta_{i,j}$ refers to the Kronecker delta. Due to the way the concept is defined, a partial Bonferroni summation is available only when two or more classes of events are under consideration.

The relation of the partial Bonferroni summation $S_{(r),u}$ (or $S_{r,(u)}$) and the Bonferroni summation $\{S_{i,j}\}$ can be expressed in the following lemma.

Lemma 2.3.2: For any positive integers r, u, the partial Bonferroni summation can be expressed as a linear combination of a set of Bonferroni summations.

$$S_{r,(u)} = \sum_{y=0}^{m-u} (-1)^y \binom{y+u}{u} S_{r,y+u} \qquad S_{(r),u} = \sum_{x=0}^{n-r} (-1)^x \binom{x+r}{r} S_{x+r,u}.$$

Proof: Notice that if $v_1 = r$, $\delta_{v_1,r} = 1$,

$$\sum_{y=0}^{0} (-1)^y \binom{v_1 - v_1}{y} = 1, \qquad \binom{v_1}{r} = \binom{v_1}{v_1} = 1.$$

If $v_1 > r$, $\delta_{v_1,r} = 0$,

$$\sum_{y=0}^{v_1-r} (-1)^y \binom{v_1 - r}{y} = (1-1)^{v_1-r} = 0.$$

Therefore

$$\delta_{v_1,r} = \sum_{y=0}^{v_1-r} (-1)^y \binom{v_1 - r}{y} \binom{v_1}{r} \qquad \text{for} \quad v_1 \geq r. \qquad (2.40)$$

From the definition of partial Bonferroni summation, we have

$$\begin{aligned}
S_{r,(u)} &= E\left(\binom{v_1}{r} \delta_{v_2,u}\right) \\
&= E\left(\binom{v_1}{r} \sum_{y=0}^{v_2-u} (-1)^y \binom{v_2 - u}{y} \binom{v_2}{u}\right),
\end{aligned}$$

from (2.40), the partial Bonferroni summation takes the form

$$\begin{aligned}
S_{r,(u)} &= E\left(\binom{v_1}{r} \sum_{y=0}^{v_2-u} (-1)^y \binom{y+u}{y} \binom{v_2}{y+u}\right) \\
&= \sum_{p=r}^{n} \sum_{q=u}^{m} \binom{p}{r} \sum_{y=0}^{q-u} (-1)^y \binom{y+u}{u} \binom{q}{y+u} P(v_1 = p, v_2 = q) \\
&= \sum_{y=0}^{m-u} (-1)^y \binom{y+u}{u} \left[\sum_{p=r}^{n} \sum_{q=y+u}^{m} \binom{p}{r} \binom{q}{y+u} P(v_1 = p, v_2 = q)\right] \\
&= \sum_{y=0}^{m-u} (-1)^y \binom{y+u}{u} S_{r,y+u},
\end{aligned}$$

by Lemma 2.3.1.

A similar argument for $S_{(r),u}$ establishes Lemma 2.3.2   □

When it is impossible to obtain all information on bivariate Bonferroni summations for Lemma 2.3.2, the next result shows how to bound the quantity of a partial Bonferroni summation with regular Bonferroni summations.

Lemma 2.3.3: For integers $n \geq 1$, $N \geq 1$, $0 \leq r \leq n$, $0 \leq u \leq N$, $0 \leq b \leq N - u - 1$, and any integer $s \geq 0$ such that $0 \leq 2s \leq n - r - 1$,

$$\sum_{x=0}^{2s+1} (-1)^x \binom{x+r}{r} S_{r+x,u+b+1} \leq S_{(r),u+b+1} \leq \sum_{x=0}^{2s} (-1)^x \binom{x+r}{r} S_{r+x,u+b+1}.$$

$$\sum_{x=0}^{2s+1} (-1)^x \binom{x+u}{u} S_{r+a+1,u+x} \leq S_{r+a+1,(u)} \leq \sum_{x=0}^{2s} (-1)^x \binom{x+u}{u} S_{r+a+1,u+x}.$$

Proof: We proceed to prove the bounds for $S_{(r),u+b+1}$. The proof for the bounds of $S_{r+a+1,(u)}$ follows similarly. For integer $0 \leq a \leq n - r - 1$, let

$$
\begin{aligned}
K &= \sum_{x=0}^{a} (-1)^x \binom{x+r}{r} S_{r+x,u+b+1} \\
&= \sum_{x=0}^{a} (-1)^x \binom{x+r}{r} \left[\sum_{i=r+x}^{n} \sum_{j=u+b+1}^{N} \binom{i}{x+r} \binom{j}{u+b+1} P(v_1 = i, v_2 = j) \right] \\
&= \sum_{j=u+b+1}^{N} \binom{j}{u+b+1} \sum_{x=0}^{a} (-1)^x \binom{x+r}{r} \sum_{i=r+x}^{n} \binom{i}{x+r} P(v_1 = i, v_2 = j) \\
&= \sum_{j=u+b+1}^{N} \binom{j}{u+b+1} \sum_{i=r}^{n} P(v_1 = i, v_2 = j) \sum_{x=0}^{T} (-1)^x \binom{x+r}{r} \binom{i}{x+r},
\end{aligned}
$$

where $T = \min(a, i-r)$. Let $y_{rj} = P(v_1 = r, v_2 = j)$. Since for $i > r$,

$$\binom{r+x}{r} \binom{i}{r+x} = \binom{i}{r} \binom{i-r}{x},$$

$$
\begin{aligned}
K &= \sum_{j=u+b+1}^{N} \binom{j}{u+b+1} \left[\sum_{i=r+1}^{n} P(v_1 = i, v_2 = j) \sum_{x=0}^{T} (-1)^x \binom{i}{r} \binom{i-r}{x} + y_{rj} \right] \\
&= \sum_{j=u+b+1}^{N} \binom{j}{u+b+1} \left[\sum_{i=r+1}^{n} P(v_1 = i, v_2 = j)(-1)^T \binom{i}{r} \binom{i-r-1}{T} + y_{rj} \right] \\
&\qquad \left(\text{from the identity } \sum_{x=0}^{T} (-1)^x \binom{i-r}{x} = (-1)^T \binom{i-r-1}{T} \right) \\
&= \sum_{j=u+b+1}^{N} \binom{j}{u+b+1} \left[(-1)^a \sum_{i=r+a+1}^{n} P(v_1 = i, v_2 = j) \binom{i}{r} \binom{i-r-1}{a} + y_{rj} \right],
\end{aligned}
$$

since $T = \min(a, i-r)$. Thus

$$K \leq \sum_{j=u+b+1}^{N} \binom{j}{u+b+1} P(v_1 = r, v_2 = j) = S_{(r),u+b+1},$$

by the definition, when the integer a is odd. And

$$K \geq \sum_{j=u+b+1}^{N} \binom{j}{u+b+1} P(v_1 = r, v_2 = j) = S_{(r),u+b+1},$$

when the integer a is even.   □

The above results on the partial Bonferroni summation will be used in deriving upper and lower Bonferroni-type bounds later in the book.

2.3.3   Decomposition of Bonferroni Summations

In Lemma 2.3.1, when the left-hand side is spelled out as the summation of all possible outcomes weighted by the corresponding probabilities, we obtain the expression of joint binomial moment for the multivariate Bonferroni summation as follows.

$$
\begin{aligned}
S_{m_1,\ldots,m_J} &= \sum_{j=1}^{J} \sum_{t_j=1}^{M_j} (\prod_{j=1}^{J} \binom{t_j}{m_j}) P(v_j = t_j, \quad 1 \leq j \leq J) \\
&= \sum \binom{t_1}{m_1} \cdots \binom{t_J}{m_J} P(v_1 = t_1 \ldots, v_J = t_J), \\
&= E(\binom{v_1}{m_1} \cdots \binom{v_J}{m_J})
\end{aligned}
\tag{2.41}
$$

The expression (2.41) shows that $S_{m_1,\ldots,m_J}$ is the joint binomial moment of $v_1 \ldots v_J$. However, the combinatorial arguments, combined with the weighted probabilities, lead to the fact that alternatively, the multivariate Bonferroni summation $S_{m_1,\ldots,m_J}$ can also be decomposed as a linear combination of $P(v_1 \geq t_1, \ldots, v_J \geq t_J)$ with the vector $(t_1, \ldots, t_J)$ summarizing over an appropriately selected index set T. We need the following basic lemma on combinations to facilitate the derivation of the decomposition theorem on multivariate Bonferroni summations.

Lemma 2.3.4 For any integer $i \geq k$, we have

$$\sum_{j=k}^{i} \binom{j-1}{k-1} = \binom{i}{k}. \tag{2.42}$$

Proof: We use the method of mathematical induction to prove the equation (2.42).
Starting from the initial value k, notice that if $i = k$,

$$\sum_{j=k}^{k} \binom{j-1}{k-1} = 1 = \binom{k}{k}.$$

So, Lemma 2.3.4 is true when $i = k$.
Now if (2.42) holds for an integer $i \geq k$, then for the next integer $i+1$, we have

$$\sum_{j=k}^{i+1} \binom{j-1}{k-1} = \sum_{j=k}^{i} \binom{j-1}{k-1} + \binom{i}{k-1}$$

$$= \binom{i}{k} + \binom{i}{k-1}$$

$$= \binom{i+1}{k}.$$

Thus, by the method of induction, (2.42) is true for any integer $i \geq k$. □

Theorem 2.3.1 (Meyer 1969) For any positive integer J and m_1, ..., m_J, the multivariate Bonferroni summation defined in (2.34) can be decomposed as follows,

$$S_{m_1...m_J} = \sum_{j=1}^{J} \sum_{t_j=1}^{M_j} \binom{t_1-1}{m_1-1} \cdots \binom{t_J-1}{m_J-1} P(v_1 \geq t_1 ..., v_J \geq t_J). \qquad (2.43)$$

Note that the decomposition structure of (2.43) is very similar to that of (2.41). In fact these two decompositions underpin the framework of derivation for multivariate upper and lower Bonferroni-type bounds in the literature. (2.43) is mainly used to derive the joint probability of *at least* t_i occurrences in the ith event set, and (2.41) is mainly used to derive the joint probability of *exactly* t_i occurrences in the ith event set.

Proof: For notational convenience, we state the proof in the case when $J = 2$, the corresponding proof for any dimension $J > 2$ is self-evident. Under this assumption, (2.43) is equivalent to

$$S_{i,j} = \sum_{t_1=i}^{M_1} \sum_{t_2=j}^{M_2} \binom{t_1-1}{i-1} \binom{t_2-1}{j-1} P(v_1 \geq t_1, v_2 \geq t_2). \qquad (2.44)$$

Recall that for any integers $N \geq 0$, $i \geq 0$ and $k \geq 0$, as shown in Figure 2.1, the order of the summation over (x,y) can be changed to (y,x) to get

$$\sum_{x=k}^{N} \sum_{y=x}^{N} f(x,y) = \sum_{y=k}^{N} \sum_{x=k}^{y} f(x,y). \qquad (2.45)$$

Now, with (2.45) we can apply Lemma 2.3.2 to decompose the multivariate Bonferroni summation as follows. Notice that

$$P(v_1 \geq t_1, v_2 \geq t_2) = \sum_{l_1=t_1}^{M_1} \sum_{l_2=t_2}^{M_2} P(v_1 = l_1, v_2 = l_2),$$

starting from the left-hand side of (2.44), we have

$$\sum_{t_1=i}^{M_1} \sum_{t_2=j}^{M_2} \binom{t_1-1}{i-1} \binom{t_2-1}{j-1} P(v_1 \geq t_1, v_2 \geq t_2)$$

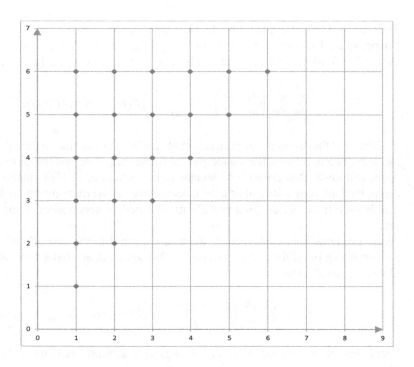

Figure 2.1 *Domain change for bivariate summation*

$$= \sum_{t_1=i}^{M_1} \sum_{t_2=j}^{M_2} \binom{t_1-1}{i-1} \binom{t_2-1}{j-1} \sum_{l_1=t_1}^{M_1} \sum_{l_2=t_2}^{M_2} P(v_1=l_1, v_2=l_2)$$

$$= \sum_{t_1=i}^{M_1} \binom{t_1-1}{i-1} \sum_{l_1=t_1}^{M_1} \sum_{t_2=j}^{M_2} \sum_{l_2=t_2}^{M_2} \binom{t_2-1}{j-1} P(v_1=l_1, v_2=l_2)$$

by (2.45)

$$= \sum_{t_1=i}^{M_1} \binom{t_1-1}{i-1} \sum_{l_1=t_1}^{M_1} \sum_{l_2=j}^{M_2} P(v_1 = l_1, v_2 = l_2) \sum_{t_2=j}^{l_2} \binom{t_2-1}{j-1}$$

$$= \sum_{t_1=i}^{M_1} \binom{t_1}{i-1} \sum_{l_1=t_1}^{M_1} \sum_{l_2=j}^{M_2} \binom{l_2}{j} P(v_1 = l_1, v_2 = l_2)$$

$$= \sum_{l_2=j}^{M_2} \binom{l_2}{j} \sum_{t_1=i}^{M_1} \sum_{l_1=t_1}^{M_1} \binom{t_1-1}{i-1} P(v_1 = l_1, v_2 = l_2)$$

$$\text{similar to (2.45) and (2.42)}$$

$$= \sum_{l_1=i}^{M_1} \sum_{l_2=j}^{M_2} \binom{l_2}{j} \binom{l_1}{i} P(v_1 = l_1, v_2 = l_2)$$

$$= S_{i,j} \quad \text{by Lemma 2.3.1.1.}$$

This completes the proof of Theorem 2.3.1. □

With the two different ways of decomposition for the multivariate Bonferroni summation, we can derive the corresponding upper and lower bounds according to the types of decomposition. Specifically, Lemma 2.3.1 leads to either upper or lower bounds for the term $P(v_i = 1, j = 1, ..., J)$, which is the probability of exactly one of the events in each of the J groups of events, occurs. While Theorem 2.3.1 results in probability inequalities for the term $P(v_1 \geq 1, v_2 \geq 1)$, which is the joint probability that at least one of the events occurs in each of the two event groups.

2.3.4 Classical Bonferroni Bounds in a Multivariate Setting

In the literature, the basic multivariate bounding theory includes the work of Meyer (1969), Galambos and Lee (1994), and Seneta and Chen (1996), among others. The first extension of the classical Bonferroni upper and lower bounds into the setting of J groups of events, the multivariate version of the classical Bonferroni bounds, is due to Meyer (1969):

Theorem 2.3.2 Let $J > 1$ be an integer representing J groups of events. Let M_j for $j = 1, ..., J$ be J integers representing the total number of events in group j. Denote v_j the number of events that occur in group j. Then for any integers m_j, $j = 1, ..., J$, we have

$$\sum_{t=\sum m_j}^{\sum m_j+2k+1} \sum_{\sum i_j = t} f(i_1 \ldots i_J; t) \leq P(v_1 = m_1; \ldots; v_J = m_J)$$

$$\leq \sum_{t=\sum m_j}^{\sum m_j+2k} \sum_{\sum i_j=t} f(i_1 \ldots i_J; t),$$

where $f(i_1 \ldots i_J;t) = (-1)^{t-\Sigma m_j} \prod_{j=1}^{J} \binom{i_j}{m_j} S_{i_1,\ldots,i_J}$, and $m_j \leq i_j \leq M_j$, $\quad 1 \leq j \leq J$.
$\square$

Proof: For convenience, we show the theorem for the case where $J = 2$. The validity of the theorem for any J sets of events is then self-evident.

First, notice that

$$S_{ij} = \sum_{t=m+n}^{N+M} \sum_{i+j=t} \binom{i}{m}\binom{j}{n} p_{ij}$$

where

$$p_{ij} = P(v_1 = i, v_2 = j).$$

Summarizing all the terms in S_{ij} gets

$$p_{ij} = \sum_{t=m+n}^{N+M} \sum_{i+j=t} (-1)^{t-(m+n)} \binom{i}{m}\binom{j}{n} S_{ij}. \tag{2.46}$$

For the upper bounds, comparing the above expression (2.46) with the upper bound, it suffices to show that

$$\sum_{t=2k+2+m+n}^{N+M} \sum_{i+j=t} (-1)^{t-(m+n)} \binom{i}{m}\binom{j}{n} S_{ij} \geq 0, \tag{2.47}$$

and

$$\sum_{t=2k+1+m+n}^{N+M} \sum_{i+j=t} (-1)^{t-(m+n)} \binom{i}{m}\binom{j}{n} S_{ij} \leq 0. \tag{2.48}$$

Multiplying variable transformations yields that (2.47) and (2.48) are, respectively, equivalent to

$$\sum_{t=2k+2}^{N+M} \sum_{i+j=t} (-1)^{t-(2k+2)} \binom{i}{m}\binom{j}{n} S_{ij} \geq 0, \tag{2.49}$$

and

$$\sum_{t=2k+1}^{N+M} \sum_{i+j=t} (-1)^{t-(2k+1)} \binom{i}{m}\binom{j}{n} S_{ij} \geq 0. \tag{2.50}$$

Thus, it suffices to show that

$$\sum_{t=r}^{N+M} \sum_{i+j=t} (-1)^{t-r} \binom{i}{m}\binom{j}{n} S_{ij} \geq 0. \tag{2.51}$$

To show (2.48), notice that

$$\sum_{t=r}^{N+M} \sum_{i+j=t} (-1)^{t-r} \binom{i}{m}\binom{j}{n} S_{ij}$$

$$= \sum_{t=r}^{N+M} \sum_{i+j=t} (-1)^{t-r} \binom{i}{m}\binom{j}{n} \sum_{y=i}^{M} \sum_{z=j}^{N} \binom{y}{i}\binom{z}{j} p_{yz}$$

$$= \sum_{i=m}^{M} \sum_{j=r-i}^{N} \sum_{y=i}^{M} \sum_{z=j}^{N} (-1)^{i+j-r} \binom{i}{m}\binom{j}{n}\binom{y}{i}\binom{z}{j} p_{yz}, \tag{2.52}$$

Now consider

$$
\sum_{j=r-i}^{N}\sum_{z=j}^{N}(-1)^{i+j-r}\binom{j}{n}\binom{z}{j}
$$

$$
= \sum_{z=r-i}^{N}\sum_{j=r-i}^{z}(-1)^{i+j-r}\binom{z-n}{z-j}\binom{z}{n}
$$

$$
= \sum_{z=r-i}^{N}\binom{z-n-1}{z-r+i}\binom{z}{n}. \tag{2.53}
$$

(2.53) can be illustrated as follows. When $z+i-r$ is odd,

$$
\sum_{j=r-i}^{z}(-1)^{i+j-r}\binom{z-n}{z-j}
$$

$$
= \binom{z-n}{z-r+i}-\binom{z-n}{z-r+i-1}+\binom{z-n}{z-r+i-2}-...+(-1)^{z+i-r}\binom{z-n}{z-z}
$$

$$
= -\binom{z-n}{0}+\binom{z-n}{1}-\binom{z-n}{2}+\binom{z-n}{3}-...+\binom{z-n}{z-r+i}
$$

$$
= -\binom{z-n-1}{0}+\binom{z-n}{1}-\binom{z-n}{2}+\binom{z-n}{3}-...+\binom{z-n}{z-r+i}
$$

$$
= \binom{z-n-1}{1}-\binom{z-n}{2}+\binom{z-n}{3}-...+\binom{z-n}{z-r+i}
$$

$$
= -\binom{z-n-1}{2}+\binom{z-n}{3}-...+\binom{z-n}{z-r+i}
$$

$$
= ...
$$

$$
= \binom{z-n-1}{z-r+i}.
$$

When $z+i-r$ is even,

$$
\sum_{j=r-i}^{z}(-1)^{i+j-r}\binom{z-n}{z-j}
$$

$$
= \binom{z-n}{z-r+i}-\binom{z-n}{z-r+i-1}+\binom{z-n}{z-r+i-2}-...+(-1)^{z+i-r}\binom{z-n}{z-z}
$$

$$
= \binom{z-n}{0}-\binom{z-n}{1}+\binom{z-n}{2}-\binom{z-n}{3}+...+\binom{z-n}{z-r+i}
$$

$$
= \binom{z-n-1}{0}-\binom{z-n}{1}+\binom{z-n}{2}-\binom{z-n}{3}+...+\binom{z-n}{z-r+i}
$$

$$
= -\binom{z-n-1}{1}+\binom{z-n}{2}-\binom{z-n}{3}+...+\binom{z-n}{z-r+i}
$$

$$
= \binom{z-n-1}{2}-\binom{z-n}{3}+...+\binom{z-n}{z-r+i}
$$

$$= \quad \dots$$

$$= \quad \binom{z-n-1}{z-r+i}.$$

Putting (2.53) into (2.52) gets

$$\sum_{t=r}^{N+M} \sum_{i+j=t} (-1)^{t-r} \binom{i}{m} \binom{j}{n} S_{ij} \geq 0.$$

This completes the proof of Theorem 2.3.2.   □

The following examples describe bounds that are included in the general setting of multivariate bounds in Theorem 2.3.2.

Example 2.3.2 Consider events A_{tj}, $1 \leq j \leq J$, $1 \leq t \leq n$ (i.e. $M_1 = \dots = M_J = n$), $k_j \geq 0$, and integer $M \geq 0$. $M_1 = \dots = M_J = n$, the upper bound of Theorem 2.3.2 becomes

$$P(v_j = k_j, \quad 1 \leq j \leq J) \leq \sum_{i=K}^{K+2M} \sum_{\mathbf{u}} f(\mathbf{u}; K; i), \tag{2.54}$$

where $\sum_{\mathbf{u}}$ signifies summation over all vectors $\mathbf{u} = (u_1 \dots u_J)$, $u_j \geq 1$ such that $\sum u_j = i$, $\sum k_j = K$, and

$$f(\mathbf{u}; K; i) = (-1)^{i-K} S_{u_1,\dots,u_J} \prod_{j=1}^{J} \binom{u_j}{k_j}. \quad \square$$

Bounds discussed in Example 2.3.2 and Theorem 2.3.2 are essentially multivariate bounds. The following example sets the connection of the multivariate bounds in Theorem 2.3.2 with univariate bounds discussed in Sections 2.1 and 2.2.

Example 2.3.3 Considering the case where $J = 1$ in Theorem 2.3.2, we obtain the classical Bonferroni-type inequalities

$$\sum_{k=0}^{2s+1} (-1)^k \binom{k+m_1}{m_1} S_{k+m_1} \quad \leq \quad P(v_1 = m_1)$$

$$\leq \quad \sum_{k=0}^{2s} (-1)^k \binom{k+m_1}{m_1} S_{k+m_1}.$$

In the case of $m_1 = 0$, the above bound takes the form:

$$\sum_{k=0}^{2s+1} (-1)^k S_k \leq P(v_1 = 0) \leq \sum_{k=0}^{2s} (-1)^k S_k.$$

Thus

$$\sum_{j=1}^{2u} (-1)^{j+1} S_j \leq P(\cup A_i) \leq \sum_{j=1}^{2u-1} (-1)^{j+1} S_j,$$

which are the classical inequalities (2.2). □

As shown in the above examples, the multivariate bound stated in Theorem 2.3.2 resembles the classical Bonferroni inequalities (2.6) and (2.7). Since (2.6) and (2.7) have refined versions of (2.8) and (2.9), it is natural to follow the spirit of the univariate refinement to seek sharpened multivariate bounds refining the result in Theorem 2.3.2. Research results in this regard include the work of Galambos and Xu (1994), and Galambos and Lee (1992), among others.

Consider the situation in which two groups of events are of interest. To bound $P(v_1 \geq r, v_2 \geq u)$ for integers $1 \leq r \leq n$ and $1 \leq u \leq m$, Galambos and Lee (1992) show that :

For any integer $a \geq 0$, $b \geq 0$ such that $r + a \leq n$, $u + b \leq m$, if both integers a and b are even, the bivariate upper bound of Theorem 2.3.2 can be improved as follows.

$$P(v_1 \geq r, v_2 \geq u) \tag{2.55}$$

$$\leq \sum_{k=0}^{a}\sum_{t=0}^{b}(-1)^{k+t}\binom{k+r-1}{r-1}\binom{t+u-1}{u-1}S_{k+r,t+u} - \frac{\frac{r}{n-r}\binom{r+a}{a}}{\binom{m}{u}}S_{r+a+1,u}$$

$$- \frac{\frac{u}{m-u}\binom{u+b}{b}}{\binom{n}{r}}S_{r,u+b+1} - \binom{b+u}{u}\binom{a+r}{r}\frac{r}{n-r}\frac{u}{m-u}S_{r+a+1,u+b+1}.$$

In Chapter 3, we will use the method of indicator functions to derive the bound (2.55). Notice that the upper bound (2.55) is limited to the condition that both integers a and b are even. If both a, b are odd, Galambos and Lee provided lower bounds improving the bivariate lower bound in Theorem 2.3.2 as follows.

$$P(v_1 \geq r, \quad v_2 \geq u)$$

$$\geq \sum_{k=0}^{a}\sum_{t=0}^{b}(-1)^{k+t}\binom{k+r-1}{r-1}\binom{t+u-1}{u-1}S_{k+r,t+u}$$

$$- \binom{b+u}{u}\binom{a+r}{r}\frac{r}{a+1}\frac{u}{b+1}S_{r+a+1,u+b+1}. \tag{2.56}$$

The upper and lower bounds listed in (2.55) and (2.56) are for the type of bivariate probabilities $P(v_1 \geq r, \quad v_2 \geq u)$ (modeling the probability of at least r events occurring from the first group and at least u events occurring from the second group). Compared with the univariate Bonferroni bounds (2.6) and (2.7), the multivariate counterpart for the exactly (r, u) events occurring was studied by Galambos and Lee (1994) as follows.

Consider two sets of events for the simultaneous occurrence of exactly certain numbers of events from each group. Assume that the first group has n events and the second group has m events, for integers r, u, a, b satisfying $0 \leq r \leq n, 0 \leq u \leq m$, $r + a \leq n, b + u \leq m$, when a and b are even, the upper bound reads,

$$P(v_1 = r, v_2 = u)$$

$$\leq \sum_{k=0}^{a}\sum_{h=0}^{b}(-1)^{k+h}\binom{k+r}{r}\binom{h+u}{u}S_{k+r,h+u}$$

$$-\frac{r+1}{n-r}\frac{u+1}{m-u}\binom{a+r+1}{a}\binom{b+u+1}{b}S_{r+a+1,u+b+1}. \qquad (2.57)$$

When both a, b are odd, the lower bound reads

$$P(v_1 = r, \quad v_2 = u)$$

$$\geq \sum_{k=0}^{a}\sum_{h=0}^{b}(-1)^{k+h}\binom{k+r}{r}\binom{t+u}{u}S_{k+r,h+u}$$

$$-\binom{b+u+1}{b}\binom{a+r+1}{a}\frac{r+1}{a+1}\frac{u+1}{b+1}S_{r+a+1,u+b+1}. \qquad (2.58)$$

From the expressions of the upper and lower bounds in (2.55) and (2.56), it follows that what is intended by (2.55) and (2.56) is essentially a multivariate generalization in the nature of the Sobel-Uppuluri-Galambos univariate bounds (2.8) refining (2.6). We now consider the following examples to illustrate the difference.

Example 2.3.4. Consider the case of $m = 1$ with $B_1 = \Omega$, for $0 \leq t \leq n$, we have

$$\begin{aligned} S_{t,1} &= \sum P(A_{i_1}...A_{i_t}\Omega) \\ &= \sum P(A_{i_1}...A_{i_t}) \\ &= S_t. \end{aligned}$$

Now, the upper bound of (2.55) becomes, for an even integer a,

$$\begin{aligned} P(v_1 \geq r) &= P(v_1 \geq r, v_2 \geq 1) \\ &\leq \sum_{k=0}^{a}(-1)^k\binom{k+r-1}{r-1}S_{k+r} - \frac{r}{n-r}\binom{r+a}{a}S_{r+a+1}. \end{aligned}$$

This is because in (2.55), b can be taken as 0 to conform with $u+b \leq m$ with $u=1$.
□

Example 2.3.5 Consider two set of events in which $m = 2$ and $B_1 = B_2 = \Omega$. To obtain a univariate bound for the first set of events when the integer a is odd, we take $b = 1$ and $u = 1$ in (2.55). In this setting, $S_{k+r,1} = 2S_{k+r}$ and $S_{k+r,2} = S_{k+r}$. Thus for an odd integer a, (2.56) reads

$$\begin{aligned} P(v_1 \geq r) &= P(v_1 \geq r, v_2 \geq 1) \\ &\geq \sum_{k=0}^{a}\sum_{t=0}^{1}(-1)^{k+t}\binom{k+r-1}{r-1}S_{k+r,t+1} \\ &= \sum_{k=0}^{a}(-1)^k(2-1)\binom{k+r-1}{r-1}S_{k+r}. \end{aligned}$$

This means that (2.56) reverts to

$$P(v_1 \geq r) = P(v_1 \geq r, v_2 \geq 1) \geq \sum_{k=0}^{a} (-1)^k \binom{k+r-1}{r-1} S_{k+r} \qquad (2.59)$$

for odd integer a satisfying $r + a \leq n$.

Now, when $r = 1$, the inequality (2.59) becomes

$$P(v_1 \geq 1) \geq \sum_{k=0}^{a} (-1)^k S_{k+1}$$

which is a special case of a classical Bonferroni lower bound (2.6). $\qquad \square$

The above two examples illustrate the bivariate bound of the type $P(v_1 \geq r, \quad v_2 \geq u)$ when the bound is reverted to a univariate situation. In what follows in the section, we examine the similar performance of the bivariate bounds for the type of joint probability $P(v_1 = r, \quad v_2 = u)$ for the upper and lower bounds (2.57) and (2.58).

Example 2.3.6 Consider two sets of events in which $B_1 = \Omega$, $B_2 = \ldots = B_m = \emptyset$. Thus for any events in the first set and for any $1 \leq t \leq n$,

$$
\begin{aligned}
S_{t,1} &= \sum_{j} \sum_{1 \leq i_1 \leq \ldots \leq i_t \leq n} P(A_{i_1} \ldots A_{i_t} B_j) \\
&= \sum_{1 \leq i_1 \leq \ldots \leq i_t \leq n} P(A_{i_1} \ldots A_{i_t} \Omega) \\
&= \sum_{1 \leq i_1 \leq \ldots \leq i_t \leq n} P(A_{i_1} \ldots A_{i_t}) \\
&= S_t(A),
\end{aligned}
$$

also, when $k \geq 2$,

$$
\begin{aligned}
S_{t,k} &= \sum_{1 \leq j_1 \leq \ldots \leq j_k \leq m} \sum_{1 \leq i_1 \leq \ldots \leq i_t \leq n} P(A_{i_1} \ldots A_{i_t} B_{j_1} \ldots B_{j_k}) \\
&= 0.
\end{aligned}
$$

In this case, for any $0 \leq r \leq n$, $a + r \leq n$, and $b + 1 \leq m$, if both a, b are even, (2.57) reads,

$$
\begin{aligned}
&P(v_1 = r) \\
=\;& P(v_1 = r, v_2 = 1) \\
\leq\;& \sum_{k=0}^{a} \sum_{h=0}^{b} (-1)^{k+h} \binom{k+r}{r} \binom{h+1}{1} S_{k+r,h+1} \\
&- \frac{r+1}{n-r} \frac{u+1}{m-u} \binom{a+r+1}{a} \binom{b+u+1}{b} S_{r+a+1,u+b+1}
\end{aligned}
$$

$$= \sum_{k=0}^{a} (-1)^k \binom{k+r}{r} \binom{1}{1} S_{k+r,0+1}$$

$$= \sum_{k=0}^{a} (-1)^k \binom{k+r}{r} S_{k+r}(A),$$

due to the special design of the events $B_1, ..., B_m$ in this example. When both a, b are odd, (2.58) correspondingly reads

$$P(v_1 = r) = P(v_1 = r, v_2 = 1)$$

$$\geq \sum_{k=0}^{a} \sum_{h=0}^{b} (-1)^{k+h} \binom{k+r}{r} \binom{h+1}{1} S_{k+r,h+1}$$

$$- \frac{r+1}{a+1} \frac{u+1}{b+1} \binom{a+r+1}{a} \binom{b+u+1}{b} S_{r+a+1,u+b+1}$$

$$= \sum_{k=0}^{a} (-1)^k \binom{k+r}{r} \binom{1}{1} S_{k+r,0+1}$$

$$= \sum_{k=0}^{a} (-1)^k \binom{k+r}{r} S_{k+r}(A),$$

for the specially selected events $B_1, ..., B_m$. Therefore, bounds for $P(v_1 = r, v_2 = u)$ in Galambos and Lee (1994) are extensions of the classical univariate bounds (2.6)-(2.7). More lower and upper bounds of this type are studied in Chapter 3. In the following section, we introduce the concept of multivariate optimality for multivariate probability bounds.

2.4   Multivariate Optimality

In Section 2.3, we discuss two types of multivariate classical Bonferroni bounds that are improved by different bounding strategies. Similar to optimality for univariate bounds, a natural task is to seek the extent to which a bound can be improved. Such a task necessitates the discussion of multivariate optimality because once an optimal bound is reached, there is no space for improvement within the same bounding domain.

For univariate optimality, Section 2.2 presents three types of optimality: Fréchet optimality, linear optimality, and optimal bound defined in linear programming. For multivariate bounds, the complete account of multivariate optimality extends from definitions of multivariate optimality to explicit forms of optimal bounds in each particular setting. Such information is tailored to specific topics of discussion in later chapters. For example, Chapter 4 elucidates the bounding method of linear programming, which connects different concepts of multivariate optimality for two sets of events. In this section, we mainly introduce definitions of multivariate optimality, and leave further discussion on bivariate upper bounds to Chapter 5, bivariate lower bounds to Chapter 7, where multivariate optimality is addressed in detail.

Without loss of generality, for convenience, we use bivariate upper and lower bounds as examples to introduce the idea of multivariate optimality. There are three concepts for multivariate bounds, namely Fréchet optimality, linear optimality, and optimality defined in linear programming.

We first discuss the concept of multivariate Fréchet optimality. As shown in Section 2.2, the formulation of Fréchet optimality essentially treats each bound as a function of the Bonferroni summations in a permissible domain, and identifies the achievable bound among all the bounds in the class under consideration. Thus, we start the discussion with the domain of a set of multivariate Bonferroni summations.

Consider two sets of events $A_1, ..., A_n$ and $B_1, ..., B_m$. Let p, q be fixed integers satisfying $1 \leq p \leq n$ and $1 \leq q \leq m$. Denote T a subset of $\{(i, j); i = 0, 1, ..., n; j = 0, 1, ..., m\}$. Let U be the class of functions $F(.)$ defined on all values of $\{S_{ij}, (i, j) \in T\}$ such that for any collections of events $A_1, ..., A_n, B_1, ..., B_m$ in any probability space

$$P(v_1 \geq p, \quad v_2 \geq q) \leq F(S_{i,j}, (i, j) \in T).$$

Note that if the values of $\{S_{ij}, (i, j) \in T\}$ are arbitrary, there may not be any probability space that can accommodate such values as the set of Bonferroni summations. This is the case especially when those given values are not generated by legitimate events in a probability space. Thus, it is necessary to set a restriction on the legitimate values for the quantities $\{S_{ij}, (i, j) \in T\}$, which is the domain of the set of bivariate Bonferroni summations under consideration. Such a set of bivariate Bonferroni summations S_{ij}'s is called *consistent* in the sequel.

Definition 2.4.1 (Consistency): For a set of values $\{s_{ij}, \quad (i, j) \in T\}$, if there exists a probability space and two sets of events $\{A_i, i = 1, ..., n\}$ and $\{B_j, j = 1, ..., m\}$ such that $S_{ij}(A, B) = s_{ij}$ for $(i, j) \in T$, then the set of values $\{s_{ij}, (i, j) \in T\}$ is called a set of consistent bivariate Bonferroni summations.

Example 2.4.1 For two sets of ten events $A_1, ..., A_{10}$ and $B_1, ..., B_{10}$ in any probability space, the set of values $\{1.2, 17, 1.3, 2\}$ is not consistent for $\{S_{11}, S_{12}, S_{21}, S_{22}\}$ because the condition of

$$S_{12} = E[\binom{v_2}{2}\binom{v_1}{1}] = E[\frac{v_2(v_2 - 1)}{2}v_1] \leq 9S_{11}$$

is violated in this set of selected values. □

For a set of consistent multivariate Bonferroni summations, the domain of the bounding function $F(.)$ can be restricted to S_{ij}'s for collections of events $A_1, ..., A_n$, $B_1, ..., B_m$ in some probability space.

Definition 2.4.2 (Bivariate Fréchet Optimality) Suppose an upper bound $F_1 \in U$ (U is a class of bounding functions defined above) satisfies the condition that for any set of consistent bivariate Bonferroni summations $S_{ij}, (i, j) \in T$, there exists a set of events $A_1^*, ..., A_n^*$, and $B_1^*, ..., B_m^*$ in a probability space such that

$$S_{ij}(A_1^*, ..., A_n^*, B_1^*, ..., B_m^*) = S_{ij}, \quad (i, j) \in T$$

and

$$P(v_1(A^*) \geq p, \quad v_2(B^*) \geq q) = F_1(S_{ij}, (i,j) \in T), \tag{2.60}$$

then F_1 is a bivariate Fréchet optimal upper bound for $P(v_1 \geq p, \quad v_2 \geq q)$. □

With Definition 2.4.2, the definition of bivariate Fréchet optimal lower bound is self-evident.

Example 2.4.2: Consider $T = \{(0,0), (1,1), (1,2), (2,1), (2,2)\}$, which involves the probabilities of intersections of any two events of each class. Bounds using Bonferroni summations in T are called bivariate degree-two optimal (upper or lower) bounds. The bivariate Fréchet optimal lower bound using Bonferroni summations with an index in set T is the following.

For any consistent set of values of S_{ij}'s for

$$(i,j) \in T = \{(0,0), (1,1), (1,2), (2,1), (2,2)\},$$

consider the set of lower bounds L

$$P(v_1 \geq p, \quad v_2 \geq q) \geq L(S_{i,j}, (i,j) \in T).$$

Assuming that one of the lower bounds $F_1 \in L$ (L is the set of all degree-two lower bounds) is such that for any set of consistent bivariate Bonferroni summations S_{ij}, $(i,j) \in \{(0,0), (1,1), (1,2), (2,1), (2,2)\}$, there exist two sets of events $A_1^*, \ldots, A_n^*$, $B_1^*, \ldots, B_m^*$ in a possibly different probability space such that

$$S_{ij}(A_1^*, \ldots, A_n^*, B_1^*, \ldots, B_m^*) = S_{ij}, \quad (i,j) \in \{(0,0), (1,1), (1,2), (2,1), (2,2)\}$$

and

$$P(v_1(A^*) \geq p, \quad v_2(B^*) \geq q) = F_1(S_{ij}, (i,j) \in T), \tag{2.61}$$

then F_1 is said to be a Fréchet optimal lower bound for $P(v_1 \geq p, \quad v_2 \geq q)$. □

As mentioned in Section 2.2, the idea of Fréchet optimality dates to work of Fréchet in 1935 for comparisons of two functions (Seneta, 1992, Section 7) in such a way that for any set of consistent Bonferroni summations S_{ij}'s,

$$F_1(S_{ij}, (i,j) \in T) = P(v_1(A^*) \geq p, v_2(B^*) \geq q) \leq F(S_{ij}, (i,j) \in T)$$

for any permissible upper bound $F \in L$. Notice that F_1 has the highest value among all the lower bounds with the same degrees.

Although there are a lot of similarities between the univariate Fréchet optimality and multivariate Fréchet optimality, the latter involves more information on the correlation between the two event-classes.

Example 2.4.3 For any integers a and b satisfying $1 \leq a \leq n$ and $1 \leq b \leq m$, we have a lower bound

$$P(v_1 \geq 1, v_2 \geq 1) \geq \frac{4S_{1,1}}{(a+1)(b+1)} - \frac{4S_{2,1}}{a(a+1)(b+1)} -$$
$$- \frac{4S_{1,2}}{b(a+1)(b+1)} + \frac{4S_{2,2}}{ab(a+1)(b+1)} \tag{2.62}$$

for $(a,b) \in D = \{(k_1, k_2): \text{ either } k_1 \geq n/2 \text{ or } k_2 \geq m/2\}$.

The lower bound in (2.62) resembles the product of two univariate Dawson-Sankoff lower bounds. The univariate versions of this bound were proved to be an optimal univariate lower bound in Section 2.2. However, as shown in Chapter 7, the product of such univariate optimal lower bounds is not an optimal bivariate lower bound. Such a paradox in bounding theory stems from the dependence between the two sets of events in the formation of a bivariate joint probability. We will discuss this type of bound in Chapter 7, where an algorithm is constructed to identify the multivariate Fréchet optimal lower bound. $\square$

The above bounds are in terms of the full combinations of $S_{11}, S_{12}, S_{21}, S_{22}$. In fact, the structure of the multivariate optimal bound consists of various forms of the bounds. For example, in terms of S_{11}, S_{12}, and S_{21} only, Galambos and Xu (1993), Seneta and Chen (2000), have investigated this type of bound.

Example 2.4.4 For any two integers a and b satisfying $1 \leq a \leq n$ and $1 \leq b \leq m$, the optimal bivariate lower bound using the three bivariate Bonferroni summations, S_{11}, S_{12}, S_{21}, can be given as follows.

$$P(v_1 \geq 1, v_2 \geq 1) \geq \frac{1}{ab}[(3 - \frac{1}{a} - \frac{1}{b})S_{11} - \frac{2}{a}S_{21} - \frac{2}{b}S_{12}]. \quad (2.63)$$

More details on the analysis of the bound (2.63) are given in Chapter 7. $\square$

The second concept of multivariate optimality is linear optimality.

Definition 2.4.3 (Bivariate Linear Optimality): Consider a collection of upper bounds taking the linear form

$$cS_{11} + dS_{12} + eS_{21} + fS_{22} \quad (2.64)$$

with coefficients $(c, d, e, f) \in D$, a set of permissible constants that make (2.64) an upper bound for the probability $P(v_1 \geq 1, v_2 \geq 1)$. If, for any set of events in any probability spaces, $\{A_i\}, i = 1, ..., n, \{B_j\}, j = 1, ..., m$, there exists an upper bound

$$P(v_1 \geq 1, v_2 \geq 1) \leq \alpha S_{11} + \beta S_{21} + \gamma S_{12} + \delta S_{22}$$

where coefficients $\alpha, \beta, \gamma, \delta$ depend on the total event sizes n, m, but not on the space or the events $\{A_i\}, \{B_j\}$, and

$$\alpha S_{11} + \beta S_{21} + \gamma S_{12} + \delta S_{22} \leq cS_{11} + dS_{12} + eS_{21} + fS_{22}$$

for any bound with $(c, d, e, f) \in D$, then the upper bound $\alpha S_{11} + \beta S_{21} + \gamma S_{12} + \delta S_{22}$ is a bivariate optimal linear upper bound. $\square$

As shown in Section 2.2, the concept of multivariate linear optimality is an extension of univariate linear optimality in the setting of multivariate bounds.

The following example introduces a bivariate optimal linear upper bound (defined in Definition 2.4.2) using the bivariate Bonferroni summations $\{S_{11}, S_{12}, S_{21}, S_{22}\}$. The complete proof of linear optimality is deferred to Chapter 5, which focuses on bivariate upper bounds.

Example 2.4.5 Let

$$U_{n,m} = S_{11} - \frac{2}{n}S_{21} - \frac{2}{m}S_{12} + \frac{4}{nm}S_{22},$$

then

$$P(v_1 \geq 1, v_2 \geq 1) \leq U_{n,m} \qquad (2.65)$$

and $U_{n,m}$ is the best upper bound among all upper bounds of the form $d_1 S_{1,1} + d_2 S_{2,1} + d_3 S_{1,2} + d_4 S_{2,2}$ where the coefficients d_j, $1 \leq j \leq 4$ are constants.   □

To illustrate the difference between linear optimality and Fréchet optimality in a bivariate setting, we use an example to show that for some special values of Bonferroni summations, the corresponding events $\{A_i^*\}$ and $\{B_j^*\}$ as specified in Definition 2.4.3 may not even exist. Thus a linear optimal bound is not necessarily a Fréchet optimal bound.

The construction of the example requires the specification of all probabilities for the elementary conjunctions of two sets of events. For a set of events A_1, ..., A_n, and B_1, ..., B_m in a probability space, the *elementary conjunctions of these two sets of event* are the 2^{n+m} different combinations of the possible occurrences of the events.

Example 2.4.6   Elementary conjunctions of two sets of events
For a set of four events A_1, A_2 and B_1, B_2 (here $n = m = 2$), the elementary conjunctions of the two sets of events consist of the following 16 joint events:

$$A_1 \cap A_2 \cap B_1 \cap B_2 \quad A_1 \cap A_2 \cap B_1 \cap B_2^c \quad A_1 \cap A_2 \cap B_1^c \cap B_2 \quad A_1 \cap A_2 \cap B_1^c \cap B_2^c$$

$$A_1 \cap A_2^c \cap B_1 \cap B_2 \quad A_1 \cap A_2^c \cap B_1 \cap B_2^c \quad A_1 \cap A_2^c \cap B_1^c \cap B_2 \quad A_1 \cap A_2^c \cap B_1^c \cap B_2^c$$

$$A_1^c \cap A_2 \cap B_1 \cap B_2 \quad A_1^c \cap A_2 \cap B_1 \cap B_2^c \quad A_1^c \cap A_2 \cap B_1^c \cap B_2 \quad A_1^c \cap A_2 \cap B_1^c \cap B_2^c$$

$$A_1^c \cap A_2^c \cap B_1 \cap B_2 \quad A_1^c \cap A_2^c \cap B_1 \cap B_2^c \quad A_1^c \cap A_2^c \cap B_1^c \cap B_2 \quad A_1^c \cap A_2^c \cap B_1^c \cap B_2^c$$

In this way, the probability of $P(A_1 B_1)$ equals the summation of the probabilities of the elementary conjunctions that are part of the event $A_1 A_2$. For instance,

$$P(A_1 B_1) = P(A_1 A_2 B_1 B_2) + P(A_1 A_2^c B_1 B_2) + P(A_1 A_2^c B_1 B_2) + P(A_1 A_2^c B_1 B_2^c),$$

because all the events are mutually exclusive.   □

The following numerical example is constructed on the basis of the idea of elementary conjunctions of two sets of events. It shows that a set of consistent bivariate Bonferroni summations may result in unachievable linear optimal upper bounds.

Example 2.4.7 Consider a univariate example with events $A_1,...,A_6$ in a probability space. To specify an actual example requires the specification of probabilities of all 2^6 elementary conjunctions involving the events A_1, ..., A_6 and their complements as follows.

With the probabilities of the elementary conjunctions of A_1, ..., A_6 specified in Tables 2.1 and 2.2, we get the pattern of relevant features as follows:

Table 2.1 *Numerical example of inconsistent optimality-I*

Event of elementary conjunction C	Probability P(C)
$A_1A_2A_3A_4A_5A_6^c$	0.0155
$A_1A_2A_3A_4A_5^cA_6$	0.0155
$A_1A_2A_3A_4A_5A_6$	0
$A_1A_2A_3A_4A_5^cA_6^c$	0.0315
$A_1A_2A_3A_4^cA_5A_6^c$	0.01563
$A_1A_2A_3A_4^cA_5^cA_6$	0.01563
$A_1A_2A_3A_4^cA_5A_6$	0
$A_1A_2A_3A_4^cA_5^cA_6^c$	0.03124
$A_1A_2A_3^cA_4A_5A_6^c$	0.005625
$A_1A_2A_3^cA_4A_5^cA_6$	0.005625
$A_1A_2A_3^cA_4A_5A_6$	0
$A_1A_2A_3^cA_4A_5^cA_6^c$	0.05125
$A_1A_2A_3^cA_4^cA_5A_6^c$	0.015125
$A_1A_2A_3^cA_4^cA_5^cA_6$	0.015125
$A_1A_2A_3^cA_4^cA_5A_6$	0
$A_1A_2A_3^cA_4^cA_5^cA_6^c$	0.03225
$A_1A_2^cA_3A_4A_5A_6^c$	0.020625
$A_1A_2^cA_3A_4A_5^cA_6$	0.020625
$A_1A_2^cA_3A_4A_5A_6$	0
$A_1A_2^cA_3A_4A_5^cA_6^c$	0.02125
$A_1A_2^cA_3A_4^cA_5A_6^c$	0.001
$A_1A_2^cA_3A_4^cA_5^cA_6$	0.001
$A_1A_2^cA_3A_4^cA_5A_6$	0
$A_1A_2^cA_3A_4^cA_5^cA_6^c$	0.0605
$A_1A_2^cA_3^cA_4A_5A_6^c$	0.005
$A_1A_2^cA_3^cA_4A_5^cA_6$	0.005
$A_1A_2^cA_3^cA_4A_5A_6$	0
$A_1A_2^cA_3^cA_4A_5^cA_6^c$	0.0615
$A_1A_2^cA_3^cA_4^cA_5A_6^c$	0.00025
$A_1A_2^cA_3^cA_4^cA_5^cA_6$	0.00025
$A_1A_2^cA_3^cA_4^cA_5A_6$	0
$A_1A_2^cA_3^cA_4^cA_5^cA_6^c$	0.062

$$P(A_1A_2) = 0.25 \quad P(A_1A_3) = 0.25 \quad P(A_1A_4) = 0.25$$

$$P(A_1A_5) - 0.0745 \quad P(A_1A_6) = 0.0745 \quad P(A_2A_3) = 0.25$$

$$P(A_2A_4) = 0.244 \quad P(A_2A_5) = 0.1085 \quad P(A_2A_6) = 0.108$$

$$P(A_3A_4) = 0.25 \quad P(A_3A_5) = 0.109 \quad P(A_3A_6) = 0.109$$

$$P(A_4A_5) = 0.089 \quad P(A_4A_6) = 0.089 \quad P(A_5A_6) = 0$$

Table 2.2 *Numerical example of inconsistent optimality-II*

Event of elementary conjunction C	Probability P(C)
$A_1^c A_2 A_3 A_4 A_5 A_6^c$	0.0156
$A_1^c A_2 A_3 A_4 A_5^c A_6$	0.0156
$A_1^c A_2 A_3 A_4 A_5 A_6$	0
$A_1^c A_2 A_3 A_4 A_5^c A_6^c$	0.0313
$A_1^c A_2 A_3 A_4^c A_5 A_6^c$	0.01
$A_1^c A_2 A_3 A_4^c A_5^c A_6$	0.01
$A_1^c A_2 A_3 A_4^c A_5 A_6$	0
$A_1^c A_2 A_3 A_4^c A_5^c A_6^c$	0.0425
$A_1^c A_2 A_3^c A_4 A_5 A_6^c$	0.00555
$A_1^c A_2 A_3^c A_4 A_5^c A_6$	0.00555
$A_1^c A_2 A_3^c A_4 A_5 A_6$	0.0257
$A_1^c A_2 A_3^c A_4 A_5^c A_6^c$	0.0257
$A_1^c A_2 A_3^c A_4^c A_5 A_6^c$	0
$A_1^c A_2 A_3^c A_4^c A_5^c A_6$	0.021
$A_1^c A_2 A_3^c A_4^c A_5 A_6$	0
$A_1^c A_2 A_3^c A_4^c A_5^c A_6^c$	0.0415
$A_1^c A_2^c A_3 A_4 A_5 A_6^c$	0.0165
$A_1^c A_2^c A_3 A_4 A_5^c A_6$	0.0165
$A_1^c A_2^c A_3 A_4 A_5 A_6$	0
$A_1^c A_2^c A_3 A_4 A_5^c A_6^c$	0.0295
$A_1^c A_2^c A_3 A_4^c A_5 A_6^c$	0.014
$A_1^c A_2^c A_3 A_4^c A_5^c A_6$	0.014
$A_1^c A_2^c A_3 A_4^c A_5 A_6$	0
$A_1^c A_2^c A_3 A_4^c A_5^c A_6^c$	0.0345
$A_1^c A_2^c A_3^c A_4 A_5 A_6^c$	0.00905
$A_1^c A_2^c A_3^c A_4 A_5^c A_6$	0.00905
$A_1^c A_2^c A_3^c A_4 A_5 A_6$	0
$A_1^c A_2^c A_3^c A_4 A_5^c A_6^c$	0.0444
$A_1^c A_2^c A_3^c A_4^c A_5 A_6^c$	0.01635
$A_1^c A_2^c A_3^c A_4^c A_5^c A_6$	0.01635
$A_1^c A_2^c A_3^c A_4^c A_5 A_6$	0
$A_1^c A_2^c A_3^c A_4^c A_5^c A_6^c$	0.0298

$$P(A_1) = 0.109 \quad P(A_2) = 0.5 \quad P(A_3) = 0.5$$
$$P(A_4) = 0.49 \quad P(A_5) = 0.187 \quad P(A_6) = 0.187$$

and the associated Bonferroni summations for two classes of events $E_i = A_i$, $H_i = A_i$ in two independent probability spaces for $i = 1, ..., 6$:

$$S_{11} = S_1(E)S_1(H) = 2.364 \times 2.364 = 5.59,$$

$$S_{12} = S_1(E)S_2(H) = 2.364 \times 2.252 = 5.32,$$

$$S_{21} = S_2(E)S_1(H) = 2.252 \times 2.364 = 5.32,$$
$$S_{22} = S_2(E)S_2(H) = 2.252 \times 2.252 = 5.07.$$

The value of the associated linear optimal bound reads (see Example 2.4.1)

$$S_{11} - \frac{2}{n}S_{12} - \frac{2}{m}S_{21} + \frac{4}{nm}S_{22}$$
$$= (S_1(E) - \frac{2}{n}S_2(E))(S_1(H) - \frac{2}{n}S_2(H))$$
$$= (2.364 - (2.252/6))(2.364 - (2.252/6))$$
$$= 3.96 > 1.$$

Such an optimal upper bound can never be achieved, even though the set of Bonferroni summations $\{5.59, 5.32, 5.32, 5.07\}$ is consistent with $\{S_{11}, S_{12}, S_{21}, S_{22}\}$. We can see from this example that linear optimality does not imply Fréchet optimality. □

We now introduce the concept of bivariate optimality defined in linear programming.

Definition 2.4.4 (Bivariate Linear Optimality): Denote $p_{ij} = P(v_1 = i, v_2 = j)$. An optimal upper bound for $P(v_1 \geq 1, v_2 \geq 1)$, given consistent bivariate Bonferroni summations S_{ij}, $i = 1, ..., t$, $j = 1, ..., k$, is defined in the solution of the optimum value for the linear programming problem:

$$\max(p_{11} + ... + p_{nm})$$

subject to $\sum_{i=0}^{n}\sum_{j=0}^{m} p_{ij} = 1$, and

$$p_{11} + 2p_{21} + ... + tp_{t1} + ... + nmp_{nm} = S_{11}$$
$$p_{21} + ... + \binom{t}{2}p_{t1} + ... + \binom{n}{2}mp_{nm} = S_{21}$$
$$..... \quad ..$$
$$p_{tk} + ... + \binom{n}{t}\binom{m}{k}p_{nm} = S_{tk}$$

for $1 \leq t \leq n$, $1 \leq k \leq m$.   □

This concept will be discussed in detail in Chapter 4, focusing on the bounding approach of linear programming, in which we also show the connection between Fréchet optimality and the optimality in linear programming.

With the fundamental issues introduced in this chapter, we can construct upper and lower bounds and discuss the corresponding optimality. In the next two chapters, we concentrate on two basic principles in the construction of Bonferroni-type bounds. One is the method of indicator functions (Chapter 3) and the other is the method of linear programming (Chapter 4).

Chapter 3

Multivariate Indicator Functions

With basic terminologies of Bonferroni-type inequalities in the preceding chapter, we can now discuss methods in the construction of Bonferroni-type inequalities, an essential component in probability bounding theory. Chapters 3 and 4 feature two discernible methods in the derivation of Bonferroni-type inequalities. The current chapter focuses on the use of indicator functions and Chapter 4 concentrates on the method of linear programming and optimality.

The method of indicator functions is useful not only in constructing new bounds, but also in bridging probability bounding theory with the theory of combinatorics. Such a connection enhances mathematical elegance in probability bounding theory. Materials in this chapter mainly include the work of Galambos and Lee (1994), Galambos and Xu (1993), Galambos and Mucci (1980), Jogdeo (1977), Chen and Seneta (1996), and Seneta and Chen (2005).

3.1 Method of Indicator Functions

The principle of the method of indicator functions is essentially the decomposition of the probability of the number of occurrences (for a set of events) into the expected value of a sequence of binomial moments. The method can be described as follows.

The indicator function of an event is usually defined as

$$I(A) = \begin{cases} 1, & \text{if } A \text{ occurs} \\ 0, & \text{if } A \text{ fails,} \end{cases}$$

where the indicator function $I(A)$ satisfies the following properties:

- The indicator function of a joint event $I(A \cap B) = I(A)I(B)$ for any two events A and B;

- The indicator function of a complementary event $I(A^c) = 1 - I(A)$.

In the introduction of univariate bounds in Section 2.1.2, we used the method of indicator functions to discuss the univariate bounds. In what follows, we will focus on the method for the setting of bivariate bounds. Similar principles of analysis are then applicable to multivariate bounds.

For any two classes of events $A_1, ..., A_n$, and $B_1, ..., B_m$, let T be a subset of the complete index set, $T \subset \{(i,j), i = 1,...,n, j = 1,...,m\}$. Let v_1 and v_2 be the numbers of the events $\{A_i, i = 1,...,n\}$ and $\{B_j, j = 1,...,m\}$ that occur, respectively, at a sample point. Assume that two functions on v_1 and v_2 exist such that

$$G_1(v_1, v_2) = \sum_{(i,j) \in T} c_{ij} \binom{v_i}{i} \binom{v_j}{j},$$

and

$$G_2(v_1, v_2) = \sum_{(i,j) \in T} d_{ij} \binom{v_i}{i} \binom{v_j}{j}.$$

Suppose that

$$G_1(v_1, v_2) \leq I(v_1 \geq 1, v_2 \geq 1) \leq G_2(v_1, v_2), \tag{3.1}$$

where c_{ij} and d_{ij} are coefficients that may depend on n and m, but not on the classes of events under consideration.

Since

$$E(I(v_1 \geq 1, v_2 \geq 1)) = P(v_1 \geq 1, v_2 \geq 1),$$

and

$$E(\binom{v_1}{i} \binom{v_2}{j}) = S_{ij}, \tag{3.2}$$

taking the expectation in (3.1), in conjunction with (3.2), yields

$$\sum_{(i,j) \in T} c_{ij} E(\binom{v_i}{i} \binom{v_j}{j}) \leq E(I(v_1 \geq 1, v_2 \geq 1)) \leq \sum_{(i,j) \in T} d_{ij} E(\binom{v_i}{i} \binom{v_j}{j}).$$

The above inequality is equivalent to the following inequality, after considering the domain of the variables v_1 and v_2,

$$\sum_{(i,j) \in T} c_{ij} S_{ij} \leq P(v_1 \geq 1, v_2 \geq 1) \leq \sum_{(i,j) \in T} d_{ij} S_{ij}. \tag{3.3}$$

Notice that in (3.3), the left-hand side $\sum_{(i,j) \in T} c_{ij} S_{ij}$ is actually a lower Bonferroni-type bound while the right-hand side $\sum_{(i,j) \in T} d_{ij} S_{ij}$ is an upper Bonferroni-type bound of the joint probability of at least one of the events occurring in each of the two classes of events $P(v_1 \geq 1, v_2 \geq 1)$. In this way, we can always obtain an upper bound or a lower bound for the joint probability of the occurrence of two sets of events. This is the *method of indicator functions* for the joint probability that at least one event occurs in both classes.

The key step toward the construction of upper and lower probability bounds is the materialization of the assumption on the two function G_1 and G_2. In what follows, we provide some examples to illustrate the construction of such "bound-to-be" functions (see, for example, Recsei and Seneta (1987), Lee (1992)).

Example 3.1.1 Consider degree-two bivariate lower Bonferroni-type bounds where the bivariate Bonferroni assumptions are S_{11}, S_{12}, S_{21} and S_{22}. The bivariate Bonferroni summations can be expressed as

$$\begin{aligned}
S_{11} &= E(v_1 v_2) \\
S_{12} &= E[v_1 v_2 (v_2 - 1)/2] \\
S_{21} &= E[v_2 v_1 (v_1 - 1)/2] \\
S_{22} &= E[v_1 (v_1 - 1) v_2 (v_2 - 1)/4].
\end{aligned}$$

Denote $h(x,y)$ the function with algebraic variables, x, y that takes the following form:

$$h(x,y) = axy + b\frac{x(x-1)y}{2} + c\frac{xy(y-1)}{2} + d\frac{x(x-1)y(y-1)}{4}$$

for $0 \le x \le n, 0 \le y \le m$, where

$$h(x,y) \le 1,$$

and constants a, b, c, and d are to be decided by the method of indicator functions. Noting that $h(0,y) = h(x,0) = 0$ and using random variables v_1, v_2, we get

$$h(v_1, v_2) \le I(v_1 \ge 1, v_2 \ge 1),$$

so

$$E(h(v_1, v_2)) \le P(v_1 \ge 1, v_2 \ge 1)$$

where

$$\begin{aligned}
&E(h(v_1, v_2)) \\
&= aE(\binom{v_1}{1}\binom{v_2}{1}) + bE(\binom{v_1}{2}\binom{v_2}{1}) + cE(\binom{v_1}{1}\binom{v_2}{2}) + dE(\binom{v_1}{2}\binom{v_2}{2}) \\
&= aS_{11} + bS_{21} + cS_{12} + dS_{22}. \tag{3.4}
\end{aligned}$$

We now discuss the determination of the constants a, b, c, and d under two different scenarios. The first scenario is discussed in Lee (1992) in which the four constants (a, b, c, and d in (3.4)) are reduced to two parameters to include all possible bounds of this type. The second scenario is found in Chen and Seneta (1995a) where the four constants are reduced to one parameter.

Example 3.1.2 Lee (1992)

For any integers k_1, k_2,　$k_1 \ge 2$　$k_2 \ge 2$, the bivariate lower bound of (3.4) takes the form

$$P(v_1 \ge 1, v_2 \ge 1) \ge \frac{2}{k_1 k_2} S_{11} - \frac{2}{k_1 k_2^2} S_{12} - \frac{2}{k_1^2 k_2} S_{21}.$$

Proof: By (3.4), as in the discussion for the method of indicator functions, it suffices to show

$$f(x,y) = xy(2k_1k_2 + k_1 + k_2 - k_1y - k_2x) \leq (k_1k_2)^2$$

for $1 \leq x \leq n$ and $1 \leq y \leq m$.
When x and y are real values in the bivariate function,

$$f(x,y) = xy(2k_1k_2 + k_1 + k_2 - k_1y - k_2x),$$

taking partial derivatives gets

$$f_x' = y(2k_1k_2 + k_1 + k_2 - k_1y - k_2x) - k_2xy$$

$$f_y' = x(2k_1k_2 + k_1 + k_2 - k_1y - k_2x) - k_1xy.$$

Letting $f_x' = 0$ and $f_y' = 0$, we have

$$x_0 = \frac{2k_1k_2 + k_1 + k_2}{3k_2} \quad \text{and} \quad y_0 = \frac{2k_1k_2 + k_1 + k_2}{3k_1}$$

for $x > 0$ and $y > 0$.
Now

$$f_{xx}''(x_0, y_0)f_{yy}''(x_0, y_0) - f_{xy}''^{2}(x_0, y_0) = \frac{1}{3}(2k_1k_2 + k_1 + k_2)^2 > 0$$

$$f_{xx}''(x_0, y_0) = -\frac{2k_2^2(2k_1k_2 + k_1 + k_2)}{3k_1k_2} < 0$$

$$f_{yy}''(x_0, y_0) = -\frac{2k_1^2(2k_1k_2 + k_1 + k_2)}{3k_1k_2} < 0.$$

So, at (x_0, y_0), $f(x,y)$ has a local maximum; this maximum value equals

$$\left(\frac{2k_1k_2 + k_1 + k_2}{3}\right)^3 \frac{1}{k_1k_2} = (k_1k_2)^2\left(\frac{2}{3} + \frac{1}{3k_2} + \frac{1}{3k_1}\right)^3 \leq (k_1k_2)^2$$

for $k_1 \geq 2$ $k_2 \geq 2$.
   Since we require $x \geq 1$, $y \geq 1$, in the area of $\{(x,y) : x \geq 1, y \geq 1\}$, no other unusual points exist such that $f_x' = 0$ and $f_y' = 0$. Thus the local maximum value at (x_0, y_0) is the absolute maximum of $f(x,y)$ for $x \geq 1$, $y \geq 1$. This implies that,

$$\max_{x \geq 1, y \geq 1} f(x,y) \leq (k_1k_2)^2. \tag{3.5}$$

The inequality in (3.5) can be written as, for $v_1 \geq 1$, $v_2 \geq 1$,

$$\frac{2}{k_1k_2}v_1v_2 - \frac{2}{k_1^2k_2}\frac{v_1(v_1-1)v_2}{2} - \frac{2}{k_2^2k_1}\frac{v_1v_2(v_2-1)}{2} \leq I(v_1 \geq 1, v_2 \geq 1).$$

Taking expectation in the above inequality yields

$$P(v_1 \geq 1, v_2 \geq 1) \geq \frac{2}{k_1 k_2} S_{11} - \frac{2}{k_1 k_2^2} S_{12} - \frac{2}{k_1^2 k_2} S_{21},$$

which gives a lower bound for $P(v_1 \geq 1, v_2 \geq 1)$, in terms of S_{11}, S_{12} and S_{21}. Note that it uses two parameters k_1 and k_2 as parameters in the bound. $\square$

As an application of the method of indicator functions, we can now use the same technique to establish another lower bound for $P(v_1 \geq 1, v_2 \geq 1)$, using the three bivariate Bonferroni summations S_{11}, S_{12}, and S_{21} again, but depending on *only one parameter t* in the following derivation. More discussion of this issue can be found in Chen and Seneta (1995a).

Example 3.1.3 Chen and Seneta (1995a)

The bivariate bound in (3.4) can be simplified as follows.

$$P(v_1 \geq 1, v_2 \geq 1) \geq \frac{2}{k_1^3 k_2^3} S_{11} - \frac{2}{k_1^3 k_2^3 (k_1 k_2 - 1)} S_{12} - \frac{2}{k_1^3 k_2^3 (k_1 k_2 - 1)} S_{21}$$

for any k_1, k_2 integers, $k_1 k_2 \geq 2$.

Notice the joint effect of the product $k_1 k_2$, we can put $k_1 k_2 = t$ to get

$$P(v_1 \geq 1, v_2 \geq 1) \geq \frac{2}{t^3} S_{11} - \frac{2}{t^3 (t-1)} S_{12} - \frac{2}{t^3 (t-1)} S_{21}.$$

Proof: Using the method of indicator functions and considering the following function that has an analogous structure to the bound in Example 3.1.2,

$$f(x,y) = xy(2(k_1 k_2 - 1) - (y-1) - (x-1)).$$

Taking partial derivatives of the above bivariate function yields

$$f_x' = y(2k_1 k_2 - y - x) - xy$$

$$f_y' = x(2k_1 k_2 - y - x) - xy.$$

Now, letting $f_x' = f_y' = 0$ results in the equilibrium $x = y$, which consequently implies that

$$y(2k_1 k_2 - 2y) - y^2 = 0,$$

$$y(2k_1 k_2 - 3y) = 0,$$

and

$$y = \frac{2}{3} k_1 k_2 \qquad (\text{for} \quad y \geq 1).$$

Therefore, the unusual point is $(\frac{2}{3} k_1 k_2, \frac{2}{3} k_1 k_2)$.

To verify the second derivatives, we have

$$f''_{xx} = -2y,$$

$$f''_{yy} = -2x,$$

and

$$f''_{xy} = 2k_1 k_2 - x - y - x - y.$$

Thus

$$f''_{xx}|_{(\frac{2}{3}k_1k_2, \frac{2}{3}k_1k_2)} = -\frac{4}{3}k_1k_2 < 0$$

$$f''_{yy}|_{(\frac{2}{3}k_1k_2, \frac{2}{3}k_1k_2)} = -\frac{4}{3}k_1k_2 < 0$$

$$f''_{xy}|_{(\frac{2}{3}k_1k_2, \frac{2}{3}k_1k_2)} = -\frac{2}{3}k_1k_2 < 0,$$

and

$$f''_{xx}f''_{yy} - (f''_{xy})^2 > 0.$$

The above verification shows that $(\frac{2}{3}k_1k_2, \frac{2}{3}k_1k_2)$ is a maximal point in $x \geq 1$ and $y \geq 1$. Therefore, we have

$$\begin{aligned} f(\frac{2}{3}k_1k_2, \frac{2}{3}k_1k_2) &= xy(2k_1k_2 - x - y)|_{(\frac{2}{3}k_1k_2, \frac{2}{3}k_1k_2)} \\ &= (\frac{2}{3}k_1k_2)^3 \\ &\leq (k_1k_2)^3. \end{aligned} \tag{3.6}$$

Putting the result of (3.6) into the indicator function yields

$$f(x,y) \leq (k_1k_2)^3 \qquad x \geq 1, \quad y \geq 1,$$

which means that

$$\frac{1}{k_1^3 k_2^3} xy(2(k_1k_2 - 1) - (y - 1) - (x - 1)) \leq 1 \quad \text{for} \quad x \geq 1, \quad y \geq 1.$$

Now, taking expectation in the above inequality leads to

$$\begin{aligned} &P(v_1 \geq 1, v_2 \geq 1) \\ &\geq (\frac{2}{k_1^3 k_2^3}S_{11} - \frac{2}{k_1^3 k_2^3(k_1k_2 - 1)}S_{12} - \frac{2}{k_1^3 k_2^3(k_1k_2 - 1)}S_{21})(k_1k_2 - 1) \\ &\geq \frac{2}{k_1^3 k_2^3}S_{11} - \frac{2}{k_1^3 k_2^3(k_1k_2 - 1)}S_{12} - \frac{2}{k_1^3 k_2^3(k_1k_2 - 1)}S_{21}, \end{aligned} \tag{3.7}$$

for $k_1 k_2 \geq 2$. Taking $t = k_1 k_2$ in (3.7) gets the conclusion in Example 3.1.3. □

We now use currently published results to further illustrate the method of indicator functions.

3.2 Moments of Bivariate Indicator Functions

In this section, we discuss the use of indicator functions in deriving high-degree multivariate upper and lower bounds. For convenience, we discuss the topic of bounding the probabilities of the occurrence of "exactly r events" and "at least r events," separately. In fact, a simplified version of these bounds was mentioned (without derivation) in Section 2.3 as an example to introduce the concept of multivariate Bonferroni type bounds. Here, we delineate the derivation aspect to show the method of generating probability bounds using indicator functions (see, for example, Galambos and Simonelli (2004), and Galambos and Xu (1995)).

3.2.1 Bounds for Joint Probability of Exactly r Occurrences

As mentioned in Chapter 2 (Section 2.3) regarding a multivariate generalization of the classical Bonferroni inequalities, the result of Galambos and Lee (1994), when applied to the univariate case, becomes

$$P(v_1 = r) \geq \sum_{i=0}^{2k-1} (-1)^i \binom{i+r}{r} S_{i+r}(A),$$

for any integer $k \leq (n+1)/2$. Taking $r = 0$ gets

$$P(v_1 = 0) \geq \sum_{i=0}^{2k-1} (-1)^i S_i(A),$$

which implies

$$P(v_1 \geq 1) \leq \sum_{i=1}^{2k-1} (-1)^{i+1} S_i.$$

The above bound is, however, weaker than the Sobel-Uppuluri-Galambos upper bound. It is just a special case of the right-hand side of the classical Bonferroni bound.

Seeking the corresponding bivariate version that refines the classical Bonferroni bound, we consider the bivariate quantity parallel to the univariate summation $A_r^n(a)$ and $B_r^n(a)$ (see (2.6) and (2.7) in Section 2.1) as follows.

For any integers $a \geq 0$, and $b \geq 0$, denote

$$A_{ru}^{nm}(a,b) = \sum_{k=0}^{a}\sum_{t=0}^{b} (-1)^{k+t} \binom{k+r-1}{r-1}\binom{t+u-1}{u-1} S_{k+r,t+u} \qquad (3.8)$$

$$B_{ru}^{nm}(a,b) = \sum_{k=0}^{a}\sum_{t=0}^{b} (-1)^{k+t} \binom{k+r}{r}\binom{t+u}{u} S_{k+r,t+u}. \qquad (3.9)$$

With the above notation, the method of indicator functions leads to the following results on the occurrence of exactly r events from the class of events $\{A_i\}$ and exactly u events from the class of events $\{B_j\}$.

Theorem 3.2.1 Let a and b be two positive even integers. For integers $r \leq n$, $u \leq m$, the bivariate upper bound of the probability of $\{v_1 = r, v_2 = u\}$ can be obtained as

$$P(v_1 = r, v_2 = u) \leq B_{ru}^{nm}(a,b) - \Delta_1 - \Delta_2 - \Delta_3 \qquad (3.10)$$

where

$$
\begin{aligned}
\Delta_1 &= & 0 \quad \text{for} \quad n = r; \\
&= & \binom{r+a}{a} \frac{r+a+1}{n-r} S_{r+a+1,(u)} \quad \text{for} \quad n > r. \\
\Delta_2 &= & 0 \quad \text{for} \quad m = u; \\
&= & \binom{u+b}{b} \frac{u+b+1}{m-u} S_{(r),u+b+1} \quad \text{for} \quad m > u. \\
\Delta_3 &= & 0 \quad \text{for} \quad n = r \quad \text{or} \quad m = u; \\
&= & \binom{r+a}{a}\binom{u+b}{b} \frac{r+a+1}{n-r} \frac{u+b+1}{m-u} S_{r+a+1,u+b+1} \quad \text{for} \quad n > r \quad \text{and} \quad m > u.
\end{aligned}
$$

Notice that the bivariate bound in the above theorem involves bivariate Bonferroni summations $S_{i,j}$ as well as the partial Bonferroni summations $S_{(i),j}$ and $S_{i,(j)}$ introduced in Chapter 2. Before starting with the derivation of the above theorem, we connect the bivariate bound in (3.10) with the associated univariate bounds as follows.

Example 3.2.1 Consider the case where $m = 1$, $B_1 = \Omega$, $u = 1$, $S_{t,(1)} = S_t(A)$, $P(v_1 = r, v_2 = 1) = P(v_1 = r)$, $S_{t,1} = S_t(A)$; and $S_{t,j} = 0$ for $j > 1$. In this case, we have

$$B_{r1}^{nm}(a,b) = \sum_{k=0}^{a} (-1)^k \binom{k+r}{r} S_{k+r}(A),$$

$$\Delta_2 = \Delta_3 = 0, \quad \Delta_1 = \binom{r+a}{a} \frac{r+a+1}{n-r} S_{r+a+1}^A = \frac{a+1}{n-r} \binom{a+r+1}{r} S_{r+a+1}^A.$$

So, the upper bound in Theorem 3.2.1 returns to the univariate Sobel-Uppuluri-Galambos upper bound. □

We now start the proof of Theorem 3.2.1.
Proof: Notice that the method of indicator functions essentially takes the expected value of the binomial moment. We start the proof with the following result on linear combinations of a set of Bonferroni summations.

For any integer $v_1 \geq r$,

$$\sum_{k=0}^{a} (-1)^k \binom{k+r}{r} \binom{v_1}{k+r} = \delta_{v_1,r} + (-1)^a \binom{v_1}{r} \binom{v_1 - r - 1}{a} I_{(v_1 > r)}. \qquad (3.11)$$

To show (3.11), notice that

$$\binom{k+r}{r}\binom{v_1}{k+r} = \binom{v_1}{r}\binom{v_1-r}{k}.$$

We have

$$\sum_{k=0}^{a}(-1)^k\binom{k+r}{r}\binom{v_1}{k+r} = \sum_{k=0}^{a}(-1)^k\binom{v_1}{r}\binom{v_1-r}{k}. \tag{3.12}$$

Now, recall

$$\binom{x}{j} = \frac{x(x-1)\ldots(x-j+1)}{j!} \quad \text{for each real} \quad x, \quad j\geq 1,$$

and

$$\binom{x}{0} = 1 \quad \text{for each real x, by definition.}$$

If $v_1 = r$,

$$\sum_{k=0}^{a}(-1)^k\binom{v_1}{r}\binom{v_1-r}{k} = 1 = \delta_{v_1,r}+(-1)^a\binom{v_1}{r}\binom{v_1-r-1}{a}I_{(v_1>r)}.$$

If $v_1 > r$,

$$\sum_{k=0}^{a}(-1)^k\binom{v_1}{r}\binom{v_1-r}{k} = (-1)^a\binom{v_1}{r}\binom{v_1-r-1}{a}$$

$$= \delta_{v_1,r}+(-1)^a\binom{v_1}{r}\binom{v_1-r-1}{a}I_{(v_1>r)},$$

because

$$\sum_{k=0}^{\alpha}(-1)^k\binom{\beta}{k} = (-1)^\alpha\binom{\beta-1}{\alpha} \tag{3.13}$$

for any β and integer $\alpha \geq 0$. The equation (3.13) can be proved by induction on the integer $\alpha \geq 0$. Thus, from (3.12) we get (3.11).

Now, with the combinatorial equation (3.11), we have, by (3.9),

$$B_{ru}^{nm} = E[\sum_{k=0}^{a}\sum_{t=0}^{b}(-1)^{k+t}\binom{k+r}{r}\binom{t+u}{u}\binom{v_1}{k+r}\binom{v_2}{t+u}]$$

$$= E[(\sum_{k=0}^{a}(-1)^k\binom{k+r}{r}\binom{v_1}{k+r})(\sum_{t=0}^{b}(-1)^t\binom{t+u}{u}\binom{v_2}{t+u})]. \tag{3.14}$$

The key to decomposing the right-hand side of the term above is to handle each summation in the product separately. To this end, notice that for any positive integer v_1,

$$(-1)^a \binom{v_1}{r} \binom{v_1-r-1}{a} I_{(v_1>r)}$$

$$= (-1)^a \binom{r+a}{a} \binom{v_1}{r+a+1} \frac{r+a+1}{v_1-r} I_{(v_1>r)}$$

$$= (-1)^a \frac{1}{(r+a)!} \binom{r+a}{a} v_1 \ldots (v_1-r+1)(v_1-r-1)\ldots(v_1-a-r) I_{(v_1>r)}.$$

If $v_1 = r$, the left-hand side equals zero and the right-hand side is taken as zero because of the factor $I_{(v_1>r)}$. Thus by (3.11) we get the following equation,

$$\sum_{k=0}^{a}(-1)^k \binom{k+r}{r}\binom{v_1}{k+r}$$

$$= \delta_{v_1,r}+(-1)^a \binom{r+a}{a}\binom{v_1}{r+a+1}\frac{r+a+1}{v_1-r}I_{(v_1>r)}, \qquad (3.15)$$

for any integer $v_1 \geq r$.

Applying a similar technique to the other terms in the product of (3.14) gets

$$\sum_{t=0}^{b}(-1)^t \binom{t+u}{u}\binom{v_2}{t+u}$$

$$= \delta_{v_2,u}+(-1)^b \binom{b+u}{b}\binom{v_2}{u+b+1}\frac{u+b+1}{v_2-u}I_{(v_2>u)}. \qquad (3.16)$$

Therefore, plugging (3.15), (3.16) into (3.14) and multiplying out the two summations yields

$$B_{ru}^{nm} = E[\delta_{v_1,r}\delta_{v_2,u}+(-1)^a \binom{r+a}{a}\binom{v_1}{r+a+1}\frac{r+a+1}{v_1-r}I_{(v_1>r)}\delta_{v_2,u}+$$

$$+(-1)^b \binom{b+u}{b}\binom{v_2}{u+b+1}\frac{u+b+1}{v_2-u}I_{(v_2>u)}\delta_{v_1,r}+(-1)^{a+b}R, \qquad (3.17)$$

where

$$R=\binom{b+u}{b}\binom{v_2}{u+b+1}\frac{u+b+1}{v_2-u}I_{(v_2>u)}\binom{r+a}{a}\binom{v_1}{r+a+1}\frac{r+a+1}{v_1-r}I_{(v_1>r)}].$$

Now, we can take expectations inside the summation and examine each term of the right-hand side of (3.17) as follows.

The first term becomes

$$E(\delta_{v_1,r}\delta_{v_2,u}) = P(v_1=r,v_2=u).$$

For the second term, using the definition of $S_{(r),u}$ in Section 2.3 in conjunction with the condition of $v_2 - u \le m - u$, yields

$$E\left(\binom{r+a}{a}\binom{v_1}{r+a+1}\frac{r+a+1}{v_1-r}I_{(v_1>r)}\delta_{v_2,u}\right)$$
$$= \begin{cases} 0, & \text{if } n = r \\ \binom{r+a}{a}\frac{r+a+1}{n-r}S_{r+a+1,(u)}, & \text{if } n > r. \end{cases} \tag{3.18}$$

For the second term, using a similar derivation as in (3.18) gets

$$E\left(\binom{b+u}{b}\binom{v_2}{u+b+1}\frac{u+b+1}{v_2-u}I_{(v_2>u)}\delta_{v_1,r}\right)$$
$$= \begin{cases} 0, & \text{if } m = u \\ \binom{u+b}{b}\frac{u+b+1}{m-u}S_{(r),u+b+1}, & \text{if } m > u. \end{cases} \tag{3.19}$$

Finally, for the last term of the summation in the right-hand side of (3.17), we have

$$E(R) = E\left[\binom{b+u}{b}\binom{v_2}{u+b+1}\frac{u+b+1}{v_2-u}I_{(v_2>u)}\binom{r+a}{a}\binom{v_1}{r+a+1}\frac{r+a+1}{v_1-r}I_{(v_1>r)}\right].$$

Obviously, $T = 0$ for $n = r$ or $m = u$. When $r < n$ and $u < m$, if both integers a and b are even (as given in the condition of the theorem), $(-1)^a = (-1)^b = (-1)^{a+b} = 1$, and

$$E(R) \ge \binom{r+a}{a}\binom{u+b}{b}\frac{r+a+1}{n-r}\frac{u+b+1}{m-u}S_{r+a+1,u+b+1}. \tag{3.20}$$

Combining the inequalities (3.20), (3.19), (3.18) with the relation (3.17) gets Theorem 3.2.1. □

As shown in the proof, the role of the indicator function is critical in constructing bivariate upper bounds. The following shows the corresponding result for bivariate lower bounds.

Theorem 3.2.2 Let a and b be two positive odd integers, $r \le n$, $u \le m$. The bivariate lower bound of the probability of $\{v_1 = r, v_2 = u\}$ can be obtained as

$$P(v_1 = r, v_2 = u) \ge B_{ru}^{nm}(a,b) + \Delta_1 + \Delta_2 - \Delta_3', \tag{3.21}$$

where the forms of Δ_1, Δ_2 are as specified in (3.10) in Theorem 3.2.1, and the term Δ_3' is given as follows.

$$\Delta_3' = \begin{cases} 0, & \text{if } n = r \text{ or } m = u \\ \binom{r+a}{a}\binom{u+b}{b}\frac{r+a+1}{a+1}\frac{u+b+1}{b+1}S_{r+a+1,u+b+1}, & \text{if } n > r \text{ and } m > u. \end{cases}$$

The lower bound in Theorem 3.2.2 reverts to the Sobel-Uppuluri-Galambos lower bound (2.9).

Proof: When $r < n$ and $u < m$, if both integers a and b are odd, $(-1)^a = (-1)^b = -1$ $(-1)^{a+b} = 1$. Also, noticing that

$$\binom{v_1}{r+a+1} = 0,$$

if $v_1 < r + a + 1$, similar to (3.20), we have

$$E(R) \leq \binom{r+a}{a}\binom{u+b}{b}\frac{r+a+1}{a+1}\frac{u+b+1}{b+1}S_{r+a+1,u+b+1}, \qquad (3.22)$$

Now, combining the inequalities (3.22), (3.19), (3.18) with the relation (3.17) gets Theorem 3.2.2. □

Theorems 3.2.1 and 3.2.2 use partial Bonferroni summations, as discussed in Section 2.3. In the scenarios where the partial Bonferroni summations are unavailable, the corresponding upper and lower bivariate Sobel-Uppuluri-Galambos bounds can be given as follows.

Theorem 3.2.3 For any even $a \geq 0$, $b \geq 0$, $r \leq n$, $u \leq m$

$$P(v_1 = r, v_2 = u) \leq B_{ru}^{nm}(a,b) - \Gamma_1 - \Gamma_2 - \Delta_3,$$

where for positive even integers a, b

$$\Gamma_1 = 0 \quad n = r; \quad = \binom{r+a}{a}\frac{r+a+1}{n-r}\sum_{y=0}^{m-u}(-1)^y S_{r+a+1,u+y}\binom{y+u}{u}, \quad n > r$$

$$\Gamma_2 = 0 \quad m = u; \quad = \binom{u+b}{b}\frac{u+b+1}{m-u}\sum_{x=0}^{n-r}(-1)^x S_{r+x,u+b+1}\binom{x+r}{r}, \quad m > u.$$

Proof: Notice that the difference between Theorem 3.2.3 and Theorem 3.2.1 is the expression of partial Bonferroni summations. Applying Lemma 2.3.2 to the partial Bonferroni summations in the right-hand side of Theorem 3.2.1 gets Theorem 3.2.3. □

Similarly, the upper bound can be expressed without partial Bonferroni summations as follows.

Theorem 3.2.4 If a, b are odd, $r \leq n$ $u \leq m$

$$P(v_1 = r, v_2 = u) \geq B_{ru}^{nm}(a,b) + \Gamma_1 + \Gamma_2 - \Delta_3'.$$

Proof: Similar to the proof of Theorem 3.2.3, applying Lemma 2.3.2 to the partial Bonferroni summations in the right-hand side of Theorem 3.2.2 leads to Theorem 3.2.4.   □

3.2.2 Bounds for Joint Probability of at Least r Occurrences

Consider two classes of events $\{A_i, i = 1, ..., n\}$ and $\{B_j, j = 1, ..., m\}$. Let v_1 and v_2 be the numbers of events in the set $\{A_i, i = 1, ..., n\}$ and $\{B_j, j = 1, ..., m\}$ that occur, respectively. For any integers $0 \le r \le n$ and $0 \le u \le m$, the previous section discusses the method bounding the probability of exactly r occurrences in the first event-class and exactly u occurrences in the second event-class. To further illustrate the method of indicator functions in this section, we concentrate on finding bounds for the probability of at least r occurrences in the first event-class and at least u occurrences in the second event-class, namely $P(v_1 \ge r, v_2 \ge u)$.

The following are two basic properties connecting the method of indicator functions, Bonferroni summations, and the joint probabilities of occurrences of the events. These properties have been discussed in, for example, Chen and Seneta (1996). The first property expresses a bivariate Bonferroni summation in the linear combination of indicator functions.

Lemma 3.2.1:

$$S_{r,u} = E[(\sum_{i=1}^{n} \binom{i-1}{r-1} I_{(v_1 \ge i)})(\sum_{j=1}^{m} \binom{j-1}{u-1} I_{(v_2 \ge j)})].$$

Proof: Consider the definition of the bivariate Bonferroni summation in Chapter 2. We have

$$
\begin{aligned}
S_{r,u} &= \sum_{i=r}^{n} \sum_{j=u}^{m} \binom{i-1}{r-1} \binom{j-1}{u-1} P(v_1 \ge i, v_2 \ge j) \quad \text{(Theorem 2.3.1)} \\
&= \sum_{i=1}^{n} \sum_{j=1}^{m} \binom{i-1}{r-1} \binom{j-1}{u-1} P(v_1 \ge i, v_2 \ge j) \\
&= \sum_{i=1}^{n} \sum_{j=1}^{m} \binom{i-1}{r-1} \binom{j-1}{u-1} E(I_{(v_1 \ge i, v_2 \ge j)}) \\
&= E[(\sum_{i=1}^{n} \binom{i-1}{r-1} I_{(v_1 \ge i)})(\sum_{j=1}^{m} \binom{j-1}{u-1} I_{(v_2 \ge j)})].
\end{aligned}
$$

The second property expresses a bivariate Bonferroni summation in a binomial moment combined with a linear combination of indicator functions.

Lemma 3.2.2:

$$S_{r,u} = E[\sum_{j=u}^{m} \binom{j-1}{u-1} I_{(v_2 \ge j)} \binom{v_1}{r}].$$

Proof: Using Lemma 3.2.1, we express a bivariate Bonferroni summation as follows.

$$
\begin{aligned}
S_{r,u} &= \sum_{i=r}^{n} \sum_{j=u}^{m} \binom{i-1}{r-1} \binom{j-1}{u-1} E(I_{(v_1 \ge i, v_2 \ge j)}) \\
&= \sum_{i=r}^{n} \sum_{j=u}^{m} \binom{i-1}{r-1} \binom{j-1}{u-1} \sum_{k=i}^{n} E(I_{(v_1 = k, v_2 \ge j)})
\end{aligned}
$$

$$= \sum_{k=r}^{n}\sum_{j=u}^{m}\binom{j-1}{u-1}E(I_{(v_1=k,v_2\geq j)})(\sum_{i=r}^{k}\binom{i-1}{r-1})$$

$$= \sum_{k=r}^{n}\sum_{j=u}^{m}\binom{j-1}{u-1}P(v_1=k,v_2\geq j)\binom{k}{r}$$

$$= E[\sum_{j=u}^{m}\binom{j-1}{u-1}I_{(v_2\geq j)}\binom{v_1}{r}],$$

by definition. □

In what follows, we use the expression of bivariate Bonferroni summations to show the method of indicator functions in the derivation of upper and lower bounds for the event of simultaneous occurrences.

Theorem 3.2.5 For any even $a, b, a \geq 0,\quad b \geq 0$ and any integer $c \geq 0,\quad d \geq 0$, the upper bound of the joint event $\{v_1 \geq r, v_2 \geq u\}$ reads

$$P(v_1 \geq r, v_2 \geq u) \leq A_{ru}^{nm}(a,b) - \frac{r}{n-r}\binom{r+a}{a}\binom{m}{u+c}^{-1}S_{r+a+1,u+c} -$$

$$-\frac{u}{m-u}\binom{u+b}{b}\binom{n}{r+d}^{-1}S_{r+d,u+b+1} -$$

$$-\frac{r}{n-r}\binom{r+a}{a}S_{r+a+1,u+b+1}\frac{u}{m-u}\binom{u+b}{b}. \qquad (3.23)$$

Proof: Noticing the expression of the term A_{ru}^{nm} in equation (3.8), we use the method of indicator functions as follows.

First, by Lemma 3.2.1, we have

$$S_{k+r,t+u} = E[\sum_{i=1}^{n}\binom{i-1}{r+k-1}\sum_{j=1}^{m}\binom{j-1}{u+t-1}I_{(v_1\geq i,v_2\geq j)}].$$

Thus,

$$A_{ru}^{nm}$$

$$= \sum_{k=0}^{a}\sum_{t=0}^{b}(-1)^{k+t}\binom{k+r-1}{r-1}\binom{t+u-1}{u-1}S_{k+r,t+u}$$

$$= \sum_{k=0}^{a}\sum_{t=0}^{b}(-1)^{k+t}\binom{k+r-1}{r-1}\binom{t+u-1}{u-1}$$

$$E[\sum_{i=1}^{n}\binom{i-1}{r+k-1}\sum_{j=1}^{m}\binom{j-1}{u+t-1}I_{(v_1\geq i,v_2\geq j)}]$$

Now, rearranging the indicator function in the above expression and denoting

$$N(i,j) = (\sum_{k=0}^{a}(-1)^k\binom{k+r-1}{r-1}\binom{i-1}{k+r-1})(\sum_{t=0}^{b}(-1)^t\binom{t+u-1}{u-1}\binom{j-1}{t+u-1}),$$

we can write the term A_{ru}^{nm} as

$$A_{ru}^{nm} = E[\sum_{i=1}^{n}\sum_{j=1}^{m} I_{(v_1 \geq i, v_2 \geq j)} N(i,j)]. \tag{3.24}$$

With the product-type form of the bivariate indicator function, we can further decompose the term A_{ru}^{nm} in the following way,

$$
\begin{aligned}
&A_{ru}^{nm} \\
&= E[\sum_{i=1}^{n}\sum_{j=1}^{m} I_{(v_1 \geq i, v_2 \geq j)}[\delta_{i,r}\delta_{j,u} + (-1)^a \binom{r+a}{a}\binom{i-1}{r+a}\frac{r}{i-r}I_{(i>r)}\delta_{j,u} + \\
&\quad +(-1)^b \binom{b+u}{b}\binom{j-1}{u+b}\frac{u}{j-u}I_{(j>u)}\delta_{i,r} + \\
&\quad +(-1)^{a+b} \binom{b+u}{b}\binom{j-1}{u+b}\frac{u}{j-u}I_{(j>u)}\binom{r+a}{a}\binom{i-1}{r+a}\frac{r}{i-r}I_{(i>r)}].
\end{aligned} \tag{3.25}
$$

Now, by the property of the indicator function $I(v_1 \geq i, v_2 \geq j)$, we can directly compute the product of the first two terms

$$E(\sum_{i=1}^{n}\sum_{j=1}^{m} I(v_1 \geq i, v_2 \geq j)\delta_{i,r}\delta_{j,u}) = P(v_1 \geq r, v_2 \geq u).$$

Recall that for any $r \geq 0, \quad d \geq 0 \quad r+d \leq n$, the indicator function satisfies

$$I_{(v_1 \geq r)} \geq \binom{v_1}{r+d}\binom{n}{r+d}^{-1}.$$

Also for any integer $j \leq m$,

$$\frac{u}{j-u} \geq \frac{u}{m-u}.$$

Thus, it follows that the second term in the summation of (3.25) becomes

$$
\begin{aligned}
&E[\sum_{i=1}^{n}\sum_{j=1}^{m} I_{(v_1 \geq i, v_2 \geq j)}\binom{b+u}{b}\binom{j-1}{u+b}\frac{u}{j-u}I_{(j>u)}\delta_{i,r}] \\
&= E(\sum_{j=u+1}^{m} \frac{u}{j-u}I_{(v_2 \geq j)}I_{(v_1 \geq r)}\binom{u+b}{b}\binom{j-1}{u+b}) \\
&\geq \frac{u}{m-u}\binom{u+b}{b}\binom{n}{r+d}^{-1} S_{r+d,u+b+1}
\end{aligned} \tag{3.26}
$$

by Lemma 3.2.2.

Due to structure-wise similarities between the second and the third terms in the summation of (3.25), we have

$$E(\sum_{i=1}^{n}\sum_{j=1}^{m}I_{(v_1\geq i,v_2\geq j)}\binom{r+a}{a}\binom{i-1}{r+a}\frac{r}{i-r}I_{(i>r)}\delta_{j,u})$$

$$\geq \frac{r}{n-r}\binom{r+a}{a}\binom{m}{u+c}^{-1}S_{r+a+1,u+c}, \qquad (3.27)$$

which is similar to (3.26).

Now, for the last term in the summation of (3.25), since

$$\frac{r}{i-r}\geq\frac{r}{n-r},$$

for $i\leq n$, we have

$$E[\sum_{i=1}^{n}\sum_{j=1}^{m}I_{(v_1\geq i,v_2\geq j)}\binom{b+u}{b}\binom{j-1}{u+b}\frac{u}{j-u}I_{(j>u)}\binom{r+a}{a}\binom{i-1}{r+a}\frac{r}{i-r}I_{(i>r)}]$$

$$\geq \frac{r}{n-r}\binom{r+a}{a}S_{r+a+1,u+b+1}\frac{u}{m-u}\binom{u+b}{b}. \qquad (3.28)$$

For even a,b, $(-1)^a = (-1)^b = (-1)^{a+b} = 1$, combining the results of (3.25), (3.26), (3.27), and (3.28) in (3.25) yields Theorem 3.2.5.   □

Theorem 3.2.5 establishes a bivariate upper bound for $P(v_1 \geq r, v_2 \geq u)$. Evaluating this joint probability necessitates a lower bound. In fact, the method of indicator functions used in the preceding theorem can be applied to derive the corresponding lower bound as follows.

Theorem 3.2.6 For any odd a,b, $a \geq 0$, $b \geq 0$ and any integer $c \geq 0$, $d \geq 0$, the lower bound of the joint event $\{v_1 \geq r, v_2 \geq u\}$ can be correspondingly obtained as follows.

$$P(v_1 \geq r, v_2 \geq u)$$

$$\geq A_{ru}^{nm}(a,b)+\frac{r}{n-r}\binom{r+a}{a}\binom{N}{u+c}^{-1}S_{r+a+1,u+c}+$$

$$+\frac{u}{m-u}\binom{u+b}{b}\binom{n}{r+d}^{-1}S_{r+d,u+b+1}-$$

$$-\frac{r}{a+1}\binom{r+a}{a}S_{r+a+1,u+b+1}\frac{u}{b+1}\binom{u+b}{b}. \qquad (3.29)$$

Proof: For odd a,b, $(-1)^a = (-1)^b = -1$, $(-1)^{a+b} = 1$, the decomposition of the term A_{ru}^{nm} into (3.25), (3.26), (3.27) is the same as in the proof for Theorem

3.2.5. However, for odd a, b, the term (3.28) can not be applied since the sign of the inequality has been changed. In this case, due to

$$\binom{i-1}{r+a} = 0$$

when $i \leq r + a$, we only need to consider the case where $i \geq r + a + 1$. Under this situation, we have

$$E[\sum_{i=1}^{n} \sum_{j=1}^{m} I_{(v_1 \geq i, v_2 \geq j)} \binom{b+u}{b} \binom{j-1}{u+b} \frac{u}{j-u} I_{(j>u)} \binom{r+a}{a} \binom{i-1}{r+a} \frac{r}{i-r} I_{(i>r)}]$$

$$\leq \frac{r}{a+1} \binom{r+a}{a} S_{r+a+1,u+b+1} \frac{u}{b+1} \binom{u+b}{b}. \tag{3.30}$$

Now Theorem 3.2.6 follows by combining the results of (3.25), (3.26), (3.27), and (3.30) in (3.25).   □

3.3   Factorization of Indicator Functions

The preceding section addressed the use of indicator functions in the derivation of refined bivariate Bonferroni-type bounds along the framework of Sobel-Uppuluri-Galambos bound. So far, we have discussed two types of bounds, one for univariate settings in Section 2.1 and another for multivariate settings in Section 3.1. However, we have not touched the topic of the relationship between the univariate and multivariate bounds. Indicator functions can also be applied to link univariate and multivariate bounds.

The device that connects univariate upper bounds with multivariate upper bounds is the factorization of indicator functions (Chen (1995), Xu (1989), Zubkov (1977)). Let A and B be two events in a probability space. The indicator functions of the joint events $A \cap B$ can be factorized into the product of the two indicator functions of A and B, respectively,

$$I_{A \cap B} = I_A I_B.$$

In other words, the multivariate indicator function of the intersection of sets of events can be factorized into the product of individual indicator functions.

In this section, we utilize the idea of factorization of multivariate indicator functions to elucidate the connection and differences between univariate and multivariate bounds. As usual, denote $S_t(A)$ the binomial moments (Bonferroni summations) for the event set $\{A_i, i = 1, ..., n\}$, and $S_t(B)$ for the event set $\{B_j, j = 1, ..., m\}$. Published work in this regard includes Galambos and Xu (1993), and Chen (1998), among others.

Theorem 3.3.1 For any given integers T and K, assume that

$$P(v_1 \geq 1) \leq \sum_{t=1}^{T} a_{t,n} S_t(A) \tag{3.31}$$

and

$$P(v_2 \geq 1) \leq \sum_{k=1}^{K} b_{k,m} S_k(B) \tag{3.32}$$

are univariate upper bounds on events $\{v_1 \geq 1\}$ and $\{v_2 \geq 1\}$, respectively. Then the joint event can be bounded as follows:

$$P(v_1 \geq 1, v_2 \geq 1) \leq \sum_{t=1}^{T} \sum_{k=1}^{K} a_{t,n} b_{k,m} S_{tk}. \tag{3.33}$$

Proof: Two different methods will be discussed for the proof of this theorem. The first approach sets the connection between univariate indicator function and bivariate indicator functions via matrix decomposition. The second method is elegant and easy to follow, courtesy of Professor F. M. Hoppe at McMaster University.

Method 1: Since $\sum_{t=1}^{T} a_{t,n} S_t(A)$ and $\sum_{k=1}^{K} b_{k,m} S_k(B)$ are upper bounds for $\{v_1 \geq 1\}$ and $\{v_2 \geq 1\}$, respectively, for any two sets of events $\{A_i\}$ and $\{B_j\}$, we have

$$P(v_1 \geq 1) \leq \sum_{t=1}^{T} a_{t,n} S_t(A)$$

and

$$P(v_2 \geq 1) \leq \sum_{k=1}^{K} b_{k,m} S_k(B).$$

Recall that the Bonferroni summations for any integer t and k can be expressed as

$$S_t(A) = \sum_{i=t}^{n} \binom{i}{t} p_i = \mathbf{a_t}' \mathbf{x}$$

and

$$S_k(B) = \sum_{j=k}^{m} \binom{j}{k} p_j = \mathbf{b_k}' \mathbf{y},$$

where $\mathbf{x}$ and $\mathbf{y}$ are vectors of probabilities such that the ith component of $\mathbf{x}$ is $x_i = P(v(A) = i)$ for $i = 1, ..., n$, and the jth component of $\mathbf{y}$ is $y_j = P(v(B) = j)$ for $j = 1, ..., m$. $\mathbf{a_t}$ is the vector of the binomial coefficients, the ith component of $\mathbf{a_t}$ is $a_t(i) = 0$ when $i < t$;

$$a_t(i) = \binom{i}{t}$$

when $i \geq t$, for $i = 1, ..., n$. Similarly, $\mathbf{b_k}$ is the vector of the binomial coefficients, the jth component of $\mathbf{b_k}$ is $b_k(j) = 0$ when $j < k$;

$$b_k(j) = \binom{j}{k}$$

when $j \geq k$, for $j = 1, ..., m$.

Denoting $\mathbf{c}' = (0, 1, ..., 1)$ and $\mathbf{d}' = (0, 1, ..., 1)$, we have

$$P(v_1 \geq 1) = \mathbf{c}'\mathbf{x}$$

and

$$P(v_2 \geq 1) = \mathbf{d}'\mathbf{y}.$$

Now, let the vectors be

$$\mathbf{a} = (a_{1,n}, ..., c_{T,n})', \qquad \mathbf{b} = (b_{1,m}, ..., b_{K,m})',$$

$$\mathbf{S^A} = (S_1^A, ..., S_T^A)', \qquad \mathbf{S^B} = (S_1^B, ..., S_K^B)'.$$

And the matrices

$$\mathbf{M^A} = (\mathbf{a}_1, ..., \mathbf{a}_T)' \qquad \mathbf{M^B} = (\mathbf{b}_1, ..., \mathbf{b}_K)'.$$

The univariate upper bounds can then be expressed in a matrix form as follows,

$$
\begin{aligned}
\sum_{t=1}^{T} a_{t,n} S_t &= \mathbf{a}'\mathbf{S^A} \\
&= \mathbf{a}'\mathbf{M^A}\mathbf{x}.
\end{aligned}
$$

Thus

$$\mathbf{c}'\mathbf{x} = P(v_1 \geq 1) \leq \sum_{t=1}^{T} a_{t,n} S_t = \mathbf{a}'\mathbf{M^A}\mathbf{x}, \tag{3.34}$$

for any vector $\mathbf{x}$ satisfying $\mathbf{x} \geq \mathbf{0}$ and $\mathbf{x}'\mathbf{1} = 1$.

Similarly, set

$$
\begin{aligned}
\sum_{k=1}^{K} b_{k,m} S_k &= \mathbf{b}'\mathbf{S^B} \\
&= \mathbf{b}'\mathbf{M^B}\mathbf{y}.
\end{aligned}
$$

So,

$$\mathbf{d}'\mathbf{y} = P(v_2 \geq 1) \leq \sum_{k=1}^{K} b_{k,m} S_k = \mathbf{b}'\mathbf{M^B}\mathbf{y}, \tag{3.35}$$

for any vector $\mathbf{y}$ satisfying $\mathbf{y} \geq \mathbf{0}$ and $\mathbf{y}'\mathbf{1} = 1$.

Taking different vectors of $\mathbf{x}$ and $\mathbf{y}$, we deduce from (3.34) and (3.35) that

$$\mathbf{a}'\mathbf{M^A} \geq \mathbf{c}',$$

and

$$\mathbf{b}'\mathbf{M^B} \geq \mathbf{d}'.$$

Now, decomposing matrix $\mathbf{M^A}$ and $\mathbf{M^B}$ according to

$$S_t(A) = \sum_{i=t}^{n} \binom{i}{t} p_i = \mathbf{a_t}'\mathbf{x}$$

and

$$S_k(B) = \sum_{j=k}^{m} \binom{j}{k} p_j = \mathbf{b_k}'\mathbf{y},$$

by the method of indicators, the two upper bounds (3.31) and (3.32) imply that

$$1 \le \sum_{t=1}^{T} a_{t,n} \binom{x}{t}, \qquad 1 \le x \le n \quad \text{integer,} \tag{3.36}$$

and

$$1 \le \sum_{k=1}^{K} b_{k,m} \binom{y}{k}, \qquad 1 \le y \le m \quad \text{integer.} \tag{3.37}$$

Multiplying the two right-hand sides of inequalities (3.36) and (3.37) leads to

$$1 \le \sum_{t=1}^{T} \sum_{k=1}^{K} a_{t,n} b_{k,m} \binom{x}{t}\binom{y}{k}, \qquad 1 \le x \le n, 1 \le y \le m,$$

where the domain of (x,y) can be extended to an area including $x = 0$ and $y = 0$ by changing the left-hand side from 1 to the indicator $I(x \ge 1, y \ge 1)$. In this way, we get

$$I(x \ge 1, y \ge 1) \le \sum_{t=1}^{T} \sum_{k=1}^{K} a_{t,n} b_{k,m} \binom{x}{t}\binom{y}{k},$$

where $0 \le x \le n, 0 \le y \le m$. Namely, if either $x = 0$ or $y = 0$, both the left-hand side and the right-hand side become zero.

Since v_1, v_2 are integers in the area of $0 \le v_1 \le n$ and $0 \le v_2 \le m$, we have

$$I(v_1 \ge 1, v_2 \ge 1) \le \sum_{t=1}^{T} \sum_{k=1}^{K} a_{t,n} b_{k,m} \binom{v_1}{t}\binom{v_2}{k}, \tag{3.38}$$

where $0 \le v_1 \le n, 0 \le v_2 \le m$. Taking expectations on both sides of (3.38) results in Theorem 3.3.1. $\square$

Method 2: Since (3.31) and (3.32) hold for arbitrary events $\{A_i, B_j\}$, for any integers x and y satisfying $1 \le x \le n$ and $1 \le y \le m$, considering the events

$$A_1 = A_2 = \ldots = A_x = \Omega, \quad A_{x+1} = A_{x+2} = \ldots = A_n = \emptyset$$

and

$$B_1 = B_2 = \ldots = B_y = \Omega, \quad B_{y+1} = B_{y+2} = \ldots = B_m = \emptyset,$$

we have, for any $t \le x$,

$$S_t(A) = \sum_{1 \le i_1 < \ldots < i_t \le n} P(A_{i_1} \ldots A_{i_t}) = \binom{x}{t},$$

and $S_t(A) = 0$ for any $t > x$. Similarly, for any $k \leq y$,

$$S_k(B) = \binom{y}{k},$$

and $S_k(B) = 0$ when $k > y$. In this case (3.31) and (3.32) become

$$1 \leq \sum_{t=1}^{T} a_{t,n} \binom{x}{t}$$

and

$$1 \leq \sum_{k=1}^{K} b_{k,m} \binom{y}{k}.$$

Therefore

$$1 \leq \sum_{t} \sum_{k} a_{t,n} b_{k,m} \binom{x}{t} \binom{y}{k}.$$

Extending the above inequality to $x = 0$ and $y = 0$, we have

$$I(v_1 \geq 1, v_2 \geq 1) \leq \sum_{t=1}^{T} \sum_{k=1}^{K} a_{t,n} b_{k,m} \binom{v_1}{t} \binom{v_2}{k}.$$

Taking expectation on the above inequality yields Theorem 3.3.1. $\square$

Theorem 3.3.1 shows the convenience of using the method of indicator functions in setting the connection between univariate upper bounds and multivariate upper bounds. Notice that this approach of factorization using the indicator function is valid for upper bounds only, in which the product of the two bounding terms is larger than 1 or zero when appropriate. However, when we consider the product of lower bounds, since the product of the corresponding forms of the indicator functions (less than 1) is not necessarily less than 1 (for example, $(-2) \times (-2) = 4 > 1$), the factorization approach cannot be similarly applied to study the bivariate lower bounds.

From Theorem 3.3.1, we know that if

$$\sum_{t=1}^{T} a_{t,n} S_t(A)$$

and

$$\sum_{k=1}^{K} b_{k,m} S_k(B)$$

are two upper bounds for $P(v_1(A) \geq 1)$ and $P(v_2(B) \geq 1)$, respectively, then by the method of factorization of indicator functions, the upper bound on the joint probability $P(v_1(A) \geq 1, v_2(B) \geq 1)$ can be bounded by $\sum_{t=1}^{T} \sum_{k=1}^{K} a_{t,n} b_{k,m} S_{t,k}$.

The above approach can be extended from bounding $P(v_1 \geq 1, v_2 \geq 1)$ to bounding $P(v_1 \geq r, v_2 \geq u)$ with $0 \leq r \leq n$, $0 \leq u \leq m$. By considering indicator functions $I(v_1 \geq r)$ and $I(v_2 \geq u)$, parallel results can be obtained for upper bounds of

$P(v_1(A) \geq r, v_2(B) \geq u)$. Such a bounding technique on the factorization of multivariate indicator functions can also be applied to obtain the upper bounds of $P(v_1(A) = r, v_2(B) = u)$ (the joint probability that exactly r and u occurrences in $\{A_i, i = 1, ..., n\}$ and $\{B_j, j = 1, ..., m\}$, respectively). Now, we further explore the effectiveness of this general bounding principle.

For events $A_1, ..., A_n$, the univariate Sobel-Uppuluri-Galambos upper bounds (2.8) read: for any integer p, $0 \leq 2p \leq n$, any integer r, $0 \leq r \leq n$

$$P(v_1(A) \geq r)$$

$$\leq \sum_{k=0}^{2p} (-1)^k \binom{k+r-1}{r-1} S_{k+r}(A) - \frac{2p+1}{n-r} \binom{2p+r}{r-1} S_{2p+r+1}(A). \quad (3.39)$$

The corresponding upper bounds for events $B_1, ..., B_m$ read: For any integer q, $0 \leq 2q \leq m$, any integer u, $0 \leq u \leq m$

$$P(v_2(B) \geq u)$$

$$\leq \sum_{t=0}^{2q} (-1)^t \binom{t+u-1}{u-1} S_{t+u}(B) - \frac{2q+1}{m-u} \binom{2q+u}{u-1} S_{2q+u+1}(B). \quad (3.40)$$

Consider the factorization of a multivariate indicator function. Theorem 3.3.1 guarantees that the corresponding upper bound for $P(v_1 \geq r, v_2 \geq u)$ takes the following form:

Theorem 3.3.2: For any integers p and q satisfying $0 \leq 2p \leq n$ and $0 \leq 2q \leq m$, and for any integers r and u satisfying $0 \leq r \leq n$ and $0 \leq u \leq m$, we use the bivariate Bonferroni summations defined in Section 2.3.1 to bound the joint probability $P(v_1 \geq r, v_2 \geq u)$ as follows.

$$P(v_1 \geq r, v_2 \geq u)$$

$$\leq \sum_{k=0}^{2p} \sum_{t=0}^{2q} (-1)^{t+k} \binom{k+r-1}{r-1} \binom{t+u-1}{u-1} S_{k+r,t+u}(A,B)$$

$$- \frac{2p+1}{n-r} \binom{2p+r}{r-1} \sum_{t=0}^{2q} (-1)^t \binom{t+u-1}{u-1} S_{2p+r+1,t+u}(A,B)$$

$$- \frac{2q+1}{m-u} \binom{2q+u}{u-1} \sum_{k=0}^{2p} (-1)^k \binom{k+r-1}{r-1} S_{k+r,2q+u+1}(A,B)$$

$$+ \frac{2p+1}{n-r} \frac{2q+1}{m-u} \binom{2p+r}{r-1} \binom{2q+u}{u-1} S_{2p+r+1,2q+u+1}(A,B). \quad (3.41)$$

From univariate Sobel-Uppuluri-Galambos type bounds (right-hand side of (2.9)), the upper bound for $P(v_1 = r, v_2 = u)$ reads as follows.

Theorem 3.3.3 For any integers p and q satisfying $0 \leq 2p \leq n$ and $0 \leq 2q \leq m$, and for any integers r and u satisfying $0 \leq r \leq n$ and $0 \leq u \leq m$, we use the bivariate

Bonferroni summations defined in Section 2.3.1 to bound the joint probability $P(v_1 = r, v_2 = u)$ as follows.

$$P(v_1 = r, v_2 = u)$$

$$\leq \sum_{k=0}^{2p} \sum_{t=0}^{2q} (-1)^{t+k} \binom{k+r}{r} \binom{t+u}{u} S_{k+r,t+u}(A,B)$$

$$- \frac{2p+1}{n-r} \binom{2p+r+1}{r} \sum_{t=0}^{2q} (-1)^t \binom{t+u}{u} S_{2p+r+1,t+r}(A,B)$$

$$- \frac{2q+1}{m-u} \binom{2q+u+1}{u} \sum_{k=0}^{2p} (-1)^k \binom{k+r}{r} S_{k+u,2q+u+1}(A,B)$$

$$+ \frac{2p+1}{n-r} \frac{2q+1}{m-u} \binom{2p+r+1}{r} \binom{2q+u+1}{u} S_{2p+r+1,2q+u+1}, \quad (3.42)$$

which is a bivariate Sobel-Uppuluri-Galambos-type bound.

3.4 A Paradox on Factorization and Binomial Moments

The preceding sections discussed two methods of bivariate bounding with indicator functions. The first one used binomial moments such that the inequalities on bino-mial moments were used to obtain bivariate upper (Theorems 3.2.1, 3.2.3, and 3.2.5) and lower bounds (Theorems 3.2.2, 3.2.4, and 3.2.6). Related references include the work of Seneta (1973) and Seneta (1986). The second approach (in Section 3.3) showed the factorization of multivariate indicator functions, in which we can gener-ate a bivariate upper bound by factorizing the joint indicator function into the product of individual indicator functions. Normally, one expects the two bounding methods to coincide with one another for the results of bivariate inequalities. However, due to the different structures of bivariate bounds, the two bivariate extensions of the Sobel-Uppuluri-Galambos bound do not concur. In this section, we provide exam-ples to illustrate the bound paradox generated from two different indicator bounding methods. More specifically, we analyze the Sobel-Uppuluri-Galambos-type upper bounds to show the difference between two different versions of the bivariate Sobel-Uppuluri-Galambos upper bounds. Similar discussion of the univariate case can be found in Example 2.1.2.

3.4.1 Upper Bound Inconsistency

This subsection focuses on comparing two types of bivariate upper bounds: one gen-erated through the factorization of indicator functions, and the other constructed through binomial moments.

First, we compare the two versions of the bivariate Sobel-Uppuluri-Galambos upper bounds for the joint probability of exactly r and u occurrences in two sets of events, $P(v_1 = r, v_2 = u)$.

Consider the bivariate bound derived in Section 3.3 (Theorem 3.3.3, Bound (3.42)). For positive integers a, b, $r < n$, and $u < m$, denote the following terms for notational convenience.

$$H_1(a,b) = 0 \quad \text{for} \quad n = r$$

$$= \frac{a+1}{n-r}\binom{a+r+1}{r}\sum_{t=0}^{b}(-1)^t\binom{t+u}{u}S_{b+r+1,t+u}(A,B),$$

$$H_2(a,b) = 0 \quad \text{for} \quad m = u$$

$$= \frac{b+1}{m-u}\binom{b+u+1}{u}\sum_{k=0}^{a}(-1)^k\binom{k+r}{r}S_{k+r,b+u+1}(A,B).$$

To conveniently compare the upper bounds of $P(v_1 = r, v_2 = u)$ with bounds that were derived in Section 3.2 (Theorem 3.2.1, Bound (3.10)), we denote

$$\Delta_1 = 0 \quad \text{for} \quad n = r;$$

$$= \binom{r+2p}{2p}\frac{r+2p+1}{n-r}S_{r+2p+1,(u)} \quad \text{for} \quad n > r.$$

$$\Delta_2 = 0 \quad \text{for} \quad m = u;$$

$$= \binom{u+2q}{b}\frac{u+2q+1}{m-u}S_{(r),u+2q+1} \quad \text{for} \quad m > u.$$

$$\Delta_3 = 0 \quad \text{for} \quad n = r \quad \text{or} \quad m = u;$$

$$= \binom{r+2p}{2p}\binom{u+2q}{2q}\frac{r+2p+1}{n-r}\frac{u+2q+1}{m-u}S_{r+2p+1,u+2q+1},$$

for $n > r$ and $m > u$.

Now, in the comparisons of the two upper bounds, for the integer $r < n$, we can decompose the first term H_1 as follows.

$$H_1 = E\left(\frac{2p+1}{n-r}\binom{2p+r+1}{r}\sum_{t=0}^{2q}(-1)^t\binom{t+u}{u}\binom{v_1}{r+2p+1}\binom{v_2}{t+u}\right)$$

$$= \frac{2p+1}{n-r}\binom{2p+r+1}{r}E\left(\sum_{t=0}^{2q}(-1)^t\binom{t+u}{u}\binom{v_1}{r+2p+1}\binom{v_2}{t+u}\right)$$

$$= \binom{r+2p}{2p}\frac{r+2p+1}{n-r}$$

$$E\left[\binom{v_1}{r+2p+1}\left(\delta_{v_2,u}+\binom{v_2}{2q+u+1}\binom{2q+u}{2q}\frac{u+2q+1}{v_2-u}I(v_2 > u)\right)\right]$$

by (3.16). Now for the corresponding term in the upper bound 3.2.1.1, we have

$$\Delta_1 = \binom{r+2p}{2p} \frac{r+2p+1}{n-r} E\left(\binom{v_1}{r+2p+1}\right) \delta_{v_2,u}).$$

Thus, denote $A = \{v_2 > u\}$, the difference in the second term of the two extended upper bounds is

$$
\begin{aligned}
&H_1 - \Delta_1 \\
&= \binom{r+2p}{2p} \frac{r+2p+1}{n-r} E\left(\binom{v_1}{r+2p+1}\right) \binom{v_2}{2q+u+1} \binom{2q+u}{2q} \frac{u+2q+1}{v_2-u} I_A) \\
&\geq \binom{r+2p}{2p} \frac{r+2p+1}{n-r} E\left(\binom{v_1}{r+2p+1}\right) \binom{v_2}{2q+u+1} \binom{2q+u}{2q} \frac{u+2q+1}{m-u} I_A) \\
&= \binom{r+2p}{2p} \binom{u+2q}{2q} \frac{r+2p+1}{n-r} \frac{u+2q+1}{m-u} S_{r+2p+1,u+2q+1} \\
&= \Delta_3.
\end{aligned}
$$

When $P(v_2 < m) > 0$ (for example, when the probability of the occurrence of v_2 is equally likely over its domain), since the event $\{v_2 < m\}$ has positive probability, the above inequality becomes a strict inequality.

Similarly, for the difference of the third terms,

$$H_2 - \Delta_2 \geq \Delta_3.$$

Therefore, the difference between the bound based on factorization (3.42) and the bound based on the binomial moment (3.10) reads

$$
\begin{aligned}
&(B_{ru}^{nm}(2p,2q) - \Delta_1 - \Delta_2 - \Delta_3) - (B_{ru}^{nm}(2p,2q) - H_1 - H_2 + \Delta_3) \\
&\geq (H_1 - \Delta_1) + (H_2 - \Delta_2) - 2\Delta_3 \\
&\geq 0.
\end{aligned}
$$

When $P(v_2 < m) > 0$ or $P(v_1 < n) > 0$ (when the probability of the occurrence of v_2 (or v_1) is equally likely over its domain), since the event $\{v_2 < m\}$ (or $\{v_1 < n\}$) has positive probability, the above inequality becomes a strict inequality, meaning the bound in (3.42) improves on the bound in (3.10).

Hence, the upper bound (3.42) is at least as good as the upper bound (3.10). Although both bounds (3.42) and (3.10) revert to the univariate Sobel-Uppuluri-Galambos bounds, the one derived from factorization (3.42) is at least as good as the one derived from binomial moments (3.10). And, under certain conditions, the improvement is positive in the bivariate case.

After comparing the two upper bounds for $P(v_1 = r, v_2 = u)$, we discuss the corresponding upper bounds for $P(v_1 \geq r, v_2 \geq u)$, to compare upper bounds (3.41) with (3.23). To this end, for the upper bound in (3.41), we denote

$$
\begin{aligned}
H_1^* &= 0 \quad \text{for} \quad n = r \\
&= \frac{2p+1}{n-r} \binom{2p+r}{r-1} \sum_{t=0}^{2q} (-1)^t \binom{t+u-1}{u-1} S_{2p+r+1,t+u}(A,B) \quad \text{if} \quad n > r.
\end{aligned}
$$

$$H_2^* = 0 \quad \text{for} \quad m = u$$

$$= \frac{2q+1}{m-u}\binom{2q+u}{u-1}\sum_{k=0}^{2p}(-1)^k\binom{k+r-1}{r-1}S_{k+r,2q+u+1}(A,B) \quad \text{if} \quad m > u.$$

For the upper bounds in (3.23), we denote

$$\Delta_1^* = 0 \quad \text{for} \quad n = r;$$

$$= \binom{r+2p}{2p}\frac{r}{n-r}\binom{m}{u+c}^{-1}S_{r+2p+1,u+c} \quad \text{for} \quad n > r.$$

$$\Delta_2^* = 0 \quad \text{for} \quad m = u;$$

$$= \binom{u+2q}{2q}\frac{u}{m-u}\binom{n}{r+d}^{-1}S_{r+d,u+2q+1} \quad \text{for} \quad m > u.$$

$$\Delta_3^* = 0 \quad \text{for} \quad n = r \quad \text{or} \quad m = u;$$

$$= \binom{r+2p}{2p}\binom{u+2q}{2q}\frac{r}{n-r}\frac{u}{m-u}S_{r+2p+1,u+2q+1},$$

for $n > r$ and $m > u$.

By (3.16), the first term H_1^* can be written as

$$H_1^* = \frac{2p+1}{n-r}\binom{2p+r}{r-1}E[\sum_{i=1}^{n}\sum_{j=1}^{m}I(v_1 \geq i, v_2 \geq j)$$

$$\binom{i-1}{2p+r}(\delta_{ju} + \binom{u+2q-1}{2q}\binom{j-1}{u+2q}\frac{u+2q}{j-u}I(j>u))]$$

$$= \frac{r}{n-r}\binom{2p+r}{2p}E[\sum_{i=1}^{n}\sum_{j=1}^{m}I(v_1 \geq i, v_2 \geq j)$$

$$\binom{i-1}{2p+r}(\delta_{j,u} + \binom{u+2q-1}{2q}\binom{j-1}{u+2q}\frac{u+2q}{j-u}I(j>u))].$$

The difficulty now is to decompose the indicator function $I_{(v_1 \geq r)}$ in terms of Δ_1^*. Notice that

$$I_{(v_1 \geq r)} \geq \binom{v_1}{r+d}\binom{n}{r+d}^{-1} \quad \text{for any} \quad r \geq 0, \quad d \geq 0 \quad r+d \leq n,$$

we have

$$\Delta_1^* = \frac{r}{n-r}\binom{r+2p}{2p}\binom{m}{u+c}^{-1}S_{r+2p+1,u+c}$$

$$\leq E(\sum_{i=r+1}^{n} \frac{r}{n-r} I_{(v_1 \geq i)} I_{(v_2 \geq u)} \binom{r+2p}{2p}\binom{i-1}{r+2p}))$$

by Lemma 3.2.1

$$= E[\sum_{i=1}^{n}\sum_{j=1}^{m} I_{(v_1 \geq i, v_2 \geq j)}\binom{2p+r}{2p}\binom{i-1}{r+2p}\frac{r}{n-r} I_{(i>r)}\delta_{j,u}].$$

Therefore, for any integers $r < n, u < m$, the difference between the first two terms in the upper bounds (3.41) and (3.23) becomes

$$
\begin{aligned}
H_1^* - \Delta_1^* &= \frac{r}{n-r}\binom{2p+r}{2p} E[\sum_{i=1}^{n}\sum_{j=1}^{m} I(v_1 \geq i, v_2 \geq j) \\
&\quad \binom{i-1}{2p+r}\binom{u+2q-1}{2q}\binom{j-1}{u+2q}\frac{u+2q}{j-u}I(j>u)] \\
&\geq \frac{r}{n-r}\binom{2p+r}{2p} E[\sum_{i=1}^{n}\sum_{j=1}^{m} I(v_1 \geq i, v_2 \geq j) \\
&\quad \binom{i-1}{2p+r}\binom{u+2q-1}{2q}\binom{j-1}{u+2q}\frac{u+2q}{m-u}I(j>u)], \quad \text{for } j \leq m \\
&= \frac{r}{n-r}\binom{2p+r}{2p}\binom{u+2q-1}{2q}\frac{u+2q}{j-u}S_{r+2p+1,u+2q+1} \\
&= \Delta_3^*.
\end{aligned}
$$

When $P(v_2 < m) > 0$, the event $\{v_2 < m\}$ has positive probability, and the above inequality becomes a strict inequality. Under this scenario, the improvement is positive.

Similarly,

$$H_2^* - \Delta_2^* \geq \Delta_3^*.$$

Combining the above yields the difference between bound (3.41) and (3.23) as follows.

$$
\begin{aligned}
&(A_{ru}^{nm}(2p,2q) - \Delta_1^* - \Delta_2^* - \Delta_3^*) - (A_{ru}^{nm}(2p,2q) - H_1^* - H_2^* + \Delta_3^*) \\
&\geq (H_1^* - \Delta_1^*) + (H_2^* - \Delta_2^*) - 2\Delta_3^* \\
&\geq 0.
\end{aligned}
$$

Therefore, upper bound (3.41) is at least as good as upper bound (3.23), and the improvement is positive when events $\{v_1 < n\}$ (or $\{v_2 < m\}$) have positive probability.

We see that, although the bivariate Sobel-Uppuluri-Galambos type upper bounds given by (3.41) and (3.42) revert to the univariate Sobel-Uppuluri-Galambos bounds as upper bounds (3.10) and (3.23), under the condition that the event $\{v_1 < n, v_2 < m\}$ has a positive probability, the two versions of bivariate bounds reach different values.

In next subsection, we present the corresponding inconsistency for lower bounds of $P(v_1 = r, v_2 = u)$ and $P(v_1 \geq r, v_2 \geq u)$.

3.4.2   Lower Bound Inconsistency

While the previous section illustrated the refinement of the bivariate bound derived from factorization of bivariate indicator functions, it did not address the issue that the method of factorization is not applicable for bivariate Bonferroni-type lower bounds. When the product of two indicator functions of upper bounds can serve as the indicator function for the upper bound of the joint probability, the product of two indicator functions of lower bounds is not a guaranteed lower bound. Consider the following example.

Example 3.4.1. For two sets of events $\{A_i\}$ and $\{B_j\}$, if the following two lower bounds are combined to form a bivariate bound,

$$P(v_1 \geq 1) \geq -S_1(A)$$

and

$$P(v_2 \geq 1) \geq -S_1(B),$$

the corresponding indicator functions become

$$I_{(v_1 \geq 1)} \geq -\binom{v_1}{1}$$

and

$$I_{(v_2 \geq 1)} \geq -\binom{v_2}{1}.$$

Multiplying the above two terms leads to

$$I_{(v_1 \geq 1, v_2 \geq 1)} = I_{(v_1 \geq 1)} I_{(v_2 \geq 1)} \geq v_1 v_2,$$

which cannot be true when $v_1 = 2$ and $v_2 = 3$. The error stems from the multiplication of two negative terms in the right-hand side of the inequality. This example shows that the method of factorization of bivariate indicator function cannot be applied directly to generate bivariate lower bounds.   □

Certainly, the above example is an extreme case, because a negative value is essentially a trivial lower bound for a probability. However, it reveals that the method of factorization cannot guarantee the validity of a bivariate lower bound. We will revisit this issue in Chapter 7.

As shown in the examples and theorems in this chapter, although the method of indicator functions is convenient and the binomial moments keep mathematical elegance and symmetry in upper and lower bounds, the method of indicator functions does not guarantee the optimality of the bound. In practice, it is more critical to consider the issue of optimality of the bound. If an upper bound or lower bound is improved, the refinement may consequently improve the power of a test procedure or accuracy of an estimation approach in statistical inference. In the next chapter, we will discuss the method of linear programming in bounding theory, leading to an optimal solution (Fréchet optimal bounds) that is mathematically more challenging.

Chapter 4

Multivariate Linear Programming Framework

Chapter 3 showed how the binomial moments and factorization bounds, derived using the method of indicator functions, are endowed with mathematical elegance (simple, symmetric, and uniform treatment on upper and lower bounds). However, the indicator function bounding approach is unable to include optimality. If the derived bound is optimal, there is no space for improvement in terms of the given information. On the other hand, if the derived bound is not optimal, the bound can be improved to increase the efficiency of statistical inference. To overcome the drawback of being unable to measure optimality when using the indicator function approach, this chapter discusses another bounding technique that takes optimality into consideration (see, for example, Boros and Prékopa (1989), and Chen (1998)).

The concept of optimality was introduced in Section 2.2, followed by three types of multivariate optimality in Section 2.4. The Fréchet optimality (Definition 2.4.2) targets achievable points where the bound hits its limit (a fundamental sense of optimality). The bivariate linear optimality (Definition 2.4.3) directly selects the best bound among all linear bounds. The linear programming version (Definition 2.4.4) selects the best value that satisfies all the constraints of multivariate Bonferroni summations. It is critical to select the appropriate type of optimality for the corresponding bounding problem of interest.

As shown in Example 2.4.5, the concept of bivariate linear optimality differs from the concept of bivariate Fréchet optimality. However, we have not yet elucidated the relationship between the optimality in linear programming and the optimality in Fréchet style (Kounias and Marin (1976) and Kwerel (1975a)). Although the method of linear programming was mentioned in several places in previous chapters, we have not thoroughly discussed details of this bounding technique. In this chapter, we concentrate on the method of linear programming, and use bivariate upper and lower bounds as examples to illustrate this bounding skill, and its relation to Fréchet optimality.

Contrasting with the method of indicator functions, the general idea of the method of linear programming (Prékopa (1990a)) is to formulate the bounding problem as the availability of a set of positive values that sum to 1. If these values satisfy the conditions for Bonferroni summations and optimize the target function, then the feasible solution is the best bound. For instance, if p_0^* is the solution to the maximization of $P(\bigcup_{i=1}^{n} A_i)$, then $P(\bigcup A_i) \leq p_0^*$ and p_0^* is the optimal upper bound, because any upper bound must be larger than p_0^* which is also one of the probability of the union of events. Similarly, for the lower bound, if q_0^* is the solution to the minimization of $P(\bigcup_{i=1}^{n} A_i)$, $P(\bigcup A_i) \geq q_0^*$ and q_0^* is the optimal lower bound. We start the discussion with optimal bivariate upper bounds.

This chapter consists of two sections. Section 4.1 discusses general framework of linear programming for multivariate upper bounds; and Section 4.2 addresses this bounding framework for multivariate lower bounds. For convenience of illustration, we elucidate the underlying principles using bivariate bounds.

4.1   Linear Programming Upper Bounds

Recall the definition of bivariate optimal upper bounds in the setting of linear programming, as introduced in Chapter 2:

For any two sets of events $\{A_i, i = 1, ..., n\}$ and $\{B_j, j = 1, ..., m\}$, let v_1 and v_2 be the number of occurrences of the two event sets, respectively. Denote $p_{ij} = P(v_1 = i, v_2 = j)$. For any integers $1 \leq t \leq n$ and $1 \leq k \leq m$, consider a set of consistent bivariate Bonferroni summations S_{ij}, $i = 1, ..., t$, $j = 1, ..., k$. An optimal upper bound for $P(v_1 \geq 1, v_2 \geq 1)$ is defined by the maximum value of the following linear programming problem:

$$\max(p_{11} + ... + p_{nm}) \tag{4.1}$$

subject to $\sum_{i=0}^{n} \sum_{j=0}^{m} p_{ij} = 1$, and

$$p_{11} + 2p_{21} + ... + tp_{t1} + ... + nmp_{nm} = S_{11}$$

$$p_{21} + ... + \binom{t}{2}p_{t1} + ... + \binom{n}{2}mp_{nm} = S_{21}$$

$$\cdots\cdots \quad \cdots\cdots \quad \cdots$$

$$p_{tk} + ... + \binom{n}{t}\binom{m}{k}p_{nm} = S_{tk}.$$

The following is the rationale of optimality behind the bounding formulation in (4.1). For any solution $\mathbf{p}^*$ to the optimization issue (4.1), denote

$$U^* = \sum_{i=1}^{n} \sum_{j=1}^{m} p_{ij}^*.$$

Now for any two sets of events characterized by $\mathbf{p}$ in any probability space,

$$P(v_1 \geq 1, v_2 \geq 1) = \sum_{i=1}^{n} \sum_{j=1}^{m} p_{ij} \leq U^*,$$

because of (4.1). Thus the feasible optimal solution leads to an upper bound (see, for instance, Galambos and Xu (1992, 1993)).

Furthermore, for any upper bound, U, based on the quantities $\{S_{ij}, i = 1, ..., t, j = 1, ..., k\}$, we have

$$P(v_1 \geq 1, v_2 \geq 1) \leq U$$

for any events satisfying the conditions that the associated Bonferroni summations equal the pre-assigned values S_{ij} with $i = 1, ..., t$ and $j = 1, ..., k$. Since the solution U^* is one of the solutions of the optimization problem (4.1), we have

$$U^* = P(v_1^* \geq 1, v_2^* \geq 1) \leq U.$$

Therefore, U^* is the best among all the upper bounds in the sense that it is lower than or equal to any bivariate upper bound depending on S_{ij} for $i = 1, ..., t$, $j = 1, ..., k$.

As mentioned above, in the setting of linear programming upper bound (4.1), the optimality is naturally endowed once the feasible solution to the optimizing problem is found. This feature resembles the idea underpinning the definition of the Fréchet optimality (Definition 2.4.2). The linear constraints for the definition of S_{ij}, $(i, j) \in T$, can be viewed as a special class of bounds L, as defined in a multivariate Fréchet optimality. Now, we shall describe such a relationship through matrix representations of an upper bound.

4.1.1 Matrix Expression of Upper Fréchet Optimality

Recall the definition of the Fréchet optimality given in Section 2.4 for the upper bound of the bivariate probability $P(v_1 \geq 1, v_2 \geq 1)$:

Denote $T = \{(i, j) : i =, ..., t, j = 1, ..., k\}$, let F_1 be an upper bound such that for any particular set of consistent Bonferroni summations S_{ij}, $(i, j) \in T$. If there exists a set of events $A_1^*, ..., A_n^*, B_1^*, ..., B_m^*$ in a probability space where

$$S_{ij}(A_1^*, ..., A_n^*, B_1^*, ..., B_m^*) = S_{ij}, \quad (i, j) \in T$$

and

$$P(v_1(A^*) \geq 1, \quad v_2(B^*) \geq 1) = F_1(S_{ij}, (i, j) \in T), \tag{4.2}$$

then F_1 is said to be a Fréchet optimal upper bound for $P(v_1 \geq 1, \quad v_2 \geq 1)$.

The optimality defined in (4.2) can be translated into the language of linear programming, in which the domain of the function class, F_1, is limited to S_{ij}, a linear combination of p_{ij} as formulated in (4.1).

For notational convenience, without loss of generality, we consider the index set of the Bonferroni summations $T = \{(1, 1), (1, 2), (2, 1), (2, 2)\}$. The argument can be analogically extended to a more general setting for any set of consistent Bonferroni summations. First, we use the idea of elementary conjunctions of two sets of events to formulate the definition of the Fréchet optimality in matrix theory.

Given any four values consistent with bivariate Bonferroni summations $\{S_{11}, S_{12}, S_{21}, S_{22}\}$, by the definition of the Fréchet optimality, there exists a probability space and events $A_1^*, ..., A_n^*, B_1^*, ..., B_m^*$ such that

$$S_{ij}(A_1^*, ..., A_n^*, B_1^*, ..., B_m^*) = S_{ij}, \quad (i, j) \in \{(1, 1), (1, 2), (2, 1), (2, 2)\}$$

and $P(v_1(A^*) \geq 1, \quad v_2(B^*) \geq 1)$ reaches the value of the upper bound. Denote such Bonferroni-type upper bounds as $F_1(S_{ij}, (i,j) \in T)$. Now, recall that for any value $1 \leq r \leq n$ and $1 \leq u \leq m$,

$$S_{ru} = \sum_{\substack{1 \leq i_1 < \ldots < i_r \leq n \\ 1 \leq j_1 < \ldots < j_u \leq m}} P(A_{i_1} \ldots A_{i_r} B_{j_1} \ldots B_{j_u}).$$

Notice that each term in the above summation, $P(A_{i_1} \ldots A_{i_r} B_{j_1} \ldots B_{j_u})$ is actually a summation of the probabilities of the elementary conjunctions that add up to the probability of the event $A_{i_1} \ldots A_{i_r} B_{j_1} \ldots B_{j_u}$ (see Example 2.4.6). For instance, when $n = m = 2$ and $r = u = 1$,

$$P(A_1 \cap B_1) = P(A_1 A_2 B_1 B_2) + P(A_1 A_2^c B_1 B_2) + P(A_1 A_2 B_1 B_2^c) + P(A_1 A_2^c B_1 B_2^c).$$

Therefore, if we denote $\mathbf{x}$ the vector of the 2^{n+m} probabilities of elementary conjunctions, for each quantity $S_{r,u}$, there exists a 2^{n+m} row vector $\mathbf{t}'(r,u)$ such that

$$S_{r,u} = \mathbf{t}'(r,u)\mathbf{x}, \tag{4.3}$$

for $(r,u) \in \{(1,1),(1,2),(2,1),(2,2)\}$. Also, by the basic property of probability assignment, for all the 2^{n+m} elementary conjunctions,

$$\mathbf{x} \geq \mathbf{0} \qquad \mathbf{1}'\mathbf{x} = 1.$$

Now, denoting $\mathbf{b}' = (1, S_{11}, S_{12}, S_{21}, S_{22})$, we have

$$\mathbf{b} = (\mathbf{1}, \mathbf{t}(1,1), \mathbf{t}(1,2), \mathbf{t}(2,1), \mathbf{t}(2,2))'\mathbf{x} \tag{4.4}$$

where $\mathbf{1}$ is the vector with length 2^{n+m} and all elements equal 1, and $\mathbf{t}(i,j)$ is the vector specified in (4.3) for any $i,j = 1,2$.

Denote the matrix $\mathbf{R} = (\mathbf{1}, \mathbf{t}(1,1), \mathbf{t}(1,2), \mathbf{t}(2,1), \mathbf{t}(2,2))'$, a $(5 \times 2^{n+m})$ matrix with structure not affected by the values of the S_{ij}'s. We have

$$\mathbf{b} = \mathbf{R}\mathbf{x}. \tag{4.5}$$

The expression in (4.5) is the matrix representation of the Fréchet optimality (existence of such a probability vector) for a set of consistent bivariate Bonferroni summations S_{ij} with $(i,j) \in \{(1,1),(1,2),(2,1),(2,2)\}$, when the function class of bounds is set to any function based on $S_{i,j}$ where $(i,j) \in T$.

4.1.2   Target Function of Linear Programming

Now, we examine the matrix representation of the optimality defined in linear programming. As usual, denote $p_{ij} = P(v_1 = i, v_2 = j)$. Then $p_{ij} \geq 0$, and

$$\sum_{i=0}^{n} \sum_{j=0}^{m} p_{ij} = 1.$$

To avoid notational confusion, recall that the vector $\mathbf{x}$ denotes the 2^{n+m} probabilities of elementary conjunctions for any two set of events A_i and B_j where $i = 1,...,n$ and $j = 1,...,m$. We introduce the notation of another vector $\mathbf{q}$ as follows.

Letting $w = (n+1)(m+1)$ and putting w elements of p_{ij} for $i = 0,1,...,n$ and $j = 0,1,...,m$ into a vector of length w, yields

$$\mathbf{q} = \left(p_{ij}\right)_{w \times 1}$$

where p_{ij} are arranged from $p_{0,0}$ to p_{nm} by the increasing order on i for each value of j, and then on j for each increasing sequence of i. Such a method of vectorization for a bivariate array can be illustrated by the following example. Letting $n = m = 2$, the vectorized outcome of a 9×1 vector $\mathbf{q}$ reads

$$\mathbf{q} = \left(p_{00}, p_{10}, p_{20}, p_{01}, p_{11}, p_{21}, p_{02}, p_{12}, p_{22}\right)'.$$

Now consider the joint probability of at least one occurrence in both sets of events,

$$P(v_1 \geq 1,\, v_2 \geq 1) = \sum_{i=1}^{n} \sum_{j=1}^{m} p_{ij},$$

which is a summation of p_{ij} for $i \geq 1$ and $j \geq 1$. For $w = (n+1)(m+1)$, denote a length w vector $\mathbf{c} = \{c(i,j)\}$ such that

$$c(i,j) = \begin{cases} 0, & \text{for } i = 0 \text{ or } j=0 \\ 1, & \text{otherwise.} \end{cases}$$

The term $P(v_1 \geq 1, v_2 \geq 1)$ can then be written as

$$P(v_1 \geq 1,\, v_2 \geq 1) = \mathbf{c}'\mathbf{q}.$$

Given four consistent Bonferroni summations, $S_{11}, S_{21}, S_{12}, S_{22}$, there is a probability allocation yielding these values of S_{kt}. Such probability allocation consequently has a set of $(n+1)(m+1)$ values p_{ij} for $i = 0,1,...,n$ and $j = 0,1...,m$. This implies that there exists a $w \times 1$ vector $\mathbf{q}$ such that

$$\mathbf{q} \geq \mathbf{0}, \qquad \mathbf{q}'\mathbf{1} = 1.$$

Furthermore, by the expression of bivariate Bonferroni summations (Lemma 2.3.1),

$$S_{kt} = \sum_{i=k}^{n} \sum_{j=t}^{m} \binom{i}{k}\binom{j}{t} p_{ij} = \mathbf{g}'_{kt}\mathbf{q}, \quad k = 1,2 \quad t = 1,2, \tag{4.6}$$

where the row vector $\mathbf{g}'_{kt}$ is the vector of coefficients specified in (4.6). Combining the row vectors $\mathbf{g}'_{k,t}$ into a matrix $\mathbf{G}$ (with the first row as $\mathbf{1}$) for the quantities $S_{11}, S_{12}, S_{21}, S_{22}$, the matrix $\mathbf{G}$ reads

$$\begin{pmatrix}
1 & \cdots & 1 & \cdots & 1 & \cdots & 1 & 1 & \cdots & 1 & \cdots & 1 \\
0 & \cdots & 1 & \cdots & j & \cdots & m & 0 & \cdots & 2m & \cdots & nm \\
0 & \cdots & 0 & \cdots & \frac{j(j-1)}{2} & \cdots & \frac{m(m-1)}{2} & 0 & \cdots & 2\frac{m(m-1)}{2} & \cdots & n\frac{m(m-1)}{2} \\
0 & \cdots & 0 & \cdots & 0 & \cdots & 0 & 0 & \cdots & m & \cdots & \frac{n(n-1)}{2}m \\
0 & \cdots & 0 & \cdots & 0 & \cdots & 0 & 0 & \cdots & \frac{m(m-1)}{2} & \cdots & \frac{nm(n-1)(m-1)}{4}
\end{pmatrix}.$$

Therefore putting $\mathbf{b}' = (1, S_{11}, S_{12}, S_{21}, S_{22})$, for $w = (n+1)(m+1)$, there exists a $5 \times w$ matrix $\mathbf{G}$ so that

$$\mathbf{b} = \mathbf{Gq}, \qquad (4.7)$$

where the first row of the matrix $\mathbf{G}$ is $\mathbf{1}'$, the first $(m+1)th$ column of $\mathbf{G}$ is $(1, 0, 0, 0, 0)'$, and the structure of $\mathbf{G}$ is not affected by the value of the bivariate Bonferroni summations $S_{i,j}$'s.

The expression in (4.7) is essentially the matrix representation of the linear programming optimality for a set of consistent bivariate Bonferroni summations S_{ij} with $(i, j) \in \{(1,1), (1,2), (2,1), (2,2)\}$. Notice that for different index sets of summations S_{ij} used in a bound, the corresponding structure of matrix $\mathbf{G}$ changes to accommodate the associated Bonferroni summations.

According to the formulation in this subsection, the vector of Bonferroni summations is determined by a vector $\mathbf{q}$ with length $w = (n+1)(m+1)$. This is different from the formulation in Subsection 4.1.1, in which the vector of Bonferroni summations is determined by a vector $\mathbf{x}$ with length $2^{(n+m)}$. In next subsection, we show that the representations of the two optimal definitions (4.5) and (4.7) are actually equivalent.

4.1.3   Linear Programming Constraints

According to what we discussed in preceding sections on Fréchet optimality, for an upper bound using the Bonferroni summations $\{S_{11}, S_{12}, S_{21}, S_{22}\}$, there exists a $5 \times (2^{n+m})$ matrix, $\mathbf{R}$ as specified in (4.5) such that

$$\mathbf{b} = \mathbf{Rx},$$

where $\mathbf{b} = (1, S_{11}, S_{12}, S_{21}, S_{22})'$, and $\mathbf{x}$ is a vector of length 2^{n+m} that represents the probabilities of the elementary conjunctions of the events $\{A_i, i = 1, ..., n\}$ and $\{B_j, j = 1, ..., m\}$. The matrix representation of a Fréchet optimal upper bound for the class of any functions of S_{ij}, is derived in (4.5).

On the other hand, the corresponding matrix representation of the optimal upper bound using linear programming takes the form (4.7),

$$\mathbf{b} = \mathbf{Gq}, \qquad (4.8)$$

where $\mathbf{G}$ is a $5 \times w$ matrix ($w = (n+1)(m+1)$), and $\mathbf{q}$ is a vector of length $(n+1)(m+1)$ that consists of the ordered values of p_{ij} for $i = 0, 1, ..., n$, then $j = 0, 1, ..., m$. The ranking of the two indexes i and j is the following: the first element of $\mathbf{c}$ corresponds to the index $\{i = 0\ j = 0\}$, and the rest of the elements of $\mathbf{c}$ are formed by ranking over $i \geq 0$ in increasing order for each fixed j, then over $j \geq 0$ in increasing order.

By appearance, the Fréchet optimality focuses on the 2^{n+m} probabilities of elementary conjunctions of the events under consideration while the linear programming focuses on the $(n+1)(m+1)$ joint probabilities p_{ij}. However, in this subsection, we show that these two versions of formulations are actually equivalent when the optimal bound exists for a set of Bonferroni summations. For convenience, recall that $\mathbf{x}$ represents the vector of 2^{n+m} probabilities of elementary conjunctions and $\mathbf{q}$ represents the vector of $w = (n+1)(m+1)$ vectorized p_{tk}, where p_{tk} is the probability of exactly t occurrences of A_i and k occurrences of B_j for $i = 0, 1, ..., n$ and $j = 0, 1, ..., m$.

Obviously, Fréchet optimality is defined for a class of function L over a set of consistent Bonferroni summations. If the given values are attainable for the class of function L, the Fréchet optimal bound exists. However, the definition of linear programming starts with a class of any functions of the linear constraints for Bonferroni-type bounds. Related reference can be found in Kounias and Marin (1976), or Tydeman and Mitchell (1980).

Theorem 4.1.1. The Fréchet optimality based on the elementary conjunctions (4.5), for the class of *all* upper bounds depending on S_{ij}, $(i, j) \in T$, is equivalent to the linear programming optimality in terms of the vectorized term $\mathbf{q}$ specified in (4.7).

Proof: This theorem is valid for any index set T; however, for notational convenience, without loss of generality, we follow the notation in the discussion in (4.8). Note that for any probability space that is associated with a given vector $\mathbf{x}$ (the vector of probabilities of elementary conjunctions arranged in a fixed order), we can sum the associated elements of $\mathbf{x}$ to obtain each element of $\mathbf{q}$. Thus, there exists a mapping from $\mathbf{q}$ to $\mathbf{x}$ so that $\mathbf{q} = \mathbf{Dx}$ for a $(2^{n+m}) \times w$ matrix $\mathbf{D}$ where $w = (n+1)(m+1)$.

For each vector of the ordered probabilities of joint events on exact occurrences, denote $\mathbf{q} = (q_i)_{w \times 1}$. By the definition of q_i, the value q_i represents a probability value p_{st}, which means exactly s occurrences for the first set of events and t occurrences for the second set of events. Now consider the corresponding elementary conjunctions that constitute the event for the probability p_{st}. We can always assign the value of each q_i to the probability of the sample point so that $x_i = p_{st}$, where x_i and q_j are the only two non-zero values, with the remaining $(2^{n+m} - w)$ sample points are assigned with probability zero. In this way, we construct a vector, $\mathbf{x}$, with length 2^{n+m}, such that $\mathbf{q} = \mathbf{Dx}$ for the pre-fixed matrix $\mathbf{D}$.

In summary, each set of elementary conjunctions of A_i's and B_j's gives a vector $\mathbf{x}$ which consequently generates a vector $\mathbf{q}$. On the other hand, each $\mathbf{q}$ can be used to construct at least one set of the corresponding elementary conjunctions of two event classes $\{A_i, i = 1, ..., n\}$, and $\{B_j, j = 1, ..., m\}$. This connects the two matrix formulations of the Fréchet optimality with the existence of the probability space. The relation between the corresponding matrix expressions is then $\mathbf{q} = \mathbf{Dx}$ for a $2^{n+m} \times w$ matrix $\mathbf{D}$. Thus under the special setting of Theorem 4.1.1, the Fréchet optimality and the linear programming optimality are equivalent. This proves Theorem 4.1.1. $\square$

Remark: It should be noted that the concept of Fréchet optimality in Theorem 4.1.1 is defined without a specification on the class of upper bounds L. When L is specified to a special group of interest, attention should be paid to the existence of a Fréchet optimality for that particular class of bounds, because the optimal bound may not exist when L is restricted to a very small group of bounds. We use the following example to illustrate the idea behind the above proof.

Example 4.1.1 According to the definition of Fréchet optimality, for S_{ij}, $(i,j) \in T = \{(1,1),(1,2),(2,1),(2,2)\}$, assume that the class of functions U is expanded to the class of any function of S_{ij}, $(i,j) \in T$. If F is an optimal upper bound, we find a probability space and events in this space, $A_1^*, ..., A_n^*, B_1^*, ..., B_m^*$, that satisfy $S_{ij}(A^*B^*) = S_{ij}$. Denote $\mathbf{q_1}$ the corresponding $\mathbf{q}$-vector associated with $\{A_i^*, i = 1,...,n\}$ and $\{B_j^*, j = 1,...,m\}$, which satisfies $\mathbf{Gq_1} = \mathbf{b}$ where $\mathbf{G}$ is defined in (4.8). Recall that the $\mathbf{q}$-vector is the vector of sorted values of the probabilities of exactly t occurrences of the A_i^* and k occurrences of the B_j^*. Now for any probability space with $\mathbf{q}$ satisfying (4.9), we have

$$
\begin{aligned}
\mathbf{c'q_1} &= P(v_1(A^*) \geq 1, v_2(B^*) \geq 1) \\
&= F(S_{ij}, (i,j) \in T) \quad \text{(by the definition of Fréchet optimality)} \\
&\geq P(v_1 \geq 1, v_2 \geq 1) \quad (F \quad \text{being an upper bound)} \\
&= \mathbf{c'q}.
\end{aligned}
$$

So, $\mathbf{q_1}$ is the solution to the optimizing problem (4.9). □

Now, recall the notation in Equation (4.8) and the vector $\mathbf{c}$: $\mathbf{c} = (c(i,j))$ with $i = 0,1,...,n$ as the first ordering index and $j = 0,1,...,m$ as the second ordering index, such that

$$
c(i,j) = \begin{cases} 0, & \text{for } i = 0 \text{ or } j=0 \\ 1, & \text{otherwise,} \end{cases}
$$

the joint probability that at least one occurrence in each event set can be expressed as

$$
P(v_1 \geq 1, \ v_2 \geq 1) = \mathbf{c'q}.
$$

With the equivalence of optimality proved in Theorem 4.1.1, it is pertinent to discuss a theorem that characterizes upper bounds using matrix representation.

Theorem 4.1.2 For matrix $\mathbf{G}$, vectors $\mathbf{c}$ and $\mathbf{b}$ as specified in the above setting, denote the vector $\mathbf{w'} = (w_0, w_1, w_2, w_3, w_4)$ which may depend on m, n, but not on the values of $S_{11}, S_{12}, S_{21}, S_{22}$ under consideration. $\mathbf{w'b}$ is an upper bound for $P(v_1 \geq 1, v_2 \geq 1)$ for all probability spaces if and only if $\mathbf{w'G} \geq \mathbf{c'}$ (each element of the vector $\mathbf{w'G}$ is not less than the corresponding element in the vector $\mathbf{c}$).

Proof: Recall that the matrix $\mathbf{G}$ is constructed such that $\mathbf{Gq} = \mathbf{b}$ for any vector $\mathbf{q}$ associated with a probability space. Suppose $\mathbf{w'G} \geq \mathbf{c'}$. Any probability space has a probability vector corresponding to elementary conjunctions, hence an associated vector $\mathbf{x}$ as in Theorem 4.1.1. By the relationship between the vectors $\mathbf{x}$ and $\mathbf{q}$, the existence of vector $\mathbf{x}$ leads to the existence of a corresponding vector $\mathbf{q}$. Multiplying the vector of probabilities $\mathbf{q}$ by the term $\mathbf{w}'\mathbf{G} \geq \mathbf{c}'$ gets $\mathbf{w}'\mathbf{Gq} \geq \mathbf{c}'\mathbf{q}$. By the setting of the matrix $\mathbf{G}$ and vector $\mathbf{c}$, the result of multiplication yields

$$\mathbf{w}'\mathbf{b} \geq P(v_1 \geq 1, v_2 \geq 1).$$

Thus $\mathbf{w}'\mathbf{b}$ is an upper bound for $P(v_1 \geq 1, v_2 \geq 1)$.

Next suppose $\mathbf{w}'\mathbf{b} \geq P(v_1 \geq 1, v_2 \geq 1)$ for any probability space; that is $\mathbf{w}'\mathbf{Gq} \geq \mathbf{c}'\mathbf{q}$ for any $\mathbf{q}$. Now we may choose a probability space such that in $\mathbf{q}$, $p_{00}=1$ and $p_{ij} = 0$ when $(i,j) \neq (0,0)$ to reach the conclusion that the first element of $\mathbf{w}'\mathbf{G}$ is not less than the first element of the vector $\mathbf{c}$. Taking another probability space in which $p_{01}=1$ and $p_{ij} = 0$ when $(i,j) \neq (0,1)$ to obtain the inequality for the second element. We repeat the process for the rest of the elements by selecting the appropriate $\mathbf{q}$. Now, recall that $\mathbf{c}$ is a vector with length $(n+1)(m+1)$. We obtain the inequality for the two vectors $\mathbf{w}'\mathbf{G} \geq \mathbf{c}'$ by considering $(n+1)(m+1)$ different probability vectors $\mathbf{q}$. This concludes the proof of Theorem 4.1.2.   □

With Theorem 4.1.2, an upper bound for a given set of S_{ij}, $(i,j) \in T$, is characterized by a vector inequality. We now establish the existence of a linear programming optimal upper bound as follows.

4.1.4   Duality Theorem and Existence of Optimality

Consider the optimal upper bound formulated in linear programming optimality: to find a vector $\mathbf{q}$, $\mathbf{q} \geq \mathbf{0}$, $\mathbf{q}'\mathbf{1} = 1$ which maximizes the term $\mathbf{c}'\mathbf{q}$ subject to the conditions,

$$\mathbf{Gq} = \mathbf{b}, \quad \text{and} \quad \mathbf{q} \geq \mathbf{0}. \tag{4.9}$$

For this linear programming problem, given consistent bivariate Bonferroni summation S_{kt}'s (that constitute the vector $\mathbf{b}$) and feasible $\mathbf{q}$, the dual problem is to find a column vector $\mathbf{d}$, $\mathbf{d}' = (d_0, d_1, d_2, d_3, d_4)$ that minimizes $\mathbf{d}'\mathbf{b}$, subject to

$$\mathbf{d}'\mathbf{G} \geq \mathbf{c}'. \tag{4.10}$$

Using (4.10), we have the following theorem which shows that an optimizing solution to the above optimization problem always exists.

Theorem 4.1.3 If the Bonferroni summations S_{11}, S_{12}, S_{21}, and S_{22} are consistent, the linear programming upper bound for $P(v_1 \geq 1, v_2 \geq 1)$ always exists.

Proof: Since the values of the Bonferroni summations $S_{i,j}$'s are consistent, there exists at least one solution to (4.9) from (4.6). Denote E the set of solutions for (4.9). E is not empty.

If there are finite $\mathbf{x}$'s satisfying (4.9), say $E = \{\mathbf{x_1}, ..., \mathbf{x_k}\}$, taking the maximum point completes the proof.

$$\max_{\mathbf{x} \in E} \mathbf{c}'\mathbf{x} = \mathbf{c}'\mathbf{x_0},$$

where $\mathbf{x}_0 \in E$ for E is finite.

If there is an infinite number of $\mathbf{x}$'s satisfying (4.9), by Edwards (1973, Chapter 1, Theorem 8.6), the set E is a subset of $R^{(n+1)(m+1)}$. In fact, E is also compact since it is both closed and bounded.

To see the closeness of the set E, for any subsequence of E, $\{\mathbf{x}_{n_1}, ..., \mathbf{x}_{n_k}, ...\}$ if

$$\lim_{k \to \infty} \mathbf{x}_{n_k} = \mathbf{x}_n$$

then, since $\mathbf{x}_{n_k}$ satisfies condition (4.9), all $\mathbf{x}_{n_k} \geq 0$ and $\mathbf{G}\mathbf{x}_{n_k} = \mathbf{b}$,

$$\mathbf{x}_n = \lim_{k \to \infty} \mathbf{x}_{n_k} \geq 0$$

and

$$\mathbf{G}\mathbf{x}_n = \mathbf{G}\lim_{k \to \infty} \mathbf{x}_{n_k} = \lim_{k \to \infty} \mathbf{G}\mathbf{x}_{n_k} = \mathbf{b}$$

so $\mathbf{x}_n \in E$. By Edwards (1973, Chapter 1, Proposition 8.1), E is closed.

To see that the set E is bounded, consider that for any $\mathbf{x} \in E$, $\mathbf{x} \geq 0$, $\mathbf{x}'\mathbf{1} = 1$ implies that

$$\mathbf{x}'\mathbf{x} \leq \mathbf{x}'\mathbf{1} = 1$$

so E is bounded. Therefore the set E is compact.

Now, consider the function $f : E \to R$, $f(\mathbf{x}) = \mathbf{c}'\mathbf{x}$ is a continuous function, by Edwards (1973, Chapter 1, Theorem 8.8) $f(\mathbf{x}) = \mathbf{c}'\mathbf{x}$ attains its maximum value at a point of E, thus there exists $\mathbf{x}_0 \in E$ such that

$$\sum_{\mathbf{x} \in E} \mathbf{c}'\mathbf{x} = \mathbf{c}'\mathbf{x}_0.$$

Therefore, $\mathbf{x}_0$ is the solution to the optimization problem in (4.10). $\square$

Example 4.1.2. The condition of consistency in Theorem 4.1.3 is critical to understanding the existence of feasible solutions serving as candidates for the optimal upper bound. Consider a simple situation where $S_1 = 1$ and $S_2 = 100$ with $n = 4$. The linear programming upper bound does not exist, because there is no probability space that satisfies $S_1 = 1$ and $S_2 = 100$ with $n = 4$:

$$S_2 = E(\frac{v(v-1)}{2}) \leq \frac{n-1}{2}E(v) = \frac{n-1}{2}S_1,$$

while for this example

$$S_2 = 100 > 1.5 = \frac{4-1}{2}S_1.$$

With Theorem 4.1.3, we can now prove the following theorem on the existence of Fréchet optimal upper bounds.

Theorem 4.1.4 For a set of consistent Bonferroni summations S_{11}, S_{12}, S_{21}, and S_{22}, when the class of bounding functions U consists of all upper bounds, the Fréchet optimal upper bound exists.

Proof: By Theorem 4.1.3, the solution to the optimizing problem (4.9) exists. Assume that x_0 is a solution to the optimizing problem (4.9). By its dual version (4.10), and the Duality Theorem of linear programming (Hadley, 1962), there exists a vector d_0 such that $d_0'b = c'x_0$ and $d_0'G \geq c'$. Now multiplying the associated probability vector yields that $d_0'b$ is an upper bound for $P(v_1 \geq 1, v_2 \geq 1)$. For the optimality, since $x_0 \geq 0$ and $x_0'1 = 1$ from (4.9), x_0 can then be used to specify probabilities of elementary conjunctions of events $\{A_i^{**}, i = 1, ..., n\}$ and $\{B_j^{**}, j = 1, ..., m\}$ by using an appropriate matrix D as in Theorem 4.1.1. (Alternatively, one can choose a suitable probability space by directly using $p_{ij} = P(v_1 = i, v_2 = j)$). Under this scenario, $G(Dx_0) = b$ implies $S_{ij}(A^{**}B^{**}) = S_{ij}$, since $d_0'b = c'x_0$, $d_0'b$ is then a Fréchet optimal bound. Thus, Theorem 4.1.4 follows. □

Remark: Due to the duality theorem, the bounding coefficients d_0 depends on the given value S_{ij}. Thus the existence of the Fréchet optimal upper bound does not guarantee existence of the linear optimal upper bound. This is because once the optimizing coefficients d_i's (say d_i^*), depends on the fixed but arbitrary consistent Bonferroni summations $S_{11}, S_{12}, S_{21}, S_{22}$, the optimal value in both the primal and dual problem becomes

$$F_1 = d_0^* + d_1^* S_{1,1} + d_2^* S_{1,2} + d_3^* S_{2,1} + d_4^* S_{2,2}. \tag{4.11}$$

Being Fréchet optimal, the bound essentially differs from the linearly optimal bound, which has a set of bounding coefficients that depend on the number of events (n or m) only, but not on the values of S_{ij}, $(i, j) \in T$. In this sense, the linear optimality is uniform for all values of the consistent Bonferroni summations S_{ij}, $(i, j) \in T$, but the Fréchet optimality is not.

We conclude this section with an example of the use of linear programming for univariate upper bounds. More detailed discussion of bivariate upper bounds can be found in Chapter 5, which is specifically devoted to bivariate upper bounds.

Example 4.1.3 Consider the upper bound via linear programming solutions with information given in Example 2.2.1 where $S_1 = 0.8$ and $S_2 = 1$ for a set of events $A_1, ..., A_5$. Denote $p_i = P(v = i)$, as usual. We are interested in finding $p_1, ..., p_n$ satisfying

$$\max(p_1 + ... + p_n)$$

subject to $\sum_{i=0}^{n} p_i = 1$, and

$$p_1 + 2p_2 + ... + tp_t + ... + np_n = S_1$$

$$p_2 + ... + \binom{t}{2}p_t + ... + \binom{n}{2}p_n = S_2.$$

Here, the matrix expression of the linear constraints becomes

$$G = \begin{pmatrix} 1 & 1 & 1 & 1 & ... & 1 & ... & 1 \\ 0 & 1 & 2 & 3 & ... & j & ... & n \\ 0 & 0 & 1 & 6 & ... & \frac{j(j-1)}{2} & ... & \frac{n(n-1)}{2} \end{pmatrix}$$

$$\mathbf{b}' = (1, S_1, S_2),$$

and

$$\mathbf{p}' = (p_0, p_1, ..., p_n),$$

such that

$$\mathbf{Gp}' = \mathbf{b}.$$

Taking three columns from $\mathbf{G}$ to form the following matrix

$$\mathbf{M} = \begin{pmatrix} 1 & 1 & 1 \\ 0 & 1 & n \\ 0 & 0 & \frac{n(n-1)}{2} \end{pmatrix},$$

we have

$$\mathbf{M}^{-1} = \begin{pmatrix} 1 & -1 & \frac{2}{n} \\ 0 & 1 & -\frac{2}{n-1} \\ 0 & 0 & \frac{2}{n(n-1)} \end{pmatrix},$$

thus

$$
\begin{aligned}
\mathbf{p}^* &= \mathbf{M}^{-1}\mathbf{b} \\
&= (1 - S_1 + \frac{2}{n}S_2, S_1 - \frac{2}{n-1}S_2, \frac{2}{n(n-1)}S_2)'.
\end{aligned}
$$

Notice that

$$S_1 - \frac{2}{n-1}S_2 = E(v - \frac{v(v-1)}{n-1}) \geq 0,$$

when $1 - S_1 + \frac{2}{n}S_2 \geq 0$, p^* is a feasible solution. By Example 2.2.1,

$$P(v \geq 1) \leq S_1 - \frac{2}{n}S_2.$$

Therefore

$$\sum_{i=1}^{n} p_i \leq \sum_{i=1}^{n} p_i^*,$$

for any probability space $\mathbf{p}$. Thus the linear programming upper bound is $S_1 - \frac{2}{n}S_2 = 0.4$ as the numerical value of Fréchet optimal bound given in Example 2.2.1.

When $1 - S_1 + \frac{2}{n}S_2 \leq 0$, such as the case where the consistent Bonferroni summations $S_1 = S_2 = 3$ and $n = 5$, $\mathbf{p}^*$ is not a feasible solution. Under this scenario, since the Bonferroni summations are consistent, the corresponding probability space is $p_i = 0$ for $i \neq 3$ and $p_3 = 1$, which is a feasible solution, and

$$P(v \geq 1) \leq 1$$

is true for any events in any probability space. The optimal upper bound for this set of Bonferroni summations ($S_1 = S_2 = 3$ and $n = 5$) is 1. $\square$

This example shows that linear programming upper bounds corresponding to different input values of the Bonferroni summations, are different. In Chapter 5, we will discuss this issue with the method of truncated bounds in detail.

The method seeking the solution for Example 4.1.3 uses a direct solution to find the explicit upper bound in a linear programming setting. However, the explicit form is not always available. In the next section, we will discuss an iteration approach to finding a linear programming bound, which addresses a different bounding technique of linear programming when the direct solution is not available.

4.2   Linear Programming Lower Bounds

In this section, we discuss linear programming in the construction of bivariate lower bounds. The method can be extended to construct multivariate lower bounds for the joint probability of the occurrence of an arbitrary number of event sets. It should be noted that although the explicit form of the univariate Fréchet optimal lower bound (the Dawson-Sankoff lower bound) exists, the corresponding product-type explicit form is not always a bivariate lower bound. This is a discernible feature of multivariate lower bounds when using linear programming (Kwerel (1975b), Chen (1998)).

Due to the nature of multivariate lower bounds, we use notations that are slightly different from those we used in the preceding section in the formulation of linear programming problems. We show that the linear programming lower bound (the Fréchet optimal lower bound) in terms of $S_{i,j}, i, j = 1, 2$, is not the product of two univariate lower bounds, and the linear programming lower bound in terms of $S_{i,j}, i, j = 1, 2$ can only be obtained by an iteration algorithm. We mainly use bivariate lower bounds in terms of $S_{i,j}, i, j = 1, 2$ as examples to show the operation of linear programming bounding techniques.

Given $S_{i,j}, i, j = 1, 2$ (fixed but arbitrary) and the number of events $n(\geq 3)$, $m(\geq 3)$, to find an optimal lower bound for $P(v_1 \geq 1, v_2 \geq 1)$ in terms of $S_{i,j}, i, j = 1, 2$, we write $P(v_1 \geq 1, v_2 \geq 1) = 1 - P(v_1 = 0 \text{ or } v_2 = 0)$. This way of finding a lower bound (the minimal value) of $P(v_1 \geq 1, v_2 \geq 1)$ is equivalent to finding the maximum value of $P(v_1 = 0 \text{ or } v_2 = 0)$ under the appropriate constraints. One constraint, for example, is to satisfy the following equations:

$$S_{ij} = E[\binom{v_1}{i}\binom{v_2}{j}] \quad i = 1, 2; \quad j = 1, 2.$$

Similar to the argument in the preceding section, to optimize the target function, we find a vector (probability space) $\mathbf{p}$ to minimize $\sum_{i=1}^{n} p_i$ subject to $\mathbf{p} \geq \mathbf{0}$ and $\mathbf{Ap} = \mathbf{b}$, where

$$\mathbf{b} = (1, S_{11}, S_{12}, S_{21}, S_{22})',$$

and matrix $\mathbf{A}$ reads

$$
\begin{pmatrix}
1 & 1 & 1 & \cdots & 1 & \cdots & 1 & 1 & \cdots & 1 & \cdots & \cdots & 1 \\
0 & 1 & 2 & \cdots & j & \cdots & m & 2 & \cdots & 2m & \cdots & \cdots & nm \\
0 & 0 & 1 & \cdots & \frac{j(j-1)}{2} & \cdots & \frac{m(m-1)}{2} & 0 & \cdots & 2\frac{m(m-1)}{2} & \cdots & \cdots & n\frac{m(m-1)}{2} \\
0 & 0 & 0 & \cdots & 0 & \cdots & 0 & 1 & \cdots & m & \cdots & \cdots & \frac{n(n-1)}{2}m \\
0 & 0 & 0 & \cdots & 0 & \cdots & 0 & 0 & \cdots & \frac{m(m-1)}{2} & \cdots & \cdots & \frac{nm(n-1)(m-1)}{4}
\end{pmatrix}.
$$

Here, $\mathbf{A}$ is a $5 \times (nm+1)$ matrix with a typical column $\mathbf{a}_t$, where the general form of $\mathbf{a}_t$ is:

$$
\mathbf{a}_t' = \left(1, ij, \frac{ij(j-1)}{2}, \frac{ij(i-1)}{2}, \frac{ij(i-1)(j-1)}{4}\right), \tag{4.12}
$$

for integers $1 \le i \le n$ and $1 \le j \le m$.

The first column of $\mathbf{A}$ corresponds to the index $\{i=0 \text{ or } j=0\}$. And the remaining columns of $\mathbf{A}$ are arranged over $i \ge 1$ in increasing order for each fixed j, and then over $j \ge 1$ in increasing order.

$$
\mathbf{p}' = (p_0, p_{11}, ..., p_{n1}, ..., ..., p_{nm})_{1 \times (nm+1)},
$$

where $p_0 = P(v_1 = 0 \text{ or } v_2 = 0)$, and $p_{ij} = P(v_1 = i, v_2 = j)$ for $i = 1, ..., n$; $j = 1, ..., m$. Notice that the construction of matrix $\mathbf{A}$ is different from the corresponding matrix $\mathbf{G}$ in the preceding section, due to different formulations of the probability vector $\mathbf{p}$.

In this section, using the four common Bonferroni summations $S_{i,j}$ for $i, j = 1, 2$, we show that the solution for bivariate lower linear programming bound exists, but cannot be factorized into the product of two univariate linear programming bounds. We then present a linear programming approach to find the optimal feasible solution.

Recall that for consistent Bonferroni summations, there exists one probability space $\mathbf{p}$ associated with this vector of Bonferroni summations. On the other hand, any vector $\mathbf{p}$ of $(nm+1)$ non-negative entries summing to unity is tantamount to the existence of a probability space when the entries of $\mathbf{p}$ are interpreted as $p_{ij} = P(v_1 = i, v_2 = j)$. Therefore, we refer to $\mathbf{p}$ as a probability space in the sequel for notational convenience.

4.2.1 Inconsistency of Linear Programming Lower Bounds

We start with an example showing that the explicit form of univariate linear programming lower bound exists, but the product of two univariate linear programming lower bounds is not a bivariate lower bound.

Example 4.2.1: Consider the linear programming lower bound with S_1 and S_2 for a set of events A_1, ..., A_n. Denote $p_i = P(v = i)$. The optimization problem now is to find a probability space $\mathbf{p}^0$ so that

$$
\sum_{i=1}^{n} p_i^0 = \min(p_1 + ... + p_n)
$$

subject to $\sum_{i=0}^{n} p_i = 1$, and

$$p_1 + 2p_2 + ... + tp_t + ... + np_n = S_1$$

$$p_2 + ... + \binom{t}{2} p_t + ... + \binom{n}{2} p_n = S_2.$$

Similar to Example 4.1.3, the matrix expression of the linear constraints becomes

$$\mathbf{A} = \begin{pmatrix} 1 & 1 & 1 & 1 & ... & 1 & ... & 1 \\ 0 & 1 & 2 & 3 & ... & j & ... & n \\ 0 & 0 & 1 & 6 & ... & \frac{j(j-1)}{2} & ... & \frac{n(n-1)}{2} \end{pmatrix}$$

$$\mathbf{b}' = (1, S_1, S_2),$$

and

$$\mathbf{p}' = (p_0, p_1, ..., p_n),$$

such that

$$\mathbf{A}\mathbf{p}' = \mathbf{b}.$$

Taking three columns from $\mathbf{A}$ to form the following matrix

$$\mathbf{B} = \begin{pmatrix} 1 & 1 & 1 \\ 0 & k & k+1 \\ 0 & \frac{k(k-1)}{2} & \frac{k(k+1)}{2} \end{pmatrix},$$

we have

$$\mathbf{B}^{-1} = \begin{pmatrix} 1 & -\frac{2}{k+1} & \frac{2}{k(k+1)} \\ 0 & 1 & -\frac{2}{k} \\ 0 & -\frac{k-1}{k+1} & \frac{2}{k+1} \end{pmatrix},$$

thus

$$\mathbf{p}^* = \mathbf{B}^{-1}\mathbf{b}$$

$$= (1 - \frac{2}{k+1}S_1 + \frac{2}{k(k+1)}S_2, S_1 - \frac{2}{k}S_2, -\frac{k-1}{k+1}S_1 + \frac{2}{k+1}S_2)'.$$

When the integer k is taken as $k^* = \lceil \frac{2S_2}{S_1} \rceil + 1$, $\mathbf{p}^*$ is a feasible solution to the constraints. Considering the lower bound in Example 2.2.2 shows that for any other feasible solution, $\mathbf{p}$,

$$\sum_{i=1}^{n} p_i = P(v \geq 1) \geq \frac{2}{k^*+1}S_1 + \frac{2}{k^*(k^*+1)}S_2 = \sum_{i=1}^{n} p_i^*.$$

Therefore, $\mathbf{p}^*$ is the degree-two univariate linear programming lower bound. □

When bivariate linear programming lower bounds are of interest, it is intuitive to combine two univariate bounds to get a bivariate bound. We can use the method of factorization to get

$$P(v_1 \geq 1, v_2 \geq 1) \geq$$
$$\frac{4S_{11}}{(a+1)(b+1)} - \frac{4S_{21}}{a(a+1)(b+1)} - \frac{4S_{12}}{b(a+1)(b+1)} + \frac{4S_{22}}{ab(a+1)(b+1)}$$

where $a \geq 1$, $b \geq 1$ are arbitrary integers. However, as shown in the following example, the product of two univariate linear programming lower bounds is not always a lower bound. More discussion on this issue can be found in Chapter 7, where we focus on bivariate lower bounds.

Example 4.2.2. For a set of events C_1, ..., C_8, consider the following probabilities corresponding to the 2^8 elementary conjunctions of intersections. First, assign two special intersections with probabilities:

$$P(C_1...C_8) = 0.8$$

and

$$P(C_1^c...C_8^c) = 0.19.$$

Then, equally assign each of the $2^8 - 2 = 254$ elementary conjunctions of the intersections with probability $(0.01/254)$.

Denote H the set of all 2^8 elementary conjunctions of intersections. We have

$$\sum_{C \in H} P(C) = 0.8 + 0.19 + 254 \times \frac{0.01}{254} = 1.$$

Now, consider two sets of events A_1, ..., A_n and B_1, ..., B_m with $A_1 = C_1$, $A_2 = C_2$, $A_3 = C_3$, $A_4 = C_4$ and $B_1 = C_5$, $B_2 = C_6$, $B_3 = C_7$, $B_4 = C_8$. This generates two sets of events, $\{A_1, ..., A_4\}$, and $\{B_1, ..., B_4\}$, for which we construct the bivariate lower bound of $P(v(A) \geq 1, v(B) \geq 1)$.

The associated bivariate Bonferroni summations S_{ij}'s for $i = 1, 2$ and $j = 1, 2$ are specified as follows.

For each pair of fixed (i, j), the intersection C_iC_j is actually the union of all of the associated elementary conjunctions, consisting of intersections of event $C_1, ..., C_8$, and their complement sets.

$$P(C_iC_{4+j}) = 0.8 + \frac{1}{254} \times (2^6 - 1), \quad 1 \leq i \leq 4, \quad 1 \leq j \leq 4$$

$$P(C_{i_1}C_{i_2}C_{4+j}) = 0.8 + \frac{1}{254} \times (2^5 - 1), \quad 1 \leq i_1 < i_2 \leq 4, \quad 1 \leq j \leq 4$$

$$P(C_iC_{4+j_1}C_{4+j_2}) = 0.8 + \frac{1}{254} \times (2^5 - 1), \quad 1 \leq i \leq 4, \quad 1 \leq j_1 < j_2 \leq 4.$$

Thus

$$S_{1,1} = \sum_{\substack{1 \leq i \leq 4 \\ 1 \leq j \leq 4}} P(A_iB_j)$$

$$= \sum_{\substack{1 \le i \le 4 \\ 1 \le j \le 4}} P(C_i C_{4+j})$$

$$= 0.8 + \frac{1}{254} \times (2^6 - 1) \sum_{\substack{1 \le i \le 4 \\ 1 \le j \le 4}} 1$$

$$= 0.8 + \frac{1}{254} \times (2^6 - 1) \times 16$$

$$= 12.839.$$

$$S_{1,2} = \sum_{\substack{1 \le i \le 4 \\ 1 \le j_1 < j_2 \le 4}} P(A_i B_{j_1} B_{j_2})$$

$$= \sum_{\substack{1 \le i \le 4 \\ 1 \le j_1 < j_2 \le 4}} P(C_i C_{4+j_1} C_{4+j_2})$$

$$= 0.8 + \frac{1}{254} \times (2^5 - 1) \sum_{\substack{1 \le i \le 4 \\ 1 \le j_1 < j_2 \le 4}} 1$$

$$= 0.8 + \frac{1}{254} \times (2^5 - 1) \times 4 \times \frac{4 \times 3}{2}$$

$$= 19.229,$$

and similarly, $S_{2,1} = 19.229$, $S_{2,2} = 28.821$.

Denote,

$$L(a,b) = \frac{4S_{11}}{(a+1)(b+1)} - \frac{4S_{21}}{a(a+1)(b+1)} - \frac{4S_{12}}{b(a+1)(b+1)} + \frac{4S_{22}}{ab(a+1)(b+1)}.$$

For $a = b = 1$,

$$S_{1,1} - S_{1,2} - S_{2,1} + S_{2,2} = 3.203.$$

Here, $L_{1,1} > 1$, which cannot serve as a lower bound. □

Although factorization fails to provide an explicit form for the bivariate linear programming lower bound, the bivariate linear programming bound does exist for a set of consistent bivariate Bonferroni summations. In what follows in this section, we introduce a theorem for the characterization of bivariate lower bounds, and then show the existence of the bivariate linear programming lower bound.

Denote the vector $\mathbf{c}$: $\mathbf{c} = (c(i,j))$, with the first element of $\mathbf{c}$ corresponding to the index $\{i = 0 \text{ or } j = 0\}$, and the rest of the elements of $\mathbf{c}$ formed by ranking over $i \ge 1$ in increasing order for each fixed j, then over $j \ge 1$ in increasing order. Also assign

$$c(i,j) = \begin{cases} 0, & \text{for } i = 0 \text{ or } j = 0 \\ 1, & \text{otherwise.} \end{cases}$$

The joint probability of at least one occurrence in both event sets can be expressed as

$$P(v_1 \ge 1, v_2 \ge 1) = \mathbf{c}' \mathbf{p}.$$

The above setting establishes a vector inequality for the characterization of bivariate lower bounds.

Theorem 4.2.1 For matrix $\mathbf{A}$, the vector of consistent Bonferroni summations $\mathbf{b}$, and the vector of coefficients $\mathbf{c}$ as specified above, denote the vector $\mathbf{w}' = (w_0, w_1, w_2, w_3, w_4)$ which may depend on m, n, but not on the values of S_{11}, S_{12}, S_{21}, S_{22}. The value $\mathbf{w}'\mathbf{b}$ is a lower bound for $P(v_1 \geq 1, \ v_2 \geq 1)$ for all probability spaces if and only if $\mathbf{w}'\mathbf{A} \leq \mathbf{c}'$ (each element of the vector $\mathbf{w}'\mathbf{A}$ is not greater than the corresponding element in the vector $\mathbf{c}$).

Proof: Recall that matrix $\mathbf{A}$ is constructed in the way such that $\mathbf{Ap} = \mathbf{b}$ for any vector $\mathbf{p}$ associated with a probability space.

Suppose $\mathbf{w}'\mathbf{A} \leq \mathbf{c}'$. Since any probability space has a probability vector $\mathbf{x}$ corresponding to its elementary conjunctions, according to Theorem 4.1.1 regarding the connection between the vectors $\mathbf{x}$ and $\mathbf{p}$, the existence of vector $\mathbf{x}$ leads to the existence of a corresponding vector $\mathbf{p}$. Now, multiplying the vector of probabilities $\mathbf{p}$ into the term $\mathbf{w}'\mathbf{A} \leq \mathbf{c}'$ gets $\mathbf{w}'\mathbf{Ap} \leq \mathbf{c}'\mathbf{p}$. By setting the matrix $\mathbf{A}$ and vector $\mathbf{c}$ as above, the result of multiplication yields

$$\mathbf{w}'\mathbf{b} \leq P(v_1 \geq 1, \ v_2 \geq 1).$$

Thus $\mathbf{w}'\mathbf{b}$ is a lower bound for $P(v_1 \geq 1, \ v_2 \geq 1)$ when $\mathbf{w}'\mathbf{A} \leq \mathbf{c}'$.

Next suppose $\mathbf{w}'\mathbf{b} \leq P(v_1 \geq 1, \ v_2 \geq 1)$ for any probability space. Spelling out the above condition gets $\mathbf{w}'\mathbf{Ap} \leq \mathbf{c}'\mathbf{p}$. We may choose a probability space such that in $\mathbf{p}$, $P(v_1 = 0$ or $v_2 = 0)=1$ and $p_{ij} = 0$ when $(i, j) \neq (0,0)$ to reach the conclusion that the first element of $\mathbf{w}'\mathbf{A}$ is not less than the first element of the vector $\mathbf{c}$. Taking a probability space in which $p_{0,1}=1$ and $p_{ij} = 0$ when $(i, j) \neq (0, 1)$, we obtain the inequality for the second element of the two vectors. We can then repeat this process in turn for the rest of the elements of the two vectors by selecting the appropriate probability space $\mathbf{p}$: $p_{i_0, j_0}=1$ and $p_{ij} = 0$ when $(i, j) \neq (i_0, j_0)$. Since $\mathbf{c}$ is a vector with length $(nm + 1)$, we obtain the inequality for the two vectors as $\mathbf{w}'\mathbf{A} \leq \mathbf{c}'$ after checking each row of elements across the two vectors. $\square$

Comparing the proofs of Theorem 4.2.1 and Theorem 4.1.2, one can see that the two proofs share the same rationale of using matrix representations for probability inequalities. For this reason, we state the following theorem for the existence of bivariate linear programming lower bounds without the explicit proof.

Theorem 4.2.2 For a set of consistent Bonferroni summations S_{11}, S_{12}, S_{21}, and S_{22}, the linear programming lower bound for $P(v_1 \geq 1, v_2 \geq 1)$ always exists.

Proof: The proof of Theorem 4.2.2 is self-evident based on the proof of Theorem 4.1.3. The only difference between the two proofs is to switch the upper bound of Theorem 4.1.3 to lower bound for Theorem 4.2.2. To save space, we do not repeat the proof here. $\square$

The above theorem shows the existence of the bivariate linear programming bounds, but does not show the approach to obtain them. The following subsection concentrates on the linear programming method toward the optimal lower bound.

4.2.2   Feasible Linear Programming Lower Bounds

Unlike the formulation of upper bounds in Section 4.1, for lower bounds, the product of two univariate lower bounds may not be the bivariate optimal bound, due to the structure of bivariate probability. This raises a question on the feasibility of multivariate lower bounds defined in the setting of linear programming. In this section, we start with an example bridging the structure of the matrix $\mathbf{A}$ in (4.12) and the existence of bivariate lower bounds for $P(v_1 \geq 1 \, v_2 \geq 1)$.

Example 4.2.3. Assume that $n \geq 3$ and $m \geq 3$, consider integers $a = \left[\frac{n}{2}\right] + 1$ and $b = \left[\frac{m}{2}\right] + 1$ such that $a \leq n, b \leq m$. Denote a matrix $\mathbf{B}_1$ formed with the first column chosen as the first column of A, and the rest of the four columns in the following way,

$$\mathbf{B}_1 = \begin{pmatrix} 1 & 1 & 1 & 1 & 1 \\ 0 & ab & (a+1)b & a(b+1) & (a+1)(b+1) \\ 0 & \frac{ab(b-1)}{2} & \frac{(a+1)b(b-1)}{2} & \frac{ab(b+1)}{2} & \frac{(a+1)(b+1)b}{2} \\ 0 & \frac{ab(a-1)}{2} & \frac{(a+1)ab}{2} & \frac{a(a-1)(b+1)}{2} & \frac{(a+1)(b+1)a}{2} \\ 0 & \frac{ab(b-1)(a-1)}{4} & \frac{(a+1)ab(b-1)}{4} & \frac{ab(a-1)(b+1)}{4} & \frac{ab(a+1)(b+1)b}{4} \end{pmatrix}. \quad (4.13)$$

For this 5×5 matrix, the associated inverse matrix $\mathbf{B}_1^{-1}$ is

$$\mathbf{B}_1^{-1} = \begin{pmatrix} 1 & \frac{-4}{(a+1)(b+1)} & \frac{4}{b(a+1)(b+1)} & \frac{4}{a(a+1)(b+1)} & \frac{-4}{ab(a+1)(b+1)} \\ 0 & 1 & \frac{-2}{b} & \frac{-2}{a} & \frac{4}{ab} \\ 0 & \frac{-(a-1)}{a+1} & \frac{2(a-1)}{(a+1)b} & \frac{2}{a+1} & \frac{-4}{b(a+1)} \\ 0 & \frac{-(b-1)}{b+1} & \frac{2}{b+1} & \frac{2(b-1)}{a(b+1)} & \frac{-4}{a(b+1)} \\ 0 & \frac{(a-1)(b-1)}{(a+1)(b+1)} & \frac{-2(a-1)}{(a+1)(b+1)} & \frac{-2(b-1)}{(a+1)(b+1)} & \frac{4}{(a+1)(b+1)} \end{pmatrix}. \quad (4.14)$$

For $i = 1, \dots, n, j = 1, \dots, m$, write

$$L(i,j) = 1 - \frac{4ij}{(a+1)(b+1)}\left(1 - \frac{i-1}{2a}\right)\left(1 - \frac{j-1}{2b}\right). \quad (4.15)$$

Let $\mathbf{s}_i'$ be the ith row of $\mathbf{B}_1^{-1}$. Notice that for $\mathbf{B}_1$, the explicit lower bound from the inverse matrix reads

$$\mathbf{s}_1' \mathbf{a_t} = L(i,j), \quad (4.16)$$

where $\mathbf{a}_t$ is defined in (4.12).

Now, for $i = 1, \dots, n, \ j = 1, \dots, m$, and any positive integers a, b,

$$\frac{4ij}{(a+1)(b+1)}\left(1 - \frac{i}{2a} \cdot \frac{1}{1}\right)\left(1 - \frac{j-1}{2b}\right)$$
$$= \left[\frac{2i}{(a+1)}\left(1 - \frac{i-1}{2a}\right)\right]\left[\frac{2j}{(b+1)}\left(1 - \frac{j-1}{2b}\right)\right]$$
$$\leq \ 1.$$

By (4.15), we have

$$
\begin{aligned}
L(i,j) &= 0 \quad \text{for} \quad i \in \{a, a+1\} \quad \text{and} \quad j \in \{b, b+1\}; \\
L(i,j) &> 0 \quad \text{for} \quad i \in \{1, ..., n\} \backslash \{a, a+1\} \quad \text{or} \quad j \in \{1, ..., m\} \backslash \{b, b+1\}.
\end{aligned}
$$

Thus, $\mathbf{1}' - \mathbf{s}_1'\mathbf{A} \leq (0, 1, ..., 1)$ for any probability space $\mathbf{p}$, meaning $\mathbf{s}_1'\mathbf{b}$ is a bivariate lower bound for $P(v_1 \geq 1, v_2 \geq 1)$.   □

Example 4.2.3 can be extended to the following general setting, which is a key step in constructing a bivariate linear programming lower bound for $P(v_1 \geq 1\ v_2 \geq 1)$.

For the matrix $\mathbf{A}$ defined in (4.12), suppose it is possible to find a 5×5 matrix, $\mathbf{B}$, formed from columns of $\mathbf{A}$ with the first column of $\mathbf{B}$ chosen to be the first column of $\mathbf{A}$, and the matrix $\mathbf{B}$ meeting the following two conditions:

- The matrix is non-singular.

$$
|\mathbf{B}| \neq 0; \tag{4.17}
$$

- The first row of $\mathbf{B}^{-1}$, denoted as $\mathbf{s}_1'$, satisfies

$$
\mathbf{s}_1'\mathbf{a_j} \geq 0 \quad \text{for all} \quad j \geq 2 \quad \text{and} \quad \mathbf{s}_1'\mathbf{a_1} = 1. \tag{4.18}
$$

Example 4.2.3 shows the existence of above specified matrix $\mathbf{B}$, leading to bivariate lower bounds as follows.

Lemma 4.2.1. With the matrix $\mathbf{B}$ defined above and satisfying the two conditions (4.17) and (4.18), $1 - \mathbf{s}_1'\mathbf{b}$ is a lower bound for $P(v_1 \geq 1, v_2 \geq 1)$.

Proof: By (4.18), we get

$$
\mathbf{1}' - \mathbf{s}_1'\mathbf{A} \leq (0, 1, ..., 1). \tag{4.19}
$$

Now, for any probability space $\mathbf{p}$ consistent with the given quantities of the Bonferroni summations $\mathbf{b}$, applying $\mathbf{p}$ to above inequality (4.19) yields

$$
1 - \mathbf{s}_1'\mathbf{b} \leq P(v_1 \geq 1, v_2 \geq 1), \tag{4.20}
$$

since $\mathbf{1}'\mathbf{p} = 1$, and $\mathbf{Ap} = \mathbf{b}$.   □

The lemma above identifies a bivariate lower bound using the linear programming approach. Furthermore, the following result shows that under certain conditions such a lower bound is achievable.

Lemma 4.2.2. Assume that the matrix $\mathbf{B}$ defined above satisfies the conditions specified by (4.17) and (4.18). If the vector $\mathbf{x_B} = \mathbf{B}^{-1}\mathbf{b} \geq \mathbf{0}$, there exists a lower bound of $P(v_1 \geq 1, v_2 \geq 1)$ for any probability space $\mathbf{p}$, and there exists a probability space where this lower bound achieves equality.

Proof: The conclusion of the lower bound follows Lemma 4.2.1 in which (4.20) holds for any probability space.

To prove the second part of Lemma 4.2.2 regarding the equality at the boundary, when

$$\mathbf{x_B} = \mathbf{B}^{-1}\mathbf{b} \geq \mathbf{0},$$

multiplying the matrix $\mathbf{B}$ on both sides gets

$$\mathbf{Bx_B} = \mathbf{b}.$$

Therefore, the following conditions are satisfied.

Condition-1, $\mathbf{x_B}'\mathbf{1} = 1$ (as the first row of $\mathbf{B}$ is $\mathbf{1}'$, and the first element of $\mathbf{b}$ is 1).

With the above condition, we specify a probability space $\mathbf{p}$, with non-zero elements assigned to elements corresponding to $\mathbf{x_B}$, because columns of the matrix $\mathbf{B}$ are chosen from columns of the matrix $\mathbf{A}$, and each column of $\mathbf{B}$ has a corresponding position in the matrix $\mathbf{A}$. This serves as an index for the the assignment of non-zero elements in $\mathbf{p}$.

Condition-2, Consider the first element of the vector $\mathbf{x_B}$, which is the first row of $\mathbf{B}^{-1}$ multiplied by the vector $\mathbf{b}$:

$$x_B(1) = \mathbf{s}_1'\mathbf{b}.$$

By the selection criteria of the matrix $\mathbf{B}$, the first column of $\mathbf{B}$ is the first column of $\mathbf{A}$. So, the first element of the vector $\mathbf{x_B}$, $x_B(1)$, is associated with $P(v_1 = 0,\ or\ v_2 = 0)$, the first element in the vector $\mathbf{p}$ specified in Condition-1.

Therefore, we have equality in (4.20) for the specified vector of probability space $\mathbf{p}$, and $1 - \mathbf{s}_1'\mathbf{b}$ is the corresponding value of $P(v_1 = 0\ or\ v_2 = 0)$. This completes the proof of Lemma 4.2.2.   □

Lemmas 4.2.1 and 4.2.2 specify conditions where the linear programming lower bound can be constructed. If the condition in Lemma 4.2.2 is satisfied, the optimal lower bound is found. However, it is not always true that $\mathbf{x_B} = \mathbf{B}^{-1}\mathbf{b} \geq \mathbf{0}$. When $\mathbf{x_B} = \mathbf{B}^{-1}\mathbf{b} \not\geq \mathbf{0}$, we need to find an alternative approach to reach the linear programming bound. This is technically more involved in linear programming. In the following, we provide details on the alternative approach (an iteration algorithm) and show that the algorithm can theoretically reach the existence condition after finite iterations.

4.2.3   A Perturbation Device in Linear Programming Optimization

When $\mathbf{B}^{-1}\mathbf{b} \not\geq \mathbf{0}$, Lemma 4.2.2 can not be applied to find the linear programming lower bound. We can update the matrix B by replacing one of the columns in $\mathbf{B}$ with a column from the set of columns of $\mathbf{A}$ to form a new $\mathbf{B}$ matrix, and denote the updated $\mathbf{B}$-matrix as $\mathbf{B}_1$. If $\mathbf{B}_1^{-1}\mathbf{b} \geq \mathbf{0}$, by Lemma 4.2.2, we find the linear programming bound. If $\mathbf{B}_1^{-1}\mathbf{b} \not\geq \mathbf{0}$, a column in $\mathbf{B}_1$ is replaced to form a new $\mathbf{B}$-matrix, denoted as $\mathbf{B}_2$. In this way, we initiate an iteration process.

The difficulty now is in selecting the proper column from $\mathbf{A}$ to form the optimal point. To this end, we use a perturbation device to show that once the column for removal/replacement is selected appropriately, we can achieve the solution for a linear programming lower bound.

First, we extend Lemma 4.2.1 in conjunction with a perturbation device (depending on ε) as follows in the iteration process to obtain the optimizing solution. The introduction of the ε below, ostensibly a perturbation device, is necessary to show that cycling doesn't occur in the iterative procedure aimed at arriving at an optimal lower bound by a linear programming implementation.

For any positive value $\varepsilon > 0$, define a vector $\mathbf{c}_B(\varepsilon)$ as follows. Let $\Gamma = \{\mathbf{c}(\varepsilon)\}$, where $\mathbf{c}(\varepsilon) = (c_1(\varepsilon), c_2(\varepsilon), ..., c_{nm+1}(\varepsilon))'_{(nm+1)\times 1}$ is any vector of ε's satisfying,

$$
\begin{aligned}
c_1(\varepsilon) &= 1 \\
c_i(\varepsilon) &= \varepsilon^{k(i)} \quad \text{for some} \quad k > 0 \quad \text{depending on} \quad i > 1 \qquad (4.21) \\
c_i(\varepsilon) &\neq c_j(\varepsilon) \quad i \neq j.
\end{aligned}
$$

For example, the function of ε can take any one of the following forms,

$$
c_i(\varepsilon) = \varepsilon^i, i \neq 1, \quad c_1(\varepsilon) = 1
$$

or

$$
c_i(\varepsilon) = \varepsilon^{i+7}, i \neq 1, \quad c_1(\varepsilon) = 1.
$$

The function $\mathbf{c}(\varepsilon)$ will be fixed through the iteration process. The possibility of different explicit forms of $\mathbf{c}(\varepsilon)$ is key in showing that it is not necessary to worry about perturbation terms in ε.

Now, denote columns from $\mathbf{A}$ forming $\mathbf{B}$ as $\mathbf{a}_1$, $\mathbf{a}_{t_2}$, ..., $\mathbf{a}_{t_5}$, then $\mathbf{c}_B(\varepsilon) = (1, c_{t_2}(\varepsilon), ..., c_{t_5}(\varepsilon))'$, a 5×1 vector formed by selecting the corresponding $c_i(\varepsilon)$'s from the $(nm+1) \times 1$ vector $\mathbf{c}(\varepsilon)$ introduced above.

With the vector $\mathbf{c}_B(\varepsilon)$ defined in (4.21), the existence condition in Lemma 4.2.1 can be extended to the following.

$$
\mathbf{c}_B(\varepsilon)'\mathbf{B}^{-1}\mathbf{a}_k \geq c_k(\varepsilon), \quad k \neq 1; \quad \text{and} \quad \mathbf{c}_B(\varepsilon)'\mathbf{B}^{-1}\mathbf{a}_1 = 1, \qquad (4.22)
$$

where $\mathbf{a}_1$, ..., $\mathbf{a}_k$ are columns of matrix $\mathbf{A}$.

Lemma 4.2.3. Assume matrix $\mathbf{B}$ defined above satisfies the conditions specified in (4.22) for any $0 < \varepsilon \leq \varepsilon_0$. If the vector $\mathbf{x}_{\mathbf{B}} = \mathbf{B}^{-1}\mathbf{b} \geq \mathbf{0}$, then there exists a lower bound of $P(v_1 \geq 1, v_2 \geq 1)$ for any probability space $\mathbf{p}$, and there exists a probability space in which this lower bound achieves equality.

Proof: If $\mathbf{x}_{\mathbf{B}} = \mathbf{B}^{-1}\mathbf{b} \geq \mathbf{0}$, since (4.22) is valid when all ε are small enough, letting $\varepsilon \to 0$, it follows that $\mathbf{B}_1$ fits the form of the matrix described in Lemma 4.2.2. Thus, letting ε go to zero in equation (4.22) gets the condition of Lemma 4.2.2. This completes the proof of the lemma. $\square$

We now show that the $\mathbf{B}$-matrix, $\mathbf{B}_1$, defined in (4.13), satisfies the condition of a bivariate lower bound.

Lemma 4.2.4. As specified in Example 4.2.3, denote

$$
\mathbf{B_1} = \begin{pmatrix}
1 & 1 & 1 & 1 & 1 \\
0 & ab & (a+1)b & a(b+1) & (a+1)(b+1) \\
0 & \frac{ab(b-1)}{2} & \frac{(a+1)b(b-1)}{2} & \frac{ab(b+1)}{2} & \frac{(a+1)(b+1)b}{2} \\
0 & \frac{ab(a-1)}{2} & \frac{(a+1)ab}{2} & \frac{a(a-1)(b+1)}{2} & \frac{(a+1)(b+1)a}{2} \\
0 & \frac{ab(b-1)(a-1)}{4} & \frac{(a+1)ab(b-1)}{4} & \frac{ab(a-1)(b+1)}{4} & \frac{ab(a+1)(b+1)b}{4}
\end{pmatrix},
$$

then there exists a positive integer ε_1 so that the matrix $\mathbf{B}_1$ satisfies condition (4.22) for all $0 < \varepsilon < \varepsilon_1$.

Proof: Consider $\mathbf{c}_{B_1}(\varepsilon)'\mathbf{B_1}^{-1}\mathbf{a_k}$. Denote $\mathbf{t}_i'$ the ith row of $\mathbf{B_1}^{-1}$ and $\mathbf{b}_1(i)$ the ith column of $\mathbf{B}_1$. To evaluate $\mathbf{c}_{B_1}(\varepsilon)'\mathbf{B_1}^{-1}\mathbf{a_k}$, we investigate the following cases.

Case 1: For $k \neq 1$, and $\mathbf{a_k}$ not being one of the columns forming $\mathbf{B}_1$ (which contains $\mathbf{a}_1$ and all the 4 columns giving $L(i,j) = 0$, see, for instance, Example 4.2.3), thus, the $L(i,j)$ resulting from $\mathbf{t}_1'\mathbf{a_k}$ is positive.

Now consider $\mathbf{c}(\varepsilon) \in \Gamma$, since $c_k(\varepsilon) \to 0$ as $\varepsilon \to 0$, recalling the structure of $\mathbf{c}_{B_1}(\varepsilon)$ yields, in general

$$
\mathbf{c}_{B_1}(\varepsilon)'\mathbf{B_1}^{-1}\mathbf{a_k} = \mathbf{t}_1'\mathbf{a_k} + \sum_{i=2}^{5} c_{t_i}(\varepsilon)\mathbf{t}_i'\mathbf{a_k},
$$

where, as in (4.21), $c_i(\varepsilon)$, is just some distinctive power of ε. Thus the second item in the expression of $\mathbf{c}_{B_1}(\varepsilon)'\mathbf{B_1}^{-1}\mathbf{a_k}$ satisfies

$$
\lim_{\varepsilon \to 0} \sum_{i=2}^{5} c_{t_i}(\varepsilon)\mathbf{t}_i'\mathbf{a_k} = 0.
$$

Therefore, there exists an $\varepsilon^* > 0$, such that for all $0 < \varepsilon < \varepsilon^*$

$$
\mathbf{c}_{B_1}(\varepsilon)'\mathbf{B_1}^{-1}\mathbf{a_k} > c_k(\varepsilon).
$$

Case 2: If $k \neq 1$ and $\mathbf{a_k}$ is one of $\mathbf{b}_i$'s, $i \neq 1$, say $\mathbf{a}_{t_i}$, $k = t_i$,

$$
\begin{aligned}
\mathbf{c}_{B_1}(\varepsilon)'\mathbf{B_1}^{-1}\mathbf{a_k} &= \mathbf{c}_{B_1}(\varepsilon)'\mathbf{B_1}^{-1}\mathbf{a}_{t_i} \\
&= \mathbf{c}_{B_1}(\varepsilon)'\mathbf{B_1}^{-1}\mathbf{b_i} \\
&= \mathbf{c}_{B_1}(\varepsilon)'\mathbf{e_i} \\
&= c_{t_i}(\varepsilon), \\
&= c_k(\varepsilon),
\end{aligned}
$$

where $\mathbf{e_i}$ is the vector with non-zero element 1 in the ith location. Thus combining Cases 1 and 2 gets

$$
\mathbf{c}_{B_1}(\varepsilon)'\mathbf{B_1}^{-1}\mathbf{a_k} \geq c_k(\varepsilon) \quad \text{when} \quad k \neq 1.
$$

Now, for $k = 1$, we have

$$\mathbf{c}_{B_1}(\varepsilon)'\mathbf{B_1}^{-1}\mathbf{a_1} = \mathbf{c}_{B_1}(\varepsilon)'\mathbf{e_1} = 1$$

Therefore, we have shown that for $\mathbf{B_1}$, condition (4.22) is satisfied, when $\varepsilon_1(\mathbf{B_1})$ is set to equal to ε^* here. $\qquad\square$

With this perturbation device, ε, we show in the next section that a matrix $\mathbf{B}$ satisfying (4.18) can always be found via an iteration process.

4.2.4 An Iteration Process in Linear Programming Optimization

With the perturbation device, $\mathbf{c}(\varepsilon)$, introduced in the preceding section, we can now concentrate on an approach to search for the linear programming solution, when the $\mathbf{B}$-matrix satisfies the condition $\mathbf{B}^{-1}\mathbf{b} \not\geq \mathbf{0}$ in the linear programming setting for bivariate lower bounds. Instead of directly solving for the explicit form of the optimization solution, we use an iteration process to find the optimal bound. In each iteration, we construct a new $\mathbf{B}$-matrix by substituting a column of the current $\mathbf{B}$-matrix with a selected column from the general matrix $\mathbf{A}$ to test the optimal condition until a feasible solution is reached. The same idea is applicable to general multivariate lower bounds. The first part of this section proves that under certain persistent conditions, the optimizing solution emerges after finite times of iterations in the iterating process. The second part endures the validity of the persistent conditions in each iteration.

For convenience, in the sequel we use the notations $\mathbf{B}$ and $\mathbf{D}$ to replace $\mathbf{B_i}$ and its replacement $\mathbf{B_{i+1}}$ respectively in the iteration process seeking the optimal bound. We need the following lemma connecting the two inverse matrices $\mathbf{B}^{-1}$ and its corresponding $\mathbf{D}^{-1}$.

Theorem 4.2.4 The inverse matrix of $\mathbf{D}$ can be decomposed as follows.

$$\mathbf{D}^{-1} = \mathbf{B}^{-1} - \begin{pmatrix} \frac{y_{1j_k}}{y_{rj_k}}\mathbf{s_r}' \\ \vdots \\ \frac{y_{5j_k}}{y_{rj_k}}\mathbf{s_r}' \end{pmatrix} + \begin{pmatrix} \mathbf{0}' \\ \vdots \\ \frac{1}{y_{rj_k}}\mathbf{s_r}' \\ \vdots \\ \mathbf{0}' \end{pmatrix}.$$

Proof: To get $\mathbf{D}^{-1}$ in terms of $\mathbf{s}_i$, $i = 1,...,5$, recall that $\mathbf{s}_i'$ is the i th row of $\mathbf{B}^{-1}$, and let

$$\mathbf{g}_i = \mathbf{s_i} - \frac{y_{ij_k}}{y_{rj_k}}\mathbf{s_r} \quad i \neq r, \quad \text{and}$$

$$\mathbf{g}_r = \frac{1}{y_{rj_k}}\mathbf{s_r}.$$

Since for any i, $\mathbf{b}_t$ is the tth column of $\mathbf{B}$,

$$\mathbf{g_i}'\mathbf{b_t} = \begin{cases} 0, & \text{if } t \neq i,r \\ 1, & \text{if } t = i \end{cases} \quad \text{for } i \neq r;$$

and $g_i'a_{j_k} = y_{ij_k} - y_{ij_k} = 0, i \neq r$, $\quad g_r'b_t = 0$, if $t \neq r$; and $\quad g_r'a_{j_k} = 1$, it follows that if we put $s_{*i} = g_i$, then $s_*(i)'$ is the ith row of D^{-1}.

Writing

$$s_*(r) = s_r - \frac{y_{rj_k}}{y_{rj_k}} s_r + \frac{1}{y_{rj_k}} s_r = \frac{1}{y_{rj_k}} s_r$$

and using the expression for $s_*(i)$ (the same as g_i) yields

$$D^{-1} = \begin{pmatrix} s_*(1)' \\ \vdots \\ s_*(5)' \end{pmatrix}$$

$$= \begin{pmatrix} s_1' \\ \vdots \\ s_5' \end{pmatrix} - \begin{pmatrix} \frac{y_{1j_k}}{y_{rj_k}} s_r' \\ \vdots \\ \frac{y_{5j_k}}{y_{rj_k}} s_r' \end{pmatrix} + \begin{pmatrix} 0' \\ \vdots \\ \frac{1}{y_{rj_k}} s_r' \\ \vdots \\ 0' \end{pmatrix}.$$

The only non-zero row in the last matrix above is the rth row. $\quad\square$

For any B satisfying (4.22), and with $x_B = B^{-1}b \not\geq 0$, we introduce the following notations for expression simplicity.

$$y_{ij} = s_i'a_j \quad i = 1, ..., 5, \quad j = 1, ..., nm + 1$$

and for each fixed $c(\varepsilon) \in \Gamma$ and $c_B(\varepsilon)$ as specified in Section 4.2.3, define

$$\theta(\varepsilon) = \max_p \{ \frac{1}{y_{rp}} [c_B(\varepsilon)'B^{-1}a_p - c_p(\varepsilon)] \}, \tag{4.23}$$

where $p \in \{j_1, ..., j_t\}$, which is the index set associated with $H(B) = \{a_{j_1}, ..., a_{j_t}\}$, a set of vectors in A, such that $s_r'a_{j_1} < 0, ..., s_r'a_{j_t} < 0$.

With the above result and notation, we can now show that there are finite iterations in the optimization process, so the iterating process will not run forever.

Theorem 4.2.5. Let n be the total number of iterations before reaching the optimal solution, if $\theta(\varepsilon) < 0$ and the condition (4.22) persists for each B matrix in the iteration process, $n < \infty$.

Proof: Put $z_{rb} = s_r' b$. By Theorem 4.2.4,

$$D^{-1}b = B^{-1}b - \begin{pmatrix} \frac{y_{1j_k}}{y_{rj_k}} z_{rb} \\ \vdots \\ \frac{y_{5j_k}}{y_{rj_k}} z_{rb} \end{pmatrix} + \begin{pmatrix} 0 \\ \vdots \\ \frac{z_{rb}}{y_{rj_k}} \\ \vdots \\ 0 \end{pmatrix}.$$

Thus,

$$\begin{aligned} \mathbf{c}_D{}'(\varepsilon)\mathbf{D}^{-1}\mathbf{b} &= \mathbf{c}_B{}'(\varepsilon)\mathbf{B}^{-1}\mathbf{b} - \theta(\varepsilon)\mathbf{s}_r{}'\mathbf{b} \\ &< \mathbf{c}_B{}'(\varepsilon)\mathbf{B}^{-1}\mathbf{b}, \end{aligned}$$

since $\theta(\varepsilon) < 0$ and $\mathbf{s}_r{}'\mathbf{b} < 0$ for $x_r < 0$.

Thus $f(\mathbf{B}) = \mathbf{c}_B(\varepsilon)'\mathbf{B}^{-1}\mathbf{b}$ decreases strictly at the next iteration, for $0 < \varepsilon < \varepsilon_0(\mathbf{B})$. The numbers of permissible $\mathbf{B}'$s is however, finite, (at most $nm(nm-1)(nm-2)(nm-3)$), so there will be a stage in the iteration process where no strict decrease is possible. But strict decrease is possible at the next stage for any $\mathbf{B}$ satisfying (4.22) and $\mathbf{B}^{-1}\mathbf{b} \not\geq \mathbf{0}$. Thus at some iteration $\mathbf{B}^{-1}\mathbf{b} \geq \mathbf{0}$. This means that after a finite number of iterations (with the first column of $\mathbf{B}$ never being changed), there exists a matrix $\mathbf{B}$ and an associated $\theta(\varepsilon)$ such that for all $0 < \varepsilon < \varepsilon_0(\mathbf{B})$

$$c_B'(\varepsilon)\mathbf{B}^{-1}\mathbf{a_j} \geq c_j(\varepsilon) \qquad c_B'(\varepsilon)\mathbf{B}^{-1}\mathbf{a_1} = 1 \quad \text{and} \quad \mathbf{B}^{-1}\mathbf{b} \geq \mathbf{0}.$$

Now, letting $\varepsilon \to 0$ leads to $\mathbf{s}_1{}'\mathbf{a_j} \geq 0$, for all $\mathbf{a_j}$ and $\mathbf{s}_1{}'\mathbf{a_1} = 1$, the condition of Lemma 4.2.3. An optimal probability space can thereby be constructed. □

We now prove that in the iteration process, when the first $\mathbf{B}$-matrix, $\mathbf{B}_1$, satisfies the condition for being a lower bound, the lower bound property (4.22) persists in the iteration process (which ends with an optimal lower bound). We need the following lemma to facilitate the discussion. First we need to show that the set of candidate vectors in the iteration process is not empty.

Lemma 4.2.5. For any $\mathbf{B}$ satisfying (4.22), and $\mathbf{x}_B = \mathbf{B}^{-1}\mathbf{b} \not\geq \mathbf{0}$, let r represent the position of the smallest negative element of $\mathbf{x_B} = \mathbf{B}^{-1}\mathbf{b}$, then $r > 1$. Furthermore, there exists a set of vectors $H(\mathbf{B}) = \{\mathbf{a}_{j_1},...,\mathbf{a}_{j_t}\}$ for some $t \geq 1$, in $\mathbf{A}$, such that $\mathbf{s}_r{}'\mathbf{a}_{j_1} < 0$, ..., $\mathbf{s}_r{}'\mathbf{a}_{j_t} < 0$, and $\mathbf{b}_i \notin H(\mathbf{B})$ where $\mathbf{b}_i$ is the ith column of $\mathbf{B}$.

Proof: For any $\mathbf{a_k}$ in $\mathbf{A}$, by (4.22), letting $\varepsilon \to 0$ gets

$$\mathbf{s}_1{}'\mathbf{a_k} \geq 0.$$

This implies

$$\mathbf{s}_1{}'\mathbf{A} \geq \mathbf{0}. \tag{4.24}$$

Since the given S_{ij}s are consistently associated with a probability space, there exists a $(mn+1)$ vector $\mathbf{p} \geq 0$ such that

$$\mathbf{A}\mathbf{p} = \mathbf{b}. \tag{4.25}$$

Applying $\mathbf{p}$ to (4.24) with consideration of (4.25) yields

$$\mathbf{s}_1{}'\mathbf{b} \geq 0 \quad \text{ie} \quad x_1 \geq 0.$$

Thus, for the negative element x_r, $r > 1$.

Now, suppose for all k, $\mathbf{s}_r{}'\mathbf{a_k} \geq 0$, ie $\mathbf{s}_r{}'\mathbf{A} \geq \mathbf{0}$. Using the probability space $\mathbf{p}$

associated with $\mathbf{b}$ would then yield $\mathbf{s}_r'\mathbf{Ap} \geq 0$, which means $\mathbf{s}_r'\mathbf{b} = x_r \geq 0$. This contradicts our definition of r being the smallest negative value x_r.

Now for any $\mathbf{b}_i$, a column of $\mathbf{B}$, since $\mathbf{s}_r'$ is the rth row of $\mathbf{B}^{-1}$, $\mathbf{s}_r'\mathbf{b}_i = \delta_{ri} \geq 0$, where $\delta_{ri} = 1$ if $i = r$; 0 if $i \neq r$. So none of $\mathbf{a}_{j_1}, ..., \mathbf{a}_{j_t}$, can be a column of the current $\mathbf{B}$. $\square$

By Lemma 4.2.5, for each $p \in \{j_1, ..., j_t\}$, $\mathbf{a}_p$ satisfies $y_{rp} = \mathbf{s}_r'\mathbf{a}_p < 0$, thus we need to maximize the following term in (4.23):

$$\frac{1}{y_{rp}}[\mathbf{c}_B(\varepsilon)'\mathbf{B}^{-1}\mathbf{a_p} - c_p(\varepsilon)]$$

$$= \frac{1}{y_{rp}}[\mathbf{s}_1'\mathbf{a_p} + \sum_{i=2}^{5} c_{t_i}(\varepsilon)\mathbf{s}_i'\mathbf{a_p} - c_p(\varepsilon)]$$

$$= \frac{1}{y_{rp}}[y_{1p} + \sum_{i=2}^{5} y_{ip}c_{t_i}(\varepsilon) - c_p(\varepsilon)]. \qquad (4.26)$$

Recalling (4.21), the term in (4.26) is a polynomial in ε for a particular p. Thus, we need to maximize t polynomials of ε. When ε belongs to a small enough interval $(0, \varepsilon_1)$, with $\varepsilon_1 = \varepsilon_1(\mathbf{B}) > 0$, we can rank all of these t polynomials. Then, $\theta(\varepsilon)$ will take just one polynomial when ε is small enough.

Using (4.26) also leads to a simpler form for $\theta(\varepsilon)$:

$$\theta(\varepsilon) = \max_p \{\frac{1}{y_{rp}}[y_{1p} + \sum_{i=2}^{5} y_{ip}c_{t_i}(\varepsilon) - c_p(\varepsilon)]\} \qquad (4.27)$$

for $\varepsilon \in (0, \varepsilon_1)$.

Theorem 4.2.3. For a matrix $\mathbf{B}$ satisfying (4.22), if $\mathbf{x}_B = \mathbf{B}^{-1}\mathbf{b} \not\geq \mathbf{0}$, the corresponding $\theta(\varepsilon) < 0$, for all $0 < \varepsilon < \varepsilon_2$, with a value $\varepsilon_2 = \varepsilon_2(\mathbf{B}) > 0$. Also, the vector $\mathbf{a}_{j_k}$, which is the maximizing vector $\mathbf{a}_p$, $p \in \{j_1, ..., j_t\}$ in (4.27), associated with $\theta(\varepsilon)$ for $0 < \varepsilon < \varepsilon_2(\mathbf{B})$, is uniquely determined.

Proof: First, we need to show that there exists $\varepsilon_2 > 0$ such that for all $0 < \varepsilon < \varepsilon_2$, $\theta(\varepsilon) < 0$.

Notice that $\mathbf{c}_B(\varepsilon)'\mathbf{B}^{-1}\mathbf{a_p} - c_p(\varepsilon) \geq 0$ for $0 < \varepsilon < \varepsilon_0(\mathbf{B})$; and since $p \in \{j_1, ..., j_t\}$, by Lemma 4.2.5, and by the definition of y_{rp}, $y_{rp} < 0$. So, by the definition of $\theta(\varepsilon)$, $\theta(\varepsilon) \leq 0$. We show that $\theta(\varepsilon) \neq 0$ for $0 < \varepsilon < \varepsilon_2$ with some appropriately chosen $\varepsilon_2 > 0$.

By Lemma 4.2.5, $t_i \neq p$ for all $i = 2, ..., 5$ since the $\mathbf{a}_{t_i}$'s are columns of $\mathbf{B}$. Taking into account (4.27), items for maximization are just polynomials of c, and there are only a finite number of zeros for such polynomials. Letting $\varepsilon_2 < \min(\varepsilon_1(\mathbf{B}), \varepsilon_0(\mathbf{B}))$ be small enough to exclude all of those positive zero points for each $p \in \{j_1, ..., j_t\}$ yields $\theta(\varepsilon) \neq 0$ for all $0 < \varepsilon < \varepsilon_2$.

So far, ε_1 ensures the selection of a single polynomial and ε_2 keeps $\theta(\varepsilon) < 0$.

For the second part of the lemma regarding the uniqueness, suppose $\mathbf{a}_{j_m} \in H$ and $\mathbf{a}_{j_n} \in H$ yield the same $\theta(\varepsilon)$, for $\varepsilon < \varepsilon_2(\mathbf{B})$, so

$$
\begin{aligned}
\theta(\varepsilon) &= \frac{1}{y_{rj_n}}[y_{1j_n} + \sum_{i=2}^{5} y_{ij_n} c_{t_i}(\varepsilon) - c_{j_n}(\varepsilon)] \\
&= \frac{1}{y_{rj_m}}[y_{1j_m} + \sum_{i=2}^{5} y_{ij_m} c_{t_i}(\varepsilon) - c_{j_m}(\varepsilon)].
\end{aligned}
$$

Let $y_{rj_n}^{-1} y_{rj_m} = q$. we have $0 < q < \infty$, thus the above equality becomes

$$
(qy_{1j_n} - y_{1j_m}) + \sum_{i=2}^{5}(qy_{ij_n} - y_{ij_m})c_{t_i}(\varepsilon) - c_{j_n}(\varepsilon)q + c_{j_m}(\varepsilon) = 0 \qquad (4.28)
$$

for all $0 < \varepsilon < \varepsilon_2$. Again, this is a polynomial of ε with different powers in items of ε, when $\varepsilon < \varepsilon_2$, $\theta(\varepsilon) < 0$. By Lemma 4.2.5, j_n, j_m are not any of t_i's, $i = 2, ..., 5$, so (4.28) can only happen when $j_n = j_m$. Otherwise, since the coefficient of $c_{j_m}(\varepsilon)$ is 1, the polynomial can't be zero for all ε in $(0, \varepsilon_2)$. $\square$

Now the iteration process may start with any fixed but arbitrary $\mathbf{c}(\varepsilon) \in \Gamma$, $\mathbf{B}_1$, and the given vector of bivariate Bonferroni summations $\mathbf{b}$.

For the matrix $\mathbf{B}_1$ defined in Lemma 4.2.4, if $\mathbf{x}_{\mathbf{B}_1} = \mathbf{B}_1^{-1}\mathbf{b} \geq \mathbf{0}$, since the equation is valid for all ε small enough, letting $\varepsilon \to 0$, $\mathbf{B}_1$ is of the form of $\mathbf{B}$. As discussed above, $\mathbf{x}_{\mathbf{B}_1}$ here can be used to construct a probability space in which linear programming optimality is reached.

If, on the other hand, $\mathbf{x}_{\mathbf{B}_1} \not\geq \mathbf{0}$, according to Lemma 4.2.4, condition (4.22) holds. By Lemma 4.2.5, there exists a non-empty set $H(\mathbf{B}_1)$ which permits us to construct a corresponding $\theta(\varepsilon)$ and an associated $\varepsilon_2(\mathbf{B}_1)$ (Theorem 4.2.3), $\varepsilon_2(\mathbf{B}_1) < \varepsilon_0(\mathbf{B}_1)$, such that $\theta(\varepsilon) < 0$ for $0 < \varepsilon < \varepsilon_2(\mathbf{B}_1)$. Further, since $\mathbf{x}_{\mathbf{B}_1} = (x_1, ..., x_5)' \not\geq \mathbf{0}$, there exists at least one element within $\mathbf{x}_{\mathbf{B}_1}$ which is negative. If x_r is the smallest negative value among the x_i's, by Lemma 4.2.5 combined with Theorem 4.2.3, $r > 1$, and we can find a vector from the columns of $\mathbf{A}$ which is not among the columns of $\mathbf{B}_1$, this vector can be used to replace the current $\mathbf{b}_r$ from $\mathbf{B}_1$ to form a new $\mathbf{B}$, say $\mathbf{B}_2$. Then either $\mathbf{B}_2^{-1}\mathbf{b} \geq \mathbf{0}$ in which case we have attained an optimal bound by Theorem 4.2.6 below. If $\mathbf{B}_2$ again satisfies condition (4.22) with $\varepsilon_0(\mathbf{B}_2) = \varepsilon_2(\mathbf{B}_1)$. The iterative process can thus proceed.

Recall that we use $\mathbf{B}$ and $\mathbf{D}$ to represent the B-matrices $\mathbf{B}_i$ and its replacement $\mathbf{B}_{i+1}$ in the iteration process.

By the definition of $\theta(\varepsilon)$, the perturbation term ε take effect only when there exist two distinct maximizing $p_1, p_2 \in \{j_1, ..., j_t\}$ such that

$$
\frac{y_{1p_1}}{y_{rp_1}} = \frac{y_{1p_2}}{y_{rp_2}}
$$

in some iteration, which then may cause degeneracy:

$$
\max_{p \in H(\mathbf{B})} \frac{y_{1p}}{y_{rp}} = 0
$$

in the iteration following it if $\mathbf{a}_{p_1}$ is selected to form the new $\mathbf{B}$, and in the newly formed $\mathbf{B}$, $y_{rp_2} < 0$ with a new r. This is because in the newly formed matrix $\mathbf{B}$, we have

$$\mathbf{s_1}'\mathbf{a}_{p_2} = (\mathbf{s_1}' - \frac{y_{1p_1}}{y_{rp_1}}\mathbf{s_r}')\mathbf{a}_{p_2} \quad \text{by definition} \tag{4.29}$$

$$= y_{1p_2} - \frac{y_{1p_1}}{y_{rp_1}}y_{rp_2}$$

$$= 0,$$

because $\frac{y_{1p_1}}{y_{rp_1}} = \frac{y_{1p_2}}{y_{rp_2}}$.

For steps without such a tie, $\frac{y_{1j_k}}{y_{rj_k}} < 0$ would ensure $\theta(\varepsilon) < 0$ for ε small enough, and the iteration procedure can continue. For those with such a tie, the choice of $\mathbf{a}_{j_k}$ should be very specific, with special attention to the power of $c_{t_i}(\varepsilon)$'s.

Theorem 4.2.6. The condition $\theta(\varepsilon) < 0$ in each iteration, which means that degeneracy does not occur in the iteration process.
Proof: It suffices to show that in any iteration during the procedure seeking an optimal lower bound, if $\mathbf{x}_B = \mathbf{B}^{-1}\mathbf{b} \not\geq \mathbf{0}$, for any two different vectors in $H(\mathbf{B})$ of Lemma 4.2.5, $\mathbf{a}_{p_1} \neq \mathbf{a}_{p_2}$ with $p_1 \neq p_2$, it's impossible to have

$$\frac{y_{ip_1}}{y_{rp_1}} = \frac{y_{ip_2}}{y_{rp_2}} \quad \text{for all} \quad i = 1, ..., 5 \tag{4.30}$$

where $y_{ij} = \mathbf{s_i}'\mathbf{a_j}$ and $\mathbf{B}$ is the associated $\mathbf{B}$-matrix in the iteration.
Now, suppose (4.30) holds. It can be rewritten as

$$y_{ip_1} = \frac{y_{rp_1}}{y_{rp_2}}y_{ip_2} \quad i = 1, ..., 5.$$

Recalling $y_{ij} = \mathbf{s_i}'\mathbf{a_j}$ yields

$$\mathbf{s_i}'\mathbf{a}_{p_1} = \frac{y_{rp_1}}{y_{rp_2}}\mathbf{s_i}'\mathbf{a}_{p_2} \quad i = 1, ..., 5.$$

This means that

$$\mathbf{B}^{-1}\mathbf{a}_{p_1} = \frac{y_{rp_1}}{y_{rp_2}}\mathbf{B}^{-1}\mathbf{a}_{p_2} \tag{4.31}$$

Applying $\mathbf{B}$ to both sides of (4.31) gets

$$\mathbf{a}_{p_1} = \frac{y_{rp_1}}{y_{rp_2}}\mathbf{a}_{p_2}. \tag{4.32}$$

Noting that the first row of $\mathbf{A}$ is $(1,...,1)$, the first element of each of $\mathbf{a}_{p_1}$ and $\mathbf{a}_{p_2}$ is 1, which implies

$$\frac{y_{rp_1}}{y_{rp_2}} = 1,$$

by (4.32), we get $\mathbf{a}_{p_1} = \mathbf{a}_{p_2}$, which is in contradiction to $\mathbf{a}_{p_1} \neq \mathbf{a}_{p_2}$. $\quad\square$

In Example 7.4.2, we discuss a situation where the occurrence of degeneracy in the iteration process is caused by an incorrect selection of the replacing vector $\mathbf{a}_{j_k}$.

Theorem 4.2.7. In the iteration process, when we sequentially reach a matrix $\mathbf{B}$ such that $\mathbf{B}^{-1}\mathbf{b} \geq \mathbf{0}$, an optimal solution is found and the process is stopped. If the associated $\mathbf{B}^{-1}\mathbf{b} \not\geq \mathbf{0}$, Condition (4.22) persists for every $\mathbf{B}$ used in each iteration, with a corresponding $\varepsilon_0(\mathbf{B}) > 0$.

Proof: We know from Lemma 4.2.4 that condition (4.22) is valid for the initial $\mathbf{B}$-matrix, $\mathbf{B}_1$, constructed in the first run of iterations. We need to show that if the current $\mathbf{B}$ satisfies condition (4.22) then, the follow-up $\mathbf{B}$-matrix, $\mathbf{D}$, also satisfies (4.22) when $\mathbf{B}^{-1}\mathbf{b} \not\geq \mathbf{0}$. To see this, note that after replacing $\mathbf{b}_r$ with $\mathbf{a}_{j_k}$, $\mathbf{B}$ changes, to $\mathbf{D}$. Thus $\mathbf{c}_{\mathbf{D}}(\varepsilon)$ is correspondingly obtained by replacing $c_{t_r}(\varepsilon)$ (the element in $\mathbf{c}_B(\varepsilon)$ which is associated with $\mathbf{b}_r$) with $c_{j_k}(\varepsilon)$ which is then located at the position of the rth element in $\mathbf{c}_{\mathbf{D}}(\varepsilon)$ (Lemma 4.2.5).

Now, using Theorem 4.2.4 and applying any $\mathbf{a}_j$ to the equation between the two inverse matrices yields

$$\mathbf{D}^{-1}\mathbf{a_j} = \mathbf{B}^{-1}\mathbf{a_j} - \begin{pmatrix} \frac{y_{1j_k}}{y_{rj_k}}y_{rj} \\ \vdots \\ \frac{y_{5j_k}}{y_{rj_k}}y_{rj} \end{pmatrix} + \begin{pmatrix} 0 \\ \vdots \\ \frac{y_{rj}}{y_{rj_k}} \\ \vdots \\ 0 \end{pmatrix}. \tag{4.33}$$

Starting the range of ε by putting

$$\varepsilon_0(\mathbf{D}) = \varepsilon_2(\mathbf{B}).$$

Therefore for $0 < \varepsilon < \varepsilon_0(\mathbf{D})$, by the definition of $\theta(\varepsilon)$ in (4.27), $\theta(\varepsilon)$ corresponding to the matrix $\mathbf{B}$ is negative ($y_{rp} < 0$), and $\mathbf{B}$ satisfies (4.22). We will only focus ε on $(0, \varepsilon_0(\mathbf{D}))$ because within this small interval, the following discussion makes sense.

Now consider the term $\mathbf{c}'_{\mathbf{D}}(\varepsilon)$. Replacing $c_{t_r}(\varepsilon)$ with $c_{j_k}(\varepsilon)$ yields

$$\mathbf{c}'_{\mathbf{D}}(\varepsilon) = \mathbf{c}'_B(\varepsilon) + (0, ..., -c_{t_r}(\varepsilon) + c_{j_k}(\varepsilon), ..., 0). \tag{4.34}$$

Therefore, by (4.33) and (4.34), recalling $\mathbf{c}_B(\varepsilon) = (1, c_{t_2}(\varepsilon), ..., c_{t_5}(\varepsilon))'$ gives the following decomposition,

$$\begin{aligned}
\mathbf{c}'_{\mathbf{D}}(\varepsilon)\mathbf{D}^{-1}\mathbf{a_j} &= \mathbf{c}'_B(\varepsilon)\mathbf{B}^{-1}\mathbf{a_j} - \frac{y_{1j_k}}{y_{rj_k}}y_{rj} - \sum_{i=2}^{5} c_{t_i}(\varepsilon)\frac{y_{ij_k}}{y_{rj_k}}y_{rj} + c_{t_r}(\varepsilon)\frac{y_{rj}}{y_{rj_k}} + \\
&\quad + (-c_{t_r}(\varepsilon) + c_{j_k}(\varepsilon))y_{rj} - (-c_{t_r}(\varepsilon) + c_{j_k}(\varepsilon))\frac{y_{rj_k}}{y_{rj_k}}y_{rj} + \\
&\quad + (-c_{t_r}(\varepsilon) + c_{j_k}(\varepsilon))\frac{y_{rj}}{y_{rj_k}}
\end{aligned}$$

$$= \mathbf{c}'_B(\varepsilon)\mathbf{B}^{-1}\mathbf{a_j} - \frac{y_{rj}}{y_{rj_k}}\left(y_{1j_k} + \sum_{i=2}^{5} c_{t_i}(\varepsilon)y_{ij_k} - c_{j_k}(\varepsilon)\right)$$

$$= \mathbf{c_B}'(\varepsilon)\mathbf{B}^{-1}\mathbf{a_j} - \theta(\varepsilon)y_{rj},$$

by the definition of $\theta(\varepsilon)$.

If $y_{rj} < 0$, precisely for any $j \in \{j_1, ..., j_t\}$ (Lemma 4.2.5), when $0 < \varepsilon < \varepsilon_0(\mathbf{D})$,

$$\theta(\varepsilon) \geq \frac{1}{y_{rj}}(\mathbf{c_B}'(\varepsilon)\mathbf{B}^{-1}\mathbf{a_j} - c_j(\varepsilon)),$$

for these j, $\theta(\varepsilon)y_{rj} \leq \mathbf{c_B}'(\varepsilon)\mathbf{B}^{-1}\mathbf{a_j} - c_j(\varepsilon)$, then, from the above decomposition,

$$\mathbf{c_D}'(\varepsilon)\mathbf{D}^{-1}\mathbf{a_j} \geq \mathbf{c_B}'(\varepsilon)\mathbf{B}^{-1}\mathbf{a_j} - \mathbf{c_B}'(\varepsilon)\mathbf{B}^{-1}\mathbf{a_j} + c_j(\varepsilon) = c_j(\varepsilon).$$

By Theorem 4.2.6, $\theta(\varepsilon) < 0$. Recall that first column of $\mathbf{B}$ is not removed in each iteration, $r \neq 1$ by Lemma 4.2.5.

Now, if $y_{rj} \geq 0$, within $(0, \varepsilon_0(\mathbf{D}))$, for $\mathbf{B}$, since $\theta(\varepsilon) < 0$ (Theorem 4.2.3), by the above decomposition ,

$$\mathbf{c_D}'(\varepsilon)\mathbf{D}^{-1}\mathbf{a_j} \geq \mathbf{c_B}'\mathbf{B}^{-1}\mathbf{a_j}$$
$$\geq c_j(\varepsilon),$$

since the matrix $\mathbf{B}$ satisfies (4.22).

Now for the last condition in (4.22),

$$\mathbf{c_D}'(\varepsilon)\mathbf{D}^{-1}\mathbf{a_1} = \mathbf{c_D}'(\varepsilon)\mathbf{D}^{-1}\mathbf{b_1}$$
$$= \mathbf{c_D}'(\varepsilon)(1,0,...,0)' \quad \text{(the first column of } \mathbf{C} \text{ is always } \mathbf{b_1})$$
$$= 1.$$

Therefore, condition (4.22) is valid for $\mathbf{D}$ with $\varepsilon \in (0, \varepsilon_0(\mathbf{D}))$ as well. Then we have shown that once (4.22) is satisfied in the first run, it persists during iterations. $\square$

Notice that by letting $\varepsilon \to 0$, the condition of Lemma 4.2.3 holds in terms of $\mathbf{D}$, thus a new bound is available for $P(v_1 \geq 1, v_2 \geq 1)$.

Since $\theta(\varepsilon)$ is a function of $\mathbf{D}$, keeping $\theta(\varepsilon)$ negative leads to $\varepsilon_2(\mathbf{D}), 0 < \varepsilon_2(\mathbf{D}) < \varepsilon_0(\mathbf{D})$, and iterations can therefore be processed.

So far, we have confirmed that if the given $\mathbf{b}$ is associated with a probability space, the bivariate linear programming optimal lower bound exists. To search for the optimal bound, Theorems 4.2.6 and 4.2.7 show that iterations are workable and Theorem 4.2.5 ensures that the optimal bound will eventually emerge after a finite number of iterations. The combination of these three theorems provides an approach to find the linear programming optimal bivariate lower bound.

In Chapter 7, we implement this approach into an algorithm and illustrate it with applications. It should be mentioned that the treatment of linear programming in this chapter is very much in the probability-oriented style which is convenient for the illustration of multivariate optimal bounds. One issue of the linear programming formulation is the restriction on the binomial moments, which significantly confines the bounding results. For example, in Section 6.2, we show that a linear programming upper bound can be improved by a hybrid bound, due to different domains for optimization as elucidated in the Section of Fréchet optimality.

When the domain of optimization is limited to binomial moments, the linear programming bound (when a feasible solution exists) is gifted with optimality. Interesting extensions in this direction include the univariate and multivariate framework of discrete moment problems by Prékopa (1988, 1990a, 1990b) and the multi-tree framework by Bukszár (2001, 2003), under various mathematical assumptions. Interested readers are referred to Bukszár, Mádi-Nagy, and Szántai (2012) for updated research results in operation research.

Chapter 5

Bivariate Upper Bounds

After introducing basic concepts in Chapter 2, and bounding approaches in Chapters 3 and 4, we now focus on bivariate Bonferroni-type upper bounds with applications in Chapters 5 and 6. The corresponding results for lower bounds are synthesized in Chapters 7 and 8.

Although upper bounds are briefly mentioned in preceding chapters as examples for different purposes, for instance, the degree-two upper bounds in Chapter 2, factorized upper bounds in Chapter 3, and optimal upper bounds in Chapter 4, we have not addressed multivariate upper bounds systematically. Toward this end, Chapters 5 and 6 are specially devoted to the theory and applications of upper bounds.

Consider two sets of events $\{A_i, i = 1, ..., n\}$ and $\{B_j, j = 1, ..., m\}$ in an arbitrary probability space with two integers $0 \leq r \leq n$ and $0 \leq u \leq m$. Let v_1 and v_2 be the numbers of the occurrences of the two sets of events, respectively. In this chapter, we summarize results regarding bivariate upper bounds for the probability of $P(v_1 \geq r, v_2 \geq u)$ and $P(v_1 = r, v_2 = u)$, and conclude the chapter with an application of upper bounds in multiple testing.

5.1 Bivariate Factorized Upper Bounds

As discussed in Chapter 3, the method of factorization bridges univariate bounds with multivariate bounds. For the construction of bivariate upper bounds for $P(v_1 \geq 1, v_2 \geq 1)$, the first type of bound that we should address is the bivariate upper bound using the method of factorization, for which some background information about direct product probability spaces is given as follows.

Suppose $A_1, ..., A_n$ are events in a probability space $(\Omega_1, \mathcal{F}_1, P_1)$, and $B_1, ..., B_m$ are events in $(\Omega_2, \mathcal{F}_2, P_2)$. Then, one can construct a direct product probability space $(\Omega_1 \times \Omega_2, \mathcal{F}_1 \times \mathcal{F}_2, P)$ where

$$P(A \times B) = P_1(A)P_2(B)$$

for events $A \times B$ in the σ-field $\mathcal{F}_1 \times \mathcal{F}_2$. It follows that

$$P(A \times \Omega_2) = P_1(A), \qquad P(\Omega_1 \times B) = P_2(B),$$

and

$$(A_1 \times \Omega_2) \cap (\Omega_1 \times B_1) = A_1 \times B_1.$$

Lemma 5.1.1 Considering events $A_1, ..., A_n$ in Ω_1, and events $B_1, ..., B_m$ in Ω_2, there exists a set of events $A_1^0, ..., A_n^0$ and $B_1^0, ..., B_m^0$ in $\Omega_1 \times \Omega_2$ such that

$$S_{ij}(A^0, B^0) = S_i(A) S_j(B).$$

Proof: Denote a new sequence of events in the product space as follows: $A_1^0 = A_1 \times \Omega_2, ..., A_n^0 = A_n \times \Omega_2$, and $B_1^0 = \Omega_1 \times B_1, ..., B_m^0 = \Omega_1 \times B_m$. For these two classes of events, we have

$$
\begin{aligned}
S_{t,k} &= \sum P([(A_{i_1} \times \Omega_2) \cap ... \cap (A_{i_t} \times \Omega_2)] \cap [(\Omega_1 \times B_{j_1}) \cap ... \cap (\Omega_1 \times B_{j_k})]) \\
&= \sum P([A_{i_1} \cap ... \cap A_{i_t}] \times [B_{j_1} \cap ... \cap B_{j_k}]) \\
&= \sum P_1(A_{i_1}...A_{i_t}) \sum P_2(B_{j_1}...B_{j_k}) \\
&= S_t(A) S_k(B).
\end{aligned}
\tag{5.1}
$$

This completes the proof of Lemma 5.1.1. □

With the above setting, we now can discuss the following theorem.

Theorem 5.1.1 For positive integers T and K, if (1) $S_{t,k}, t = 1, ..., T, \quad k = 1, ..., K$ have a simultaneous product form

$$S_{t,k} = S_t(A) S_k(B), \quad t = 1, ..., T, \quad k = 1, ..., K. \tag{5.2}$$

(2) $\sum_{t=1}^{T} a_{t,n} S_t(A)$ and $\sum_{k=1}^{K} b_{k,m} S_k(B)$ are Fréchet optimal upper bounds for $P(v_1 \geq 1)$ and $P(v_2 \geq 1)$, respectively, then

$$\sum_{t=1}^{T} \sum_{k=1}^{K} a_{t,n} b_{k,m} S_{t,k}$$

is a Fréchet optimal bound for $P(v_1 \geq 1, v_2 \geq 1)$.

Proof: Since $\sum_{t=1}^{T} a_{t,n} S_t(A)$ and $\sum_{k=1}^{K} b_{k,m} S_k(B)$ are Fréchet optimal upper bounds for $P(v_1 \geq 1)$ and $P(v_2 \geq 1)$, by Definition 2.2.1, there exist $A_1^*, ..., A_n^*$ in some probability space $(\Omega_1, \mathcal{F}_1, P_1)$, and $B_1^*, ..., B_m^*$ in $(\Omega_2, \mathcal{F}_2, P_2)$, such that

$$P(v_1(A^*) \geq 1) = \sum_{t=1}^{T} a_{t,n} S_t(A)$$

$$P(v_2(B^*) \geq 1) = \sum_{k=1}^{K} b_{k,m} S_k(B)$$

By Theorem 3.3.1, $\sum_{t=1}^{T} \sum_{k=1}^{K} a_{t,n} b_{k,m} S_{t,k}$ is an upper bound. Now we can construct a direct product probability space $(\Omega_1 \times \Omega_2, \mathcal{F}_1 \times \mathcal{F}_2, P)$ and two classes of events in it, as follows.

$$A_1^{**} = A_1^* \times \Omega_2, \quad ..., \quad A_n^{**} = A_n^* \times \Omega_2$$

$$B_1^{**} = \Omega_1 \times B_1^*, \quad ..., \quad B_m^{**} = \Omega_1 \times B_m^*,$$

such that

$$\sum_{t=1}^{T} \sum_{k=1}^{K} a_{t,n} b_{k,m} S_{t,k}(A_1^* \times \Omega_2, ..., \Omega_1 \times B_m^*)$$

$$= \sum_{t=1}^{T} \sum_{k=1}^{K} a_{t,n} b_{k,m} S_t(A^*) S_k(B^*)$$

$$= \sum_{t=1}^{T} a_{t,n} S_t(A^*) \sum_{k=1}^{K} b_{k,m} S_k(B^*)$$

$$= P_1(v_1(A^*) \geq 1) P_2(v_2(B^*) \geq 1)$$

$$= P(\text{ at least one } \{A_i^* \times \Omega_2\} \quad \text{and one} \quad \{\Omega_1 \times B_j^*\} \quad \text{occur})$$

$$= P(v_1(A^{**}) \geq 1, v_2(B_1^{**}) \geq 1).$$

By Definition 2.4.1, Theorem 5.1.1 follows. $\square$

From the theorem, if we have univariate optimal bounds, then the bivariate optimal events $\{A_i^{**}\}$ and $\{B_j^{**}\}$ can be constructed from the corresponding univariate optimal events A_i^* and B_j^* in two original spaces separately. This gives the associated bivariate Bonferroni summations S_{ij}'s expressed in a product form. However, not all $\{S_{ij}\}$ can be expressed in a product form. The condition (5.2) is one of the necessary conditions for the validity of the theorem. The result of Theorem 5.1.1 can only be applied to a subset of consistent $\{S_{ij}\}$ that has a simultaneous product form (5.2). This point is illustrated by the following example.

Example 5.1.1 Consider A_1, A_2, A_3, B_1, B_2, B_3 ($n = m = 3$) with corresponding non-zero probabilities specified below. Note that this is equivalent to defining two sets of events in a probability space.

$$P(v_1 = 0, v_2 = 0) = 0.5 \qquad P(v_1 = 1, v_2 = 1) = 0.2$$

$$P(v_1 = 1, v_2 = 2) = 0.1 \qquad P(v_1 = 2, v_2 = 1) = 0.1$$

$$P(v_1 = 3, v_2 = 3) = 0.1.$$

Then, Bonferroni summations for the above specified events read

$$S_{11} = 1.5, \quad S_{12} = S_{21} = 1, \quad S_{22} = 0.9.$$

To show that there are no events in any probability space that satisfy (5.2), suppose that there exists a sequence C_1^*, ..., C_n^*, D_1^*, ..., D_m^* in some probability space, such that, under this scenario,

$$S_1(C^*)S_1(D^*) = S_{11} = 1.5, \tag{5.3}$$

$$S_1(C^*)S_2(D^*) = S_{12} = 1, \tag{5.4}$$

$$S_2(C^*)S_1(D^*) = S_{21} = 1, \tag{5.5}$$

$$S_2(C^*)S_2(D^*) = S_{22} = 0.9, \tag{5.6}$$

by (5.3) and (5.4) we get

$$\lambda(D^*) = \frac{S_1(D^*)}{S_2(D^*)} = 1.5$$

while by (5.5) and (5.6) we have

$$\lambda(D^*) = \frac{S_1(D^*)}{S_2(D^*)} = \frac{1}{0.9} \neq 1.5.$$

The contradiction of the value of $\lambda(D^*)$ implies that the assumption on the existence of the event set $\{C_i^*, i = 1, ..., n\}$, $\{D_j^*, j = 1, ..., m\}$ is not reasonable.

In fact, for the example considered, the marginal Bonferroni summations are

$$S_1(A) = \binom{1}{1}0.3 + \binom{2}{1}0.1 + \binom{3}{1}0.1 = 0.8,$$

$$S_2(A) = \binom{2}{2}0.1 + \binom{3}{2}0.1 = 0.4.$$

Similarly, $S_1(B) = 0.8$, $S_2(B) = 0.4$. So, $S_{11}(A, B) \neq S_1(A)S_1(B)$. □

As indicated by the following example, the conditions in Theorem 5.1.1 are sufficient; however, there are other situations in which the conclusion of Theorem 5.1.1 is still valid.

Example 5.1.2 With the numerical example discussed in Example 5.1.1, we get $\{S_{ij}\} = \{1.5, 1, 1, 0.9\}$. Now, by the univariate Fréchet optimality of upper bound, there exist two sets of events $\{A_i^*, i = 1, 2, 3\}$, $\{B_j^*, j = 1, 2, 3\}$ such that

$$P(\bigcup_i A_i^*) = \min(S_1(A) - \frac{2}{n}S_2(A), 1) = S_1(A) - \frac{2}{n}S_2(A) = 0.8 - \frac{2}{3} \times 0.4$$

$$P(\bigcup_j B_j^*) = \min(S_1(B) - \frac{2}{m}S_2(B), 1) = S_1(B) - \frac{2}{n}S_2(B) = 0.8 - \frac{2}{3} \times 0.4.$$

However, because condition (5.2) of Theorem 5.1.1 is broken, we are unable to conclude immediately from Theorem 5.1.1 that there exist events $\{A_i^{**}\}$, $\{B_j^{**}\}$ such that

$$P(\cup_i A_i^{**}, \cup_j B_j^{**}) = S_{11} - \frac{2}{n}S_{21} - \frac{2}{m}S_{12} + \frac{4}{nm}S_{22} \tag{5.7}$$

by simply letting $A_i^{**} = A_i^* \times \Omega_2$, and $B_j^{**} = \Omega_1 \times B_j^*$. In Section 5.3, we show that the right-hand side of (5.7) is still a Fréchet optimal upper bound. Thus (5.2) is sufficient but not an if-and-only-if condition for Theorem 5.1.1.

As discussed in Section 3.4, bounds derived from the method of factorization, although elegant, are not necessary optimal. Any non-optimal bound has bounding space to be improved. For example, increasing the degree of an inequality normally sharpens the inequality. In Section 5.2, we briefly discuss two theoretical results on bivariate Bonferroni-type upper bounds with the use of high degree intersection among the events.

5.2   Bivariate High-degree Upper Bounds

As discussed in Chapter 3, when more information on the intersection of n events is available, the degree-r Bonferroni summations can be computed. Consequently, the degree-r upper bounds for $P(v_1 \geq k, v_2 \geq t)$ can be constructed. This type of bound is discussed in Chapter 3 as an introduction to the method of indicator functions. For completeness and concreteness, in this section, we only summarize this type of bivariate upper bound without repeating the proofs.

As shown in Chapter 2, for any positive integers a and b such that $r + 2a \leq n$, $u + 2b \leq m$, the bivariate upper bound for $P(v_1 \geq r, v_2 \geq u)$ reads

$$P(v_1 \geq r, v_2 \geq u)$$
$$\leq \sum_{k=0}^{2a} \sum_{t=0}^{2b} (-1)^{k+t} \binom{k+r-1}{r-1} \binom{t+u-1}{u-1} S_{k+r,t+u} - \frac{\frac{r}{n-r} \binom{r+2a}{2a}}{\binom{m}{u}} S_{r+2a+1,u}$$
$$- \frac{\frac{u}{m-u} \binom{u+2b}{2b}}{\binom{n}{r}} S_{r,u+2b+1} - c(r,u,a,b) S_{r+2a+1,u+2b+1}, \tag{5.8}$$

where
$$c(r,u,a,b) = \binom{2b+u}{u} \binom{2a+r}{r} \frac{r}{n-r} \frac{u}{m-u}.$$

The inequality (5.8) is a bivariate Bonferroni-type upper bound for the occurrence of at least r events in $\{A_i, i = 1, ..., n\}$ and u events in $\{B_j, j = 1, ..., m\}$. The corresponding bivariate upper bound for the probability of exact numbers of occurrences is given as follows.

Consider positive integers r, u, a, b satisfying $2a + r \leq n$, $2b + u \leq m$. We have

$$P(v_1 = r, v_2 = u)$$
$$\leq \sum_{k=0}^{2a} \sum_{h=0}^{2b} (-1)^{k+h} \binom{k+r}{r} \binom{h+u}{u} S_{k+r,h+u}$$
$$- \frac{r+1}{n-r} \frac{u+1}{m-u} \binom{2a+r+1}{2a} \binom{2b+u+1}{2b} S_{r+2a+1,u+2b+1}. \tag{5.9}$$

Galambos and Lee (1994) proposed bounds extending (5.9) to a bivariate setting ($d = 2$). Their bound can be uniformly improved as follows, where u_1 and u_2 are non-negative integers:

$$P(v_1 = r_1, v_2 = r_2) \leq \sum_{k_1=0}^{2u_1} \sum_{k_2=0}^{2u_2} (-1)^{k_1+k_2} \binom{k_1+r_1}{r_1} \binom{k_2+r_2}{r_2} S_{k_1+r_1,k_2+r_2} -$$
$$- H_1 - H_2 + H_3, \tag{5.10}$$

where

$$H_1 = 0 \quad \text{for} \quad n_1 = r_1;$$

$$= \binom{r_1 + 2u_1}{r_1} \frac{r_1 + 2u_1 + 1}{n_1 - r_1} \sum_{y=0}^{2u_2} (-1)^y S_{r_1 + 2u_1 + 1, r_2 + y} \binom{2u_2 + y}{2u_2} \quad \text{for} \quad n_1 > r_1.$$

$$H_2 = 0 \quad \text{for} \quad n_2 = r_2;$$

$$= \binom{r_2 + 2u_2}{r_2} \frac{r_2 + 2u_2 + 1}{n_2 - r_2} \sum_{x=0}^{2u_1} (-1)^x S_{r_1 + x, r_2 + 2u_2 + 1} \binom{2u_1 + x}{2u_1} \quad \text{for} \quad n_2 > r_2.$$

$$H_3 = 0 \quad \text{for} \quad n_1 = r_1 \quad \text{or} \quad n_2 = r_2;$$

$$= \binom{r_1 + 2u_1}{r_1} \binom{r_2 + 2u_2}{r_2} \frac{r_1 + 2u_1 + 1}{n_1 - r_1} \frac{r_2 + 2u_2 + 1}{n_2 - r_2} S_{r_1 + 2u_1 + 1, r_2 + 2u_2 + 1}$$

$$\text{for} \quad n_1 > r_1 \quad \text{and} \quad n_2 > r_2.$$

The bounds (5.9) and (5.10) are two examples showing the idea using high degree bivariate bounds, an integral part of bivariate bounding theory. More high degree bivariate upper bounds can be found in Chapter 3 where we discussed the bounding method of indicator functions. We conclude this section with an example to show the operation of bivariate high degree bounds.

Example 5.2.1 Consider two sets of events, A_1, A_2, A_3, A_4 and B_1, B_2, B_3, B_4 with non-zero probabilities of elementary conjunctions assigned as follows

$$P(A_1 A_2 A_3^c A_4 B_1^c B_2^c B_3^c B_4) = 0.18 \qquad P(A_1 A_2 A_3 A_4^c B_1^c B_2 B_3^c B_4^c) = 0.12$$

$$P(A_1 A_2 A_3 A_4 B_1 B_2 B_3 B_4^c) = 0.31 \qquad P(A_1 A_2 A_3 A_4 B_1 B_2^c B_3 B_4) = 0.19$$

$$P(A_1 A_2 A_3 A_4^c B_1 B_2 B_3 B_4) = 0.01 \qquad P(A_1^c A_2^c A_3^c A_4^c B_1^c B_2^c B_3^c B_4^c) = 0.19.$$

Hence we have the corresponding non-zero probabilities of the exact number of events A's and B's which occur as

$$P(v_1 = 3, v_2 = 1) = 0.3 \quad P(v_1 = 3, v_2 = 4) = 0.01,$$

$$P(v_1 = 4, v_2 = 3) = 0.50 \quad P(v_1 = 0, v_2 = 0) = 0.19,$$

and the associated Bonferroni summations read

$$S_{31} = 6.34 \quad S_{32} = 6.06 \quad S_{33} = 2.04 \quad S_{34} = 0.01,$$

$$S_{41} = 1.5 \quad S_{42} = 1.5 \quad S_{43} = 0.5 \quad S_{44} = 0.$$

Letting $u_1 = 1$, $u_2 = 0$, $n_1 = n_2 = 4$, $r_1 = 3$, $r_2 = 1$ the upper bound in (5.9) becomes

$$
\begin{aligned}
P(v_1 = 3, v_2 = 1) \;\leq\; & \sum_{k=0}^{2}\sum_{t=0}^{0}(-1)^{k+t}\binom{k+3}{3}\binom{t+1}{1}S_{k+3,t+1} - H_2 \\
\leq\; & S_{31} - 4S_{41} - \frac{2}{3}(S_{32} - 4S_{42}) \\
=\; & 0.3,
\end{aligned}
$$

which exactly hits the evaluated value $P(v_1 = 3, v_2 = 1)$. □

With the use of higher degree Bonferroni summations, the bound (5.9) (hitting the target value, 0.3) improves any upper bound of $P(v_1 = 3, v_2 = 1)$ for the above example. However, (5.9) is not always an optimal bound. The following example shows the scenario in which upper bound (5.9) becomes a trivial bound.

Example 5.2.2. Consider two sets of events A_1, A_2, A_3, A_4 and B_1, B_2,B_3, B_4 with non-zero probabilities of elementary conjunctions specified as follows:

$$P(A_1A_2A_3A_4^cB_1B_2B_3^cB_4^c) = 0.1 \qquad P(A_1^cA_2A_3A_4B_1^cB_2^cB_3B_4) = 0.1$$

$$P(A_1A_2^cA_3A_4B_1^cB_2B_3B_4^c) = 0.1 \qquad P(A_1A_2A_3^cA_4B_1B_2B_3^cB_4^c) = 0.1$$

$$P(A_1A_2A_3^cA_4B_1B_2^cB_3^cB_4^c) = 0.2 \qquad P(A_1^cA_2^cA_3^cA_4^cB_1^cB_2^cB_3^cB_4^c) = 0.1$$

$$P(A_1A_2A_3A_4^cB_1^cB_2^cB_3^cB_4) = 0.3$$

Such an assignment gives the corresponding non-zero probabilities of the number of exactly A's and B's as

$$P(v_1 = 3, v_2 = 2) = 0.4 \quad P(v_1 = 3, v_2 = 1) = 0.5 \quad P(v_1 = 0, v_2 = 0) = 0.1,$$

and the corresponding Bonferroni summations can be calculated as

$$S_{31} = 1.3 \quad S_{32} = 0.4 \quad S_{33} = 0 \quad S_{34} = 0$$

$$S_{41} = 0 \quad S_{42} = 0 \quad S_{43} = 0 \quad S_{44} = 0.$$

In this case, the upper bound (5.9) gives the bound

$$P(v_1 = 3, v_2 = 1) \leq 1.3 - \frac{2}{3} \times 0.4 = 1.03,$$

which is a trivial upper bound even though higher degree Bonferroni summations are used. □

Although high degree upper bounds are normally sharper than upper bounds with relatively lower degrees, as shown in the above two examples, there is no guarantee that these types of bounds are optimal. In the next section, we start the discussion on optimal bivariate upper bounds.

5.3   Bivariate Optimal Upper Bounds

By definition, there are three types of optimality: linear optimality, Fréchet optimality, and linear programming optimality. Among these definitions of optimality, the linear optimal bound is optimal uniformly for all events in any probability space, which puts on more restrictions than the Fréchet optimal upper bound (depending on the values of the Bonferroni summations). The value of a linear optimal upper bound appears weaker than that of a Fréchet optimal upper bound. Also, the linear optimal bound may not exist for some special situations. For example, in the setting of one set of events, the Fréchet optimal Sobel-Uppuluri upper bound (Example 4.1.3) reads,

$$P(v_1 \geq 1) \leq \min(S_1 - \frac{2}{n}S_2, 1),$$

which is an adjusted version of the linearly optimal bound $(S_1 - \frac{2}{n}S_2)$. In this section, we address the bivariate linear optimal upper bound and the bivariate Fréchet optimal upper bounds in two subsections.

5.3.1   Linear Optimal Upper Bounds

For the upper bound of $P(v_1 \geq 1, v_2 \geq 1)$, when linear combinations of the quantities S_{11}, S_{12}, S_{21}, and S_{22} are considered, Theorem 3.3.1 in conjunction with Examples 2.4.1 and 2.4.7 forms the following bivariate upper bound

$$P(v_1 \geq 1, v_2 \geq 1) \leq S_{11} - \frac{2}{n}S_{21} - \frac{2}{m}S_{12} + \frac{4}{nm}S_{22}. \tag{5.11}$$

In what follows, we show that this upper bound can be generalized in the scenario for any index set $T = \{(t,k), (t,k+1), (t+1,k), (t+1,k+1)\}$ for bivariate Bonferroni summations S_{ij} with $(i,j) \in T$. Related references can be found in, for example, Margaritescu (1990a), or Tydeman and Mitchell (1980).

We need the following lemmas to facilitate the discussion. The first lemma essentially relates two univariate Bonferroni summations.

Lemma 5.3.1. For a set of events $A_1, ..., A_n$ in any probability space,

$$S_k - \frac{k+1}{n-k}S_{k+1} \geq 0.$$

Proof: Notice that

$$S_k = \sum_{i=k}^{m} \binom{i}{k} P(v_1 = i)$$

$$= P(v_1 = k) + \sum_{i=k+1}^{m} \frac{i!}{k!(i-k)!} P(v_1 = i).$$

The second term in the right-hand side of the equality above can be written as

$$\sum_{i=k+1}^{m} \frac{i!}{(k+1)!(i-k-1)!} \frac{k+1}{i-k} P(v_1 = i) = \sum_{i=k+1}^{m} \binom{i}{k+1} \frac{k+1}{i-k} P(v_1 = i)$$

which, for all $i \leq m$, is no smaller than

$$\sum_{i=k+1}^{m} \binom{i}{k+1} \frac{k+1}{m-k} P(v_1 = i) = \frac{k+1}{m-k} S_{k+1},$$

thus,

$$S_k \geq P(v_1 = k) + \frac{k+1}{m-k} S_{k+1}$$

$$\geq \frac{k+1}{m-k} S_{k+1}.$$

This completes the proof of Lemma 5.3.1.   □

Now, the relationship of two univariate Bonferroni summations can be extended to any two bivariate Bonferroni summations as follows.

Lemma 5.3.2. For any two sets of events $A_1, ..., A_n$ and $B_1, ..., B_m$, we have

$$S_{t+1,k} \geq \frac{k+1}{m-k} S_{t+1,k+1}. \tag{5.12}$$

$$S_{t,k+1} \geq \frac{t+1}{n-t} S_{t+1,k+1}. \tag{5.13}$$

Proof: We prove (5.12); then the inequality (5.13) follows similarly. Consider

$$S_{t+1,k} \geq \binom{m}{k} \binom{m}{k+1}^{-1} S_{t+1,k+1}$$

$$= \frac{k+1}{m-k} S_{t+1,k+1}. \tag{5.14}$$

The result of Lemma 5.3.2 can be seen as follows:

$$S_{t+1,k} - \frac{k+1}{m-k} S_{t+1,k+1}$$

$$= \sum_{1 \leq i_1 < ... < i_{t+1} \leq n} \left(\sum_{1 \leq j_1 < ... < j_k \leq m} P(A_{i_1}...A_{i_{t+1}} B_{j_1}...B_{j_k}) - \right.$$

$$\left. - \frac{k+1}{m-k} \sum_{1 \leq j_1 < ... < j_{k+1} \leq m} P(A_{i_1}...A_{i_{t+1}} B_{j_1}...B_{j_{k+1}}) \right)$$

$$= \sum_{1 \leq i_1 < ... < i_{t+1} \leq n} \left(\sum_{1 \leq j_1 < ... < j_k \leq m} P(B'_{j_1} B_{j_2}...B_{j_k}) - \right.$$

$$\left. - \frac{k+1}{m-k} \sum_{1 \leq j_1 < ... < j_{k+1} \leq m} P(B'_{j_1} B_{j_2}...B_{j_{k+1}}) \right)$$

$$\geq 0, \tag{5.15}$$

by Lemma 5.3.1 combined with the notation $B'_{j_1} = A_{i_1}...A_{i_{t+1}}B_{j_1}$. This completes the proof of Lemma 5.3.2. $\square$

Lemma 5.3.3. For any two sets of events A_1, ..., A_n, and B_1, ..., B_m, we have

$$S_{t,k} - \frac{t+1}{n-t}S_{t+1,k} - \frac{k+1}{m-k}S_{t,k+1} + \frac{t+1}{n-t}\frac{k+1}{m-k}S_{t+1,k+1} \geq 0.$$

Proof: Consider the binomial moments of the bivariate Bonferroni summations,

$$S_{t,k} = E[\binom{v_1}{t}\binom{v_2}{k}].$$

Then we have

$$S_{t,k} - \frac{t+1}{n-t}S_{t+1,k} - \frac{k+1}{m-k}S_{t,k+1} + \frac{t+1}{n-t}\frac{k+1}{m-k}S_{t+1,k+1}$$

$$= E[\binom{v_1}{t}\binom{v_2}{k} - \frac{t+1}{n-t}\binom{v_1}{t+1}\binom{v_2}{k} -$$

$$- \frac{k+1}{m-k}\binom{v_1}{t}\binom{v_2}{k+1} + \frac{t+1}{n-t}\frac{k+1}{m-k}\binom{v_1}{t+1}\binom{v_2}{k+1}]$$

$$= E[\binom{v_1}{t}\binom{v_2}{k}(1 - \frac{v_1-t}{n-t} - \frac{v_2-k}{m-k} + \frac{v_2-k}{m-k}\frac{v_1-t}{n-t})]$$

$$= E[\binom{v_1}{t}\binom{v_2}{k}(1 - \frac{v_1-t}{n-t})(1 - \frac{v_2-k}{m-k})]$$

$$\geq 0,$$

because

$$\frac{v_1-t}{n-t} \leq 1, \quad \frac{v_2-k}{m-k} \leq 1.$$

This completes the proof for Lemma 5.3.3. $\square$

Note that Lemmas 5.3.1–5.3.3 can also be proved by observing that

$$\frac{S_k}{\binom{n}{k}} \geq \left(\frac{S_{k+1}}{\binom{n}{k+1}}\right)$$

because the probability average of all $(k+1)$-fold intersections cannot exceed the probability average of all k-fold intersections of events (see Fréchet (1940). Thanks to Professor F. M. Hoppe for bringing Fréchet's (1940) work to my attention. Now we can state a theorem on the generalized bivariate Bonferroni upper bounds.

Theorem 5.3.1: For $0 \leq t \leq n$, and $0 \leq k \leq m$

$$P(v_1 \geq t, v_2 \geq k)$$

$$\leq S_{t,k} - (\frac{t+1}{n-t} - \binom{n}{t+1}^{-1})S_{t+1,k} - (\frac{k+1}{m-k} - \binom{m}{k+1}^{-1})S_{t,k+1} +$$

$$+ (\frac{t+1}{n-t} - \binom{n}{t+1}^{-1})(\frac{k+1}{m-k} - \binom{m}{k+1}^{-1})S_{t+1,k+1}. \tag{5.16}$$

Bound (5.16) is at least as good as any linear combination $d_1 S_{t,k} + d_2 S_{t+1,k} + d_3 S_{t,k+1} + d_4 S_{t+1,k+1}$ (where the d_i are constants that depend purely on t, k, n, m, but not on the events being considered) that forms an upper bound for $P(v_1 \geq t, v_2 \geq k)$. *Proof*: Notice that when $t = k = 1$, the upper bound (5.16) becomes (5.11). By Theorem 3.3.1 in conjunction with the univariate bounds of Margaritescu (1986),

$$P(v_1 \geq t) \leq S_t - [\frac{t+1}{n-t} - \binom{n}{t+1}^{-1}]S_{t+1},$$

we know that (5.16) is a bivariate upper bound.

Now we focus on the optimality of (5.16) while generalizing the univariate approach of Margaritescu(1987). Suppose

$$d_1 S_{t,k} + d_2 S_{t+1,k} + d_3 S_{t,k+1} + d_4 S_{t+1,k+1} \geq P(v_1 \geq t, v_2 \geq k),$$

where the d_i's are constants that depend purely on t, k, n, m, but not on the events being considered. We choose

$$A_i = \Omega, i = 1, 2, ..., t, \qquad A_j = \phi, j = t+1, ..., n.$$

$$B_i - \Omega, i = 1, 2, ..., k, \qquad B_j - \phi, j = k+1, ..., m.$$

Then $S_{t,k} = 1, S_{t+1,k} = S_{t,k+1} = S_{t+1,k+1} = 0$, and $P(v_1 \geq t, v_2 \geq k) = 1$. So

$$d_1 \geq 1. \tag{5.17}$$

If we choose the following set of events for the inequality,

$$A_i = \Omega, i = 1, 2, ..., n, \quad B_i = \Omega, i = 1, 2, ..., a_2, \quad B_j = \phi, j = k+1, ..., m,$$

we get

$$S_{t,k} = \binom{n}{t}, \quad S_{t+1,k} = \binom{n}{t+1}, \quad S_{t,k+1} = S_{t+1,k+1} = 0,$$

and $P(v_1 \geq t, v_2 \geq k) = 1$. So

$$d_1 \binom{n}{t} + d_2 \binom{n}{t+1} \geq 1,$$

and

$$d_2 \geq \binom{n}{t+1}^{-1}(1 - d_1 \binom{n}{t}). \tag{5.18}$$

Similarly, when we choose the following sets of events,

$$A_i = \Omega, i = 1, 2, ..., t, \quad A_j = \phi, j = t+1, ..., n, \quad B_i = \Omega, i = 1, 2, ..., m,$$

we get d_3 satisfies

$$d_3 \geq \binom{m}{k+1}^{-1} [1 - d_1 \binom{m}{k}]. \tag{5.19}$$

Finally, when we choose

$$A_i = \Omega, i = 1, 2, ..., n, \quad B_i = \Omega, i = 1, 2, ..., m,$$

we have

$$d_4 \geq \binom{n}{t+1}^{-1} \binom{m}{k+1}^{-1} [1 - d_1 \binom{n}{t} \binom{m}{k} - $$
$$- d_2 \binom{n}{t+1} \binom{m}{k} - d_3 \binom{n}{t} \binom{m}{k+1}]. \tag{5.20}$$

Therefore appropriately applying (5.17) through (5.20) to

$$d_1 S_{t,k} + d_2 S_{t+1,k} + d_3 S_{t,k+1} + d_4 S_{t+1,k+1}$$

and taking into account of all the associated terms attached, yields

$$
\begin{aligned}
T &= d_1 S_{t,k} + d_2 S_{t+1,k} + d_3 S_{t,k+1} + d_4 S_{t+1,k+1} \\
&\geq d_1 S_{t,k} + d_2 S_{t+1,k} + d_3 S_{t,k+1} + \binom{n}{t+1}^{-1} \binom{m}{k+1}^{-1} [1 - d_1 \binom{n}{t} \binom{m}{k} - \\
&\quad - d_2 \binom{n}{t+1} \binom{m}{k} - d_3 \binom{n}{t} \binom{m}{k+1}] S_{t+1,k+1} \\
&\qquad \text{from (5.18) and } S_{t+1,k+1} \geq 0 \\
&= d_1 (S_{t,k} - \frac{k+1}{m-k} \frac{t+1}{n-t} S_{t+1,k+1}) + d_2 (S_{t+1,k} - \frac{k+1}{m-k} S_{t+1,k+1}) + \\
&\quad + d_3 (S_{t,k+1} - \frac{t+1}{n-t} S_{t+1,k+1}) + \binom{n}{t+1}^{-1} \binom{m}{k+1}^{-1} S_{t+1,k+1} \\
&\geq d_1 (S_{t,k} - \frac{k+1}{m-k} \frac{t+1}{n-t} S_{t+1,k+1}) + \\
&\quad + [\binom{n}{t+1}^{-1} (1 - d_1 \binom{n}{t})] (S_{t+1,k} - \frac{k+1}{m-k} S_{t+1,k+1}) + \\
&\quad + [\binom{m}{k+1}^{-1} (1 - d_1 \binom{m}{k})] (S_{t,k+1} - \frac{t+1}{n-t} S_{t+1,k+1}) + \\
&\quad + \binom{n}{t+1}^{-1} \binom{m}{k+1}^{-1} S_{t+1,k+1}.
\end{aligned}
$$

It follows from Lemma 5.3.2 that

$$T = d_1 (S_{t,k} - \frac{k+1}{m-k} \frac{t+1}{n-t} S_{t+1,k} - \frac{k+1}{m-k} S_{t,k+1} + \frac{t+1}{n-t} \frac{k+1}{m-k} S_{t+1,k+1}) +$$

$$+\binom{n}{t+1}^{-1}S_{t+1,k}+\binom{m}{k+1}^{-1}S_{t,k+1}-\binom{n}{t+1}^{-1}\frac{k+1}{m-k}S_{t+1,k+1}-$$

$$-\binom{m}{k+1}^{-1}\frac{t+1}{n-t}S_{t+1,k+1}+\binom{n}{t+1}^{-1}\binom{m}{k+1}^{-1}S_{t+1,k+1}. \tag{5.21}$$

Now, only d_1 remains on the right-hand side of the inequality. Lemma 5.3.3 leads to

$$d_1 S_{t,k}+d_2 S_{t+1,k}+d_3 S_{t,k+1}+d_4 S_{t+1,k+1}$$

$$\geq \quad S_{t,k}-\frac{t+1}{n-t}S_{t+1,k}-\frac{k+1}{m-k}S_{t,k+1}+\frac{t+1}{n-t}\frac{k+1}{m-k}S_{t+1,k+1}+$$

$$\binom{n}{t+1}^{-1}S_{t+1,k}+\binom{m}{t+1}^{-1}S_{t,k+1}-\binom{n}{t+1}^{-1}\frac{k+1}{m-k}S_{t+1,k+1}-$$

$$-\binom{m}{k+1}^{-1}\frac{t+1}{n-t}S_{t+1,k+1}+\binom{n}{t+1}^{-1}\binom{m}{k+1}^{-1}S_{t+1,k+1}$$

$$= \quad S_{t,k}-[\frac{t+1}{n-t}-\binom{n}{t+1}^{-1}]S_{t+1,k}-$$

$$-[\frac{k+1}{m-k}-\binom{m}{k+1}^{-1}]S_{t,k+1}+$$

$$+[\frac{t+1}{n-t}-\binom{n}{t+1}^{-1}][\frac{k+1}{m-k}-\binom{m}{k+1}^{-1}]S_{t+1,k+1}.$$

This establishes the optimality part of Theorem 5.3.1. □

Using Theorem 5.3.1, we have that (5.11) is a bivariate linear optimal upper bound. We now switch the discussion to bivariate Fréchet optimal upper bounds.

5.3.2 Bivariate Fréchet Optimal Upper Bounds

Recall Example 2.4.7 where we showed that the upper bound

$$S_{11}-\frac{2}{n}S_{12}-\frac{2}{m}S_{21}+\frac{4}{nm}S_{22}$$

is not Fréchet optimal. This reveals a gap between Fréchet optimality and linear optimality. To this end, we will first find the Fréchet optimal upper bound using the linear programming approach, then set the connection between the two bivariate optimal upper bounds (see, for example, Margaritescu (1990b)).

Theorem 5.3.2 For any two sets of events $A_1, ..., A_n$ and $B_1, ..., B_m$, if the Bonferroni summations S_{11}, S_{12}, S_{21}, and S_{22} are consistent, and

$$S_{11}-\frac{2}{n}S_{12}-\frac{2}{m}S_{21}+\frac{4}{nm}S_{22}\leq 1,$$

the Fréchet optimal upper bound is

$$P(v_1\geq 1, v_2\geq 1)\leq S_{11}-\frac{2}{n}S_{12}-\frac{2}{m}S_{21}+\frac{4}{nm}S_{22}.$$

Proof: Considering the relation between Fréchet optimality and linear programming as discussed in Chapter 4, it suffices to show that the bound

$$S_{11} - \frac{2}{n}S_{12} - \frac{2}{m}S_{21} + \frac{4}{nm}S_{22}$$

is a solution to the following optimization problem:

$$\max(p_{11} + \ldots + p_{nm}) \tag{5.22}$$

subject to $\sum_{i=0}^{n}\sum_{j=0}^{m} p_{ij} = 1$, and

$$
\begin{aligned}
p_{11} + 2p_{21} + \ldots + tp_{t1} + \ldots + nmp_{nm} &= S_{11} \\
p_{21} + \ldots + \binom{t}{2}p_{t1} + \ldots + \binom{n}{2}mp_{nm} &= S_{21} \\
p_{12} + \ldots + \binom{k}{2}p_{1k} + \ldots + \binom{m}{2}np_{nm} &= S_{12} \\
p_{22} + \ldots + \binom{n}{2}\binom{m}{2}p_{nm} &= S_{22}.
\end{aligned}
$$

Here, the linear constraints can be expressed in a matrix as follows. As discussed in (4.7), by putting $\mathbf{b}' = (1, S_{11}, S_{12}, S_{21}, S_{22})$, a $5 \times w$ (where $w = (n+1)(m+1)$) matrix $\mathbf{G}$ exists so that

$$\mathbf{b} = \mathbf{Gq}, \tag{5.23}$$

where the matrix $\mathbf{G}$ reads

$$
\begin{pmatrix}
1 & \ldots & 1 & \ldots & 1 & \ldots & 1 & 1 & \ldots & 1 & \ldots & 1 \\
0 & \ldots & 1 & \ldots & j & \ldots & m & 0 & \ldots & 2m & \ldots & nm \\
0 & \ldots & 0 & \ldots & \frac{j(j-1)}{2} & \ldots & \frac{m(m-1)}{2} & 0 & \ldots & 2\frac{m(m-1)}{2} & \ldots & n\frac{m(m-1)}{2} \\
0 & \ldots & 0 & \ldots & 0 & \ldots & 0 & 0 & \ldots & m & \ldots & \frac{n(n-1)}{2}m \\
0 & \ldots & 0 & \ldots & 0 & \ldots & 0 & 0 & \ldots & \frac{m(m-1)}{2} & \ldots & \frac{nm(n-1)(m-1)}{4}
\end{pmatrix}.
$$

Taking the following five columns from $\mathbf{G}$ to form the following matrix

$$
\mathbf{M} = \begin{pmatrix}
1 & 1 & 1 & 1 & 1 \\
0 & 1 & n & m & nm \\
0 & 0 & \frac{n(n-1)}{2} & 0 & \frac{n(n-1)}{2}m \\
0 & 0 & 0 & \frac{m(m-1)}{2} & \frac{m(m-1)}{2}n \\
0 & 0 & 0 & 0 & \frac{n(n-1)}{2}\frac{n(n-1)}{2}
\end{pmatrix}.
$$

We have

$$
\mathbf{M}^{-1} = \begin{pmatrix}
1 & -1 & \frac{2}{n} & \frac{2}{m} & -\frac{4}{nm} \\
0 & 1 & -\frac{2}{n-1} & -\frac{2}{m-1} & \frac{4}{(m-1)(n-1)} \\
0 & 0 & \frac{2}{n(n-1)} & 0 & -\frac{4}{n(n-1)(m-1)} \\
0 & 0 & 0 & \frac{2}{m(m-1)} & -\frac{4}{(n-1)(m-1)m} \\
0 & 0 & 0 & 0 & \frac{4}{(n-1)(m-1)mn}
\end{pmatrix}.
$$

Thus the tentative feasible solutions are

$$\mathbf{p}^* = \mathbf{M}^{-1}\mathbf{b}$$

$$= \begin{pmatrix} 1 - S_{11} + \frac{2}{n}S_{21} + \frac{2}{m}S_{12} - \frac{4}{nm}S_{22} \\ S_{11} - \frac{2}{n-1}S_{21} - \frac{2}{m-1}S_{12} + \frac{4}{(n-1)(m-1)}S_{22} \\ \frac{2}{n(n-1)}S_{21} - \frac{4}{n(n-1)(m-1)}S_{22} \\ \frac{2}{m(m-1)}S_{12} - \frac{4}{m(n-1)(m-1)}S_{22} \\ \frac{4}{n(n-1)(m-1)m}S_{22} \end{pmatrix}.$$

Notice that when

$$S_{11} - \frac{2}{n}S_{21} - \frac{2}{m}S_{12} + \frac{4}{nm}S_{22} \le 1,$$

$$1 - S_{11} + \frac{2}{n}S_{21} + \frac{2}{m}S_{12} - \frac{4}{nm}S_{22} \ge 0,$$

thus $\mathbf{p}^*$ is a feasible solution. By being an upper bound (as in Theorem 5.3.1), the optimal solution for $1 - p_0^*$ is

$$S_{11} - \frac{2}{n}S_{21} - \frac{2}{m}S_{12} + \frac{4}{nm}S_{22}.$$

This completes the proof of Theorem 5.3.2. □

Note: In the case where

$$S_{11} - \frac{2}{n}S_{21} - \frac{2}{m}S_{12} + \frac{4}{nm}S_{22} \ge 1,$$

$$1 - S_{11} + \frac{2}{n}S_{21} + \frac{2}{m}S_{12} - \frac{4}{nm}S_{22} \le 0,$$

$\mathbf{p}^* = \mathbf{M}^{-1}\mathbf{b}$ is not a feasible solution. However, for any probability, the maximum upper bound is 1. On the boundary, any $p_{ij}^* = 1$ achieves the upper bound 1. Thus, the truncated bound is Fréchet optimal:

$$P(v_1 \ge 1, v_2 \ge 1) \le \min(S_{11} - \frac{2}{n}S_{21} - \frac{2}{m}S_{12} + \frac{4}{nm}S_{22}, 1). \qquad (5.24)$$

Example 5.3.1 Consider the situation where $n = 5$, let $p_i = 0$ when $i \ne 3$ and $p_3 = 1$ for a set of event $A_1, ..., A_5$, and $B_1 = \Omega$, $m = 1$. The bound

$$P(v_1 \ge 1, v_2 \ge 1) \le S_{11} - \frac{2}{n}S_{21} - \frac{2}{m}S_{12} + \frac{4}{nm}S_{22}$$

becomes

$$P(v_1 \ge 1) \le S_1(A) - \frac{2}{5}S_2(A),$$

because $S_{11} = S_1(A)$, $S_{21} = S_2(A)$, $S_{12} = 0$ and $S_{22} = 0$.

In this case, the matrix expression of the linear constraints (similar to Example 4.1.3) becomes

$$G = \begin{pmatrix} 1 & 1 & 1 & 1 & \cdots & 1 & \cdots & 1 \\ 0 & 1 & 2 & 3 & \cdots & j & \cdots & n \\ 0 & 0 & 1 & 6 & \cdots & \frac{j(j-1)}{2} & \cdots & \frac{n(n-1)}{2} \end{pmatrix}$$

$$b' = (1, S_1, S_2),$$

and

$$p' = (p_0, p_1, ..., p_n),$$

such that

$$Gp' = b.$$

If we follow the train of thought in the proof of Theorem 5.3.2, taking three columns from G to form the following matrix

$$M = \begin{pmatrix} 1 & 1 & 1 \\ 0 & 1 & n \\ 0 & 0 & \frac{n(n-1)}{2} \end{pmatrix},$$

we have

$$M^{-1} = \begin{pmatrix} 1 & -1 & \frac{2}{n} \\ 0 & 1 & -\frac{2}{n-1} \\ 0 & 0 & \frac{2}{n(n-1)} \end{pmatrix},$$

and

$$\begin{aligned} p^* &= M^{-1}b \\ &= (1 - S_1 + \frac{2}{n}S_2, S_1 - \frac{2}{n-1}S_2, \frac{2}{n(n-1)}S_2)'. \end{aligned}$$

Notice that

$$S_1 - \frac{2}{n-1}S_2 = E(v - \frac{v(v-1)}{n-1}) \geq 0,$$

but,

$$1 - S_1 + \frac{2}{n}S_2 = 1 - 3 + \frac{6}{5} \leq 0.$$

Thus p^* now is not a feasible solution. In fact, taking

$$M = \begin{pmatrix} 1 & 1 & 1 \\ 0 & 1 & 3 \\ 0 & 0 & 3 \end{pmatrix},$$

we have

$$M^{-1} = \begin{pmatrix} 1 & -1 & \frac{2}{3} \\ 0 & 1 & -1 \\ 0 & 0 & \frac{1}{3} \end{pmatrix},$$

Thus the feasible solution is

$$\mathbf{p}^* = (0,0,1)',$$

and for this example, the optimal upper bound is 1.   □

The idea of the truncated bivariate optimal upper bound can be further extended to the following theorem that bridges Fréchet optimality with linear optimality when a set of bivariate consistent Bonferroni summations are given. It shows that the truncated version of a linear optimal bound improves the original optimal bound, when they are not identical.

Theorem 5.3.3 Let $T = \{(1,1),(1,2),(2,1),(2,2)\}$. Suppose that F_2 is a linear optimal upper bound for $P(v_1 \geq 1\ v_2 \geq 1)$:

$$F_2 = a_0 + a_1 S_{1,1} + a_2 S_{1,2} + a_3 S_{2,1} + a_4 S_{2,2}, \tag{5.25}$$

where the constants a_j, $j = 0, 1...4$ depend on the numbers of events n and m only. Then, the truncated upper bound $\min(F_2,1)$ improves F_2 when $\min(F_2,1) \neq F_2$.

Proof: Denote $\mathbf{a} = (a_0, a_1, a_2, a_3, a_4)'$. We know from (4.10) that $\mathbf{a}'\mathbf{G} \geq \mathbf{c}'$, where the matrix $\mathbf{G}$ and vector $\mathbf{c}$ are described in Section 4.1.4.

Recall

$$S_{kt} = \sum_{i=k}^{n}\sum_{j=t}^{m} \binom{i}{k}\binom{j}{t} p_{ij} = \mathbf{g}'_{k,t}\mathbf{x}, \tag{5.26}$$

and

$$P(v_1 \geq 1, v_2 \geq 1) = \sum_{i=1}^{n}\sum_{j=1}^{m} p_{ij} = \mathbf{c}'\mathbf{x}, \tag{5.27}$$

where $c(i,j)=0$ for any $(i,j) \in K$ with $K = \{(i,j) : i = 0\ or\ j = 0\}$, and $\mathbf{x}$ is the vector of probability space described in Section 4.1.4.

For $(i_0, j_0) \in K$, let $\mathbf{x_0}$ be a corresponding probability vector with $p_{i_0,j_0}=1$ and other p_{ij}'s=0. Thus in the inequality $\mathbf{a}'\mathbf{G} \geq \mathbf{c}'$, the element of $\mathbf{a}'\mathbf{G}$, which corresponds to the element $c(i_0,j_0)$ in the vector $\mathbf{c}$, is $\mathbf{a}'\mathbf{Gx_0}$. Using a probability space described by $\mathbf{x_0}$, we see from the left-hand side of (5.26) that $S_{kt}=0$ for all $(k,t) = (1,1),(1,2),(2,1),(2,2)$, which implies that $\mathbf{a}'\mathbf{Gx_0} = a_0$. Thus for any element of $\mathbf{c}'$ that is 0, we can apply the same procedure to show that the left-hand side of $\mathbf{a}'\mathbf{G} \geq \mathbf{c}'$ is a_0. Under this scenario, $\mathbf{a}'\mathbf{Gc}$ can be decomposed as:

$$a_0 \geq 0 \quad \text{and} \quad a_0\mathbf{c}' + (a_1,a_2,a_3,a_4)\mathbf{B} \geq \mathbf{c}', \tag{5.28}$$

where $\mathbf{B}$ is a $4 \times (n+1)(m+1)$ matrix of the 2nd, 3th, 4th, 5th rows of the matrix $\mathbf{G}$. Thus $\mathbf{b}^* = \mathbf{Bx} = (S_{1,1}, S_{1,2}, S_{2,1}, S_{2,2})'$ and (5.28) implies that

$$(a_1,a_2,a_3,a_4)\mathbf{B} \geq (1-a_0)\mathbf{c}'.$$

Now, if $a_0 \geq 1$, $1 - a_0 \leq 0$. Using an arbitrary probability vector $\mathbf{x}$, (5.28) implies

$$
\begin{aligned}
(a_1,a_2,a_3,a_4)'\mathbf{b}^* &= a_1 S_{1,1} + a_2 S_{1,2} + a_3 S_{2,1} + a_4 S_{2,2} \\
&\geq (1-a_0)P(v_1 \geq 1,\ v_2 \geq 1) \\
&\geq 1 - a_0,
\end{aligned}
$$

since $P(v_1 \geq 1, v_2 \geq 1) \leq 1$ and $1 - a_0 \leq 0$. Thus when $a_0 \geq 1$, we have

$$a_0 + a_1 S_{1,1} + a_2 S_{1,2} + a_3 S_{2,1} + a_4 S_{2,2} \geq a_0 + (1 - a_0) = 1, \qquad (5.29)$$

which is an unachievable trivial upper bound.

If $0 \leq a_0 < 1$, applying

$$\frac{1}{1 - a_0}(a_1, a_2, a_3, a_4)\mathbf{B} \geq \mathbf{c}'$$

to any probability space, described by $\mathbf{x}$, yields

$$\frac{1}{1 - a_0}(a_1 S_{1,1} + a_2 S_{1,2} + a_3 S_{2,1} + a_4 S_{2,2}) \geq P(v_1 \geq 1, v_2 \geq 1),$$

which means

$$\frac{1}{1 - a_0}(a_1 S_{1,1} + a_2 S_{1,2} + a_3 S_{2,1} + a_4 S_{2,2}) \geq F_2,$$

since F_2 is a linear optimal upper bound. Thus

$$a_1 S_{1,1} + a_2 S_{1,2} + a_3 S_{2,1} + a_4 S_{2,2} \geq (1 - a_0)F_2,$$

and

$$\begin{aligned} a_0 + a_1 S_{1,1} + a_2 S_{1,2} + a_3 S_{2,1} + a_4 S_{2,2} &\geq a_0 + (1 - a_0)F_2 \\ &\geq \min(1, F_2), \qquad (5.30) \end{aligned}$$

since $0 \leq a_0 < 1$. Combining (5.29) and (5.30), we obtain the conclusion of the theorem. $\square$

In the next section, we describe a method using a basic probability inequality to improve the power of a stepwise testing procedure. More powerful procedures with sharpened inequalities can be obtained using the same testing mechanism. In the literature, methods studying multiple testing include the closure principle and the partitioning principle. For example, see Domitrienko, Tamhane and Bretz (2009) or Liu (2011). In the next section, we will also introduce the third approach, a probability inequality principle, that opens a path to improving the power of multiple testing procedures. Related references on the probability inequality principle in applications include Chen and Glaz (2005), Dunn (1959, 1961), and Falk (1989), among others.

5.4 Applications in Multiple Testing

Consider a multiple testing problem for n hypotheses $H_1, H_2, \cdots, H_n$. For each hypothesis H_i, $i = 1, ..., n$, denote the p-value R_i for the test statistics X_i corresponding to a data set. For convenience, assume that the distribution of each test statistic is continuous. The issue is to test all n hypotheses simultaneously with strong control of familywise error rate.

For every non-null subset $I \subseteq \{1, 2, \cdots n\}$, denote m the size of the set I, so we have $1 \leq m \leq n$. A multiple testing procedure is said to *strongly control the family-wise error rate* (Hochberg and Tamhane, 1987) if

$$P(\text{At least one of } H_s, \ s \in I \text{ is rejected } | H_s, s \in I, \text{ true}) \leq \alpha, \qquad (5.31)$$

holds for any index set I and for a pre-specified size of error rate α.

A multiple testing procedure is said to *weakly control the familywise error rate* (Hochberg and Tamhane, 1987) if

$$P(\text{At least one of } H_s, \ s \in \{1,2,...,n\} \text{ is rejected } | H_s, s \in \{1,2,...,n\}, \text{ true}) \leq \alpha, \tag{5.32}$$

holds, for a pre-specified size of error rate α.

Obviously, (5.31) includes (5.32) because the set $\{1,2,...,n\}$ is also a subset of itself. Strong control of the familywise error rate is critical in multiple testing because the condition that all hypotheses H_i, $i = 1,...,n$ are true is a very narrow scenario. For example, when testing n doses of a drug for efficacy, when higher doses are efficacious and lower doses are not, weak controlling of the familywise error rate has no meaning because the event that all doses are inefficacious is empty. Also, the strong control of the familywise error rate can be interpreted as controlling the error of incorrectly claiming any inefficacious doses as efficacious.

5.4.1   Bonferroni Procedure

Denote $R_{(1)}, R_{(2)}, \cdots, R_{(n)}$ the ordered p-values of $R_1, \cdots, R_n$, and $H_{(1)}, H_{(2)}, \cdots, H_{(n)}$ the corresponding hypotheses. The conventional *Bonferroni* multiple test procedure rejects the composite hypothesis defined by

$$H_0 = \cap_{i=1}^{n} H_i = \{H_i, i = 1,...,n\},$$

when

$$R_{(1)} \leq \frac{\alpha}{n};$$

and does not reject any hypothesis, otherwise.

Obviously the *conventional* one-step Bonferroni procedure can only weakly control the familywise error rate because

$$P(\text{At least one of } H_s, \ s \in \{1,2,...,n\} \text{ is rejected } | H_s, s \in \{1,2,...,n\}, \text{ true})$$

$$= \ P(R_{(1)} \leq \frac{\alpha}{n})$$

$$= \ P(\bigcup_{i=1}^{n} \{R_i \leq \frac{\alpha}{n}\})$$

$$\leq \ \sum_{i=1}^{n} P(R_i \leq \frac{\alpha}{n})$$

$$= \ \alpha,$$

since each p-value R_i follows a uniform distribution.

In a stepwise style, the *stepwise* Bonferroni procedure works as follows. It strongly controls the familywise error rate.

1: Examine whether $R_{(1)} \leq \alpha/n$; if not, do not reject any $H_{(i)}, i = 1, \cdots n$, stop; if so, reject $H_{(1)}$ and go to Step 2.

2: Examine whether $R_{(2)} \leq \alpha/n$ if not, do not reject $H_{(i)}, i = 2, \cdots n$, stop; if so, reject $H_{(2)}$ and go to Step 3.

$\vdots$

k: Examine whether $R_{(k)} \leq \alpha/n$ if not, do not reject $H_{(i)}, i = k, \cdots n$, stop; if so, reject $H_{(k)}$ and go to Step $k+1$.

$\vdots$

n: Examine whether $R_{(n)} \leq \alpha/n$ if not, do not reject $H_{(n)}$, stop; if so, reject $H_{(n)}$ and stop.

Through an application of Boole's (the first degree first degree Bonferroni inequality Bonferroni) inequality, the stepwise Bonferroni procedure satisfies (5.31). For any non-null subset $I \subset \{1, 2, \cdots n\}$, denote the size of I as m. Since any hypothesis with index $j \in I$ is rejected only when the associated p-value $R_j \leq \frac{\alpha}{n}$. Denote the rank of R_j as r_j when ordering all the p-values of the n hypotheses, $R_j = R_{(r_j)}$. When $R_j \leq \frac{\alpha}{n}$, the stepwise Bonferroni procedure does not stop before examining $R_{(r_j)}$ because any p-value ranking before $R_{(r_j)}$ is less than $R_{(r_j)} < \frac{\alpha}{n}$. Thus

$$P(\text{At least one of } H_j, \ j \in I \text{ is rejected} \,|H_j, j \in I, \text{ true})$$
$$= \ P(R_j \leq \frac{\alpha}{n}, \ j \in I)$$
$$= \ P(\bigcup_{j \in I} \{R_j \leq \frac{\alpha}{n}\})$$
$$\leq \ \sum_{j \in I} P(R_j \leq \frac{\alpha}{n})$$
$$= \ \frac{m}{n}\alpha$$
$$\leq \ \alpha.$$

The stepwise Bonferroni procedure can be improved by a procedure described in the following subsection.

5.4.2 Holm Step-down Procedure

Holm (1979) refined the stepwise Bonferroni procedure. Related references include Rom (1990), Westfall, Johnson, and Utts (1997), Guo (1995), Holland and Copenhaver (1987), Hommel (1989), Hommel and Bernhard (1996), among others.

1: Examine whether $R_{(1)} \leq \alpha/n$; if not, do not reject any $H_{(i)}, i = 1, \cdots n$, stop; if so, reject $H_{(1)}$ and go to Step 2.

2: Examine whether $R_{(2)} \leq \alpha/(n-1)$ if not, do not reject $H_{(i)}, i = 2, \cdots n$, stop; if so, reject $H_{(2)}$ and go to Step 3.

$\vdots$

k: Examine whether $R_{(k)} \leq \alpha/(n-k+1)$ if not, do not reject $H_{(i)}, i = k, \cdots n$, stop; if so, reject $H_{(k)}$ and go to Step $k+1$.

$\vdots$

n: Examine whether $R_{(n)} \leq \alpha$ if not, do not reject $H_{(n)}$, stop; if so, reject $H_{(n)}$ and stop.

Holm's procedure strongly controls the familywise error rate in the following way. For any set $I \subset \{1, ..., n\}$, denote m the number of elements in I. If the procedure stops before touching any index in I, none of the hypothesis in I is rejected. A hypothesis H_i, $i \in I$ is rejected only when the corresponding $R_i < \alpha/m$ and all the p-values prior to R_i are less than their associated cut-off thresholds. Thus

$$P(\text{ At least one of } H_j, \ j \in I \text{ is rejected } |H_j, j \in I, \text{ true})$$
$$\leq \quad P(R_j \leq \frac{\alpha}{m}, \ j \in I)$$
$$= \quad P(\bigcup_{j \in I} \{R_j \leq \frac{\alpha}{m}\})$$
$$\leq \quad \sum_{j \in I} P(R_j \leq \frac{\alpha}{m})$$
$$= \quad \frac{m}{m}\alpha$$
$$\leq \quad \alpha.$$

Following Holm's procedure, efforts have be devoted to obtain more powerful procedures (such as Hochberg (1988)). Unfortunately, none of the procedures is able to strongly control the familywise error rate (Samel-Cahn, 1996), regardless of modifications on model conditions (such as the MTP_2 condition of Sarkar (1998)). In what follows, we use the degree-two Bonferroni upper bound to introduce a new multiple testing procedure that strongly controls the familywise error rate without making any model assumption.

5.4.3 Improved Holm Procedure

For illustration convenience, we introduce a step-down procedure using degree-two Sobel-Uppuluri-Galambos upper bounds in this subsection. Refined procedures with improved univariate or multivariate upper bounds can be generated similarly (Chen and Seneta (1999), Seneta and Chen (1997, 2005)).

For a set of events $\{A_1, ..., A_n\}$, let S_1 and S_2 be the first and second degree Bonferroni summations, respectively. As discussed in Chapter 2,

$$P(\bigcup_{i=1}^{n} A_i) \leq S_1 - \frac{2}{n} S_2. \tag{5.33}$$

Now, for each integer m, $m = 1, 2, ..., n$, denote $\Psi(m)$ the set of all index subsets with size m and

$$
\begin{aligned}
\gamma(m) &= \min_{I \in \Psi(m)} \frac{2}{m} S_2(I) \\
&= \min_{I \in \Psi(m)} \frac{2}{m} \sum_{i \in I, j \in I} P(R_i \leq \frac{\alpha}{m}, R_j \leq \frac{\alpha}{m} | H_i \text{ and } H_j \text{ are true}).
\end{aligned}
$$

Thus, $\gamma(1) = 0$ and

$$
\begin{aligned}
\gamma(2) &= \min_{(i,j) \in \Psi(2)} P(R_i \leq \frac{\alpha}{2}, R_j \leq \frac{\alpha}{2}) \\
&\leq P(R_1 \leq \frac{\alpha}{2}, R_2 \leq \frac{\alpha}{2}) \\
&\leq P(R_1 \leq \frac{\alpha}{2}) \\
&\leq \alpha/2.
\end{aligned}
$$

For $k = 1, ..., n$, denote $C(k)$ the condition

$$R_{(k)} \leq \min(\frac{\alpha + \gamma(n-k+1)}{n-k+1}, \frac{\alpha}{n-k}).$$

With the above notations, we have the following multiple testing procedure:

We now show that the new procedure strongly controls the familywise error rate, which necessitates the following lemma.

Lemma 5.4.1 For any integer m, $m > 1$, let I be a set with size m, we have

$$\bigcap_{i \in I} \left\{ R_i > \min\left(\frac{(\alpha + \gamma(m))}{m}, \frac{\alpha}{m-1} \right) \right\} \subseteq \{H_s, s \in I, \text{ are not rejected}\}.$$

Proof: Denote

$$\gamma(I) = \max_{i \in I} \sum_{j \in I - \{i\}} P(R_i \leq \frac{\alpha}{m}, R_j \leq \frac{\alpha}{m} | H_s, s \in I, \text{true}).$$

For any $I \subseteq \{1, ..., n\}$ with $|I| = m$, we consider the following two cases if the sample point ω satisfies the condition:

$$\omega \in \bigcap_{i \in I} \left\{ R_i > \min\left(\frac{(\alpha + \gamma(m))}{m}, \frac{\alpha}{m-1} \right) \right\}. \tag{5.34}$$

Figure 5.1 *Degree-two stepwise procedure*

Step 1: Examine whether $C(1)$ is true; if not, do not reject any $H_{(i)}, i = 1, \cdots n$, stop; if so, reject $H_{(1)}$ and go to Step 2.

Step 2: Examine whether $C(2)$ is true; if not, do not reject $H_{(i)}, i = 2, \cdots, n$, stop; if so, reject $H_{(2)}$ and go to Step 3.

$\vdots$

Step k: Examine whether $C(k)$ is true; if not, do not reject $H_{(i)}, i = k, \cdots n$, stop; if so, reject $H_{(k)}$ and go to Step $k+1$.

$\vdots$

Step n: Examine whether $C(n)$ is true; if not, do not reject $H_{(n)}$, stop; if so, reject $H_{(n)}$ and stop.

Note: Since $\gamma(1) = 0$, Step n does not contain the term γ.

Case I. If the associated p-values $R_i, i \in I$, are the largest m values among all the p-values, $R_{(i)}, i = n - m + 1, \cdots, n$, then $H_i, i \in I$, are not to be rejected by the procedure because in this case, the procedure stops before or at the $(n - m + 1)$th step.

Case II. If the p-values of the set I, $R_i, i \in I$, are not the largest m values, denote the smallest value of $R_i, i \in I$ as $R_{(n-m+1-k)}$ for some k with $1 \leq k \leq n - m$. By (5.34) we have either $R_{(n-m+1-k)} > (\alpha + \gamma)/m;$ or $R_{(n-m+1-k)} > \alpha/(m-1)$.

For the case where $R_{(n-m+1-k)} > (\alpha + \gamma)/m$, we have

$$
\begin{aligned}
R_{(n-m+1-k)} \quad &> \quad \alpha/m \\
&\geq \quad \alpha/(m+k-1) \\
&= \quad \alpha/(n-(n-m-k+1)) \\
&\geq \quad \min\left(\frac{\alpha+\gamma}{n-(n-m-k+1)+1}, \frac{\alpha}{n-(n-m-k+1)}\right).
\end{aligned}
$$

So the procedure stops before or at the $(n-m-k+1)^{th}$ step with $H_{(i)}$ not being rejected for $n - m - k + 1 \leq i \leq n$. Therefore, in this case, $H_i, i \in I$, are not rejected.

For the case where $R_{(n-m+1-k)} > \alpha/(m-1)$, we have

$$
\begin{aligned}
R_{(n-m+1-k)} \quad &> \quad \alpha/(m-1) \\
&> \quad \alpha/m \\
&\geq \quad \alpha/(m+k-1) \\
&= \quad \alpha/(n-(n-m-k+1)) \\
&\geq \quad \min\left(\frac{\alpha+\gamma}{n-(n-m-k+1)+1}, \frac{\alpha}{n-(n-m-k+1)}\right).
\end{aligned}
$$

So the procedure stops before or at the $(n-m-k+1)^{th}$ step and $H_i, i \in I$, are not rejected. This completes the proof of Lemma 5.4.1.

Theorem 5.4.1 The degree-two multiple testing procedure described in this subsection strongly controls the familywise error rate.

Proof: For notational convenience, denote $H = \{H_s, \, s \in I, \, \text{true}\}$. By Lemma 5.4.1, we have

$$P(H_s, \, s \in I, \, \text{are not rejected}|H)$$

$$\geq \; P\left(\bigcap_{i \in I}\left\{R_i > \min\left(\frac{(\alpha+\gamma(m))}{m}, \frac{\alpha}{m-1}\right)\right\}|H\right)$$

$$= \; 1 - P\left(\bigcup_{i \in I}\left\{R_i \leq \min\left(\frac{(\alpha+\gamma(m))}{m}, \frac{\alpha}{m-1}\right)\right\}|H\right)$$

$$\geq \; 1 - \sum_{i \in I} P\left(R_i \leq \min\left(\frac{(\alpha+\gamma(m))}{m}, \frac{\alpha}{m-1}\right)|H\right)$$

$$+ \frac{2}{m}\sum_{j \in I} P\left(R_i \leq \min\left(\frac{(\alpha+\gamma)}{m}, \frac{\alpha}{m-1}\right), R_j \leq \min\left(\frac{(\alpha+\gamma)}{m}, \frac{\alpha}{m-1}\right)|H\right),$$

by (5.33).

Now notice that

$$P(R_i < \min(\frac{\alpha+\gamma(m)}{m}, \frac{\alpha}{m-1}), R_j < \min(\frac{\alpha+\gamma(m)}{m}, \frac{\alpha}{m-1})) \geq P(R_i \leq \frac{\alpha}{m}, R_j \leq \frac{\alpha}{m})$$

in conjunction with the uniform distribution of each R_i, $i \in I$, we have

$$P(H_s, \, s \in I, \, \text{are not rejected}|H)$$

$$\geq \; 1 - \min((\alpha+\gamma(m)), m\alpha/(m-1)) + \frac{2}{m}\sum_{j \in I} P(R_i \leq \frac{\alpha}{m}, R_j \leq \frac{\alpha}{m}|H)$$

$$= \; 1 - \min((\alpha+\gamma(m)), m\alpha/(m-1)) + \gamma(m)$$

$$\geq \; 1 - (\alpha+\gamma(m)) + \gamma(m)$$

$$= \; 1 - \alpha. \quad \square$$

The procedure in Figure 5.1 uses a degree-two upper bound to improve the Holm's procedure. It can be further refined by using sharper upper bounds discussed in this chapter. When each hypothesis consists of two components, we can use bivariate upper bounds to improve the testing procedure. The improvement using the principle of probability inequality can be very substantial. For example, Seneta and Chen (2005) adapted Kounias' inequality (1969) to the mechanism of the procedure in Figure 5.1, and used simulations to show that the power of the procedure originating from a second degree inequality is compatible with the step-up Hochberg's procedure.

Chapter 6

Multivariate and Hybrid Upper Bounds

The previous chapter described upper bounds for the probability of the occurrence of two sets of events. In this chapter, we extend the bounding methods in two new directions. The first focus is on multivariate inequalities, upper bounds of the probability of the occurrence for any number of event sets. If two sets of events are of interest, the bounds return to bivariate upper bounds, as discussed in the previous chapter.

The second focus of this chapter is hybrid type inequalities, bounding beyond the setting of Bonferroni-type bounds. Different from bounding approaches discussed in previous chapters, the hybrid type bounding method consists of a bivariate upper bound based on the probabilities of intersections among certain events, which is different from the use of all joint probabilities for bivariate Bonferroni summations. This type of bound is hybrid, in the sense that it is not completely within the domain of Bonferroni-type bounds. We use numerical examples to show that in some cases, hybrid bounds are better than optimal bounds involving merely Bonferroni summations.

This chapter synthesizes basic ideas of the work of Hoppe (1993), Chen and Seneta (2000), Dunn (1958), and Glaz and Pozdnykov (2005), among others.

6.1 High Dimension Upper Bounds

This section extends the bounding domain from bivariate upper bounds to multivariate upper bounds, a higher dimension for evaluating the probability of occurrences among more than two sets of events.

For J groups of events, let M_j be the integer representing the total number of events in group j for $j = 1, ..., J$. Denote v_j the number of events that occur in group j. Consider multivariate upper bounds for the joint probability of exactly m_1, ..., m_J occurrences in the event sets 1, ..., J, respectively. Then, for any integers m_j, $j = 1, ..., J$, we have

$$P(v_1 = m_1 \ldots v_J = m_J) \leq \sum_{t=\sum m_j}^{\sum m_j + 2k} \sum_{\sum i_j = t} (-1)^{t - \sum m_j} \prod_{j=1}^{J} \binom{i_j}{m_j} S_{i_1, \ldots, i_J}, \qquad (6.1)$$

where $m_j \leq i_j \leq M_j$, $1 \leq j \leq J$, and

$$S_{\mathbf{k}} = E(\prod_{i=1}^{J} \binom{v_i}{k_i}) = E(\binom{\mathbf{v}}{\mathbf{k}}),$$

for J sets of events A_{ij}, $i = 1, ..., n_j$, $j = 1, ..., J$, respectively, $\mathbf{v} = (v_1, ..., v_J)'$, and $\mathbf{r} = (r_1, ..., r_J)'$.

Notice that this way of constructing multivariate inequalities involves a degree p bound with the term,

$$d_p = \sum_{|\mathbf{k}|=p} \binom{\mathbf{k}}{\mathbf{r}} S_{\mathbf{k}},$$

where $|\mathbf{k}| = \sum_{i=1}^{J} k_i$, $\mathbf{k} = (k_1, ..., k_J)'$ and

$$\binom{\mathbf{k}}{\mathbf{r}} = \prod_{i=1}^{J} \binom{k_i}{r_i}, \qquad S_{\mathbf{k}} = S_{k_1, ..., k_J},$$

in the sense that p-dimensional joint probabilities are calculated.

The multivariate bound (6.1) can be derived from the following multivariate identity (for example, Meyer (1969)),

$$P(\mathbf{v} = \mathbf{r}) = \sum_{i=|\mathbf{r}|}^{|\mathbf{n}|} (-1)^i \sum_{|\mathbf{k}|=i} \binom{\mathbf{k}}{\mathbf{r}} S_{\mathbf{k}} \qquad (6.2)$$

and its inverse form

$$S_{\mathbf{k}} = \sum_{p=|\mathbf{k}|}^{|\mathbf{n}|} \sum_{|\mathbf{s}|=p} \binom{\mathbf{s}}{\mathbf{k}} P(\mathbf{v} = \mathbf{s}),$$

which is of degree $|\mathbf{n}|$.

Note that bounds generalizing the Bonferroni bounds (2.2) by using the multivariate degree p quantities d_p are a natural way of extension, although each direction of generalization has its own advantages under different circumstances. We illustrate this point with numerical examples in this section.

Generalizing univariate inequalities into the p dimensional multivariate bounds first appeared in Meyer (1969), where univariate Bonferroni bounds were extended to a vector form, which is refined by Galambos and Xu (1993),

$$P(\mathbf{v} = \mathbf{r})$$
$$\leq \sum_{i=0}^{2t} (-1)^i \sum_{|\mathbf{k}|=i} \binom{\mathbf{k}+\mathbf{r}}{\mathbf{r}} S_{\mathbf{k}+\mathbf{r}} - \frac{2t+1}{|\mathbf{n}|-|\mathbf{r}|} \sum_{|\mathbf{k}|=2t+1} \binom{\mathbf{k}+\mathbf{r}}{\mathbf{r}} S_{\mathbf{k}+\mathbf{r}}, \qquad (6.3)$$

extending the univariate Sobel-Uppuluri-Galambos bounds into the multivariate setting. For simplicity, we discuss two main results to illustrate the idea of vector-type (p dimension) inequalities. More detail can be found in Chen and Seneta (1995b), Galambos and Xu (1993, 1996), and Hoppe and Seneta (1990).

Table 6.1 *Numerical example of high dimension upper bounds*

Non-zero probability	value
$P(A_1A_2A_3^cA_4B_1^cB_2^cB_3^cB_4)$	0.18
$P(A_1A_2A_3A_4^cB_1^cB_2B_3^cB_4^c)$	0.12
$P(A_1A_2A_3A_4B_1B_2B_3B_4^c)$	0.31
$P(A_1A_2A_3A_4B_1B_2^cB_3B_4)$	0.19
$P(A_1A_2A_3A_4^cB_1B_2B_3B_4)$	0.01
$P(A_1^cA_2^cA_3^cA_4^cB_1^cB_2^cB_3^cB_4^c)$	0.19

Theorem 6.1.1: Consider d sets of events in a probability space. For any integer $t \geq 0$ with $2t + 1 \leq |\mathbf{n}| - |\mathbf{r}|$,

$$P(\mathbf{v} = \mathbf{r})$$

$$\leq \sum_{i=0}^{2t}(-1)^i \sum_{|\mathbf{k}|=i}\binom{\mathbf{k}+\mathbf{r}}{\mathbf{r}}S_{\mathbf{k}+\mathbf{r}} - \max_i A_i(2t), \qquad (6.4)$$

where

$$A_i(x) = \sum_{|\mathbf{k}|=x} S_{k_1+r_1,\dots,k_i+r_i+1\dots k_d+r_d}\binom{k_1+r_1}{r_1}\cdots\binom{k_i+r_i+1}{r_i}\cdots\binom{k_d+r_d}{r_d}\frac{k_i+1}{n_i-r_i},$$

$i = 1,\dots,d$, and the quantity $A_i(x)$ has degree $|\mathbf{r}| + x + 1$.

We use an example to explain this bound before discussing its derivation.

Example 6.1.1. This example is similar to the numerical example discussed in Section 5.2. Consider two sets of events A_1, A_2, A_3, A_4 and B_1, B_2, B_3, B_4 with non-zero probabilities of elementary conjunctions assigned as follows.

Based on the above information, the corresponding non-zero probabilities of the number of exact occurrence among events A and B can be computed as follows.

$$P(v_1 = 3, v_2 = 1) = 0.3 \quad P(v_1 = 3, v_2 = 4) = 0.01$$

$$P(v_1 = 4, v_2 = 3) = 0.50 \quad P(v_1 = 0, v_2 = 0) = 0.19,$$

with the associated Bonferroni summations

$$S_{31} = 6.34 \quad S_{32} = 6.06 \quad S_{33} = 2.04 \quad S_{34} = 0.01$$

$$S_{41} = 1.5 \quad S_{42} = 1.5 \quad S_{43} = 0.5 \quad S_{44} - 0.$$

For the setting in Theorem 6.1.1, $d = 2$, $n_1 = n_2 = 4$, $r_1 = 3, r_2 = 1$ and $t = 1$, we have

$$P(v_1 = 3, v_2 = 1) \leq S_{31} - (2S_{32} + 4S_{41}) + (3S_{33} + 8S_{42}) - 12S_{43} = 0.34$$

and

$$P(v_1 = 3, v_2 = 1) \leq S_{31} - (2S_{32} + 4S_{41}) + (3S_{33} + 8S_{42}) - (4S_{34} + 8S_{43}) = 2.3.$$

Thus, the upper bound from Theorem 6.1.1 reads

$$P(v_1 = 3, v_2 = 1) \leq \min(0.34, 2.3) = 0.34$$

while the upper bound of (6.3) reads

$$P(v_1 = 3, v_2 = 1) \leq S_{31} - (2S_{32} + 4S_{41}) + (3S_{33} + 8S_{42}) - (3S_{34} + 9S_{43}) = 1.81.$$

Here (6.3) gives a trivial upper bound (greater than 1). The actual value of the bivariate probability is $P(v_1 = 3, v_2 = 1) = 0.30$. $\square$

We need the following lemmas in the proof of Theorem 6.1.1.

Lemma 6.1.1 For any integer $T \geq 0$ and non-negative integers $x_1, ..., x_d$,

$$\sum_{i=0}^{T} \sum_{|\mathbf{k}|=i} (-1)^i \binom{x_1}{k_1} \cdots \binom{x_d}{k_d} = (-1)^T \sum_{|\mathbf{k}|=T} \binom{x_1 - 1}{k_1} \binom{x_2}{k_2} \cdots \binom{x_d}{k_d}. \qquad (6.5)$$

Since all (x_i, k_i), $i = 1, ..., d$ are interchangeable on the left, the right-hand side can be rewritten in d different ways.

Proof: We use the method of mathematical induction to prove Lemma 6.1.1.

When $T = 0$, left=1=right. When $T = 1$, the left becomes $1 - x_1 - x_2 - ... - x_d$ while the right equals $(-1)(x_1 - 1 + x_2 + ... + x_d)$, the same as the left. Supposing that (6.5) is true for any integer $T \geq 0$. In the case of $T + 1$ we have

$$\sum_{i=0}^{T+1} \sum_{|\mathbf{k}|=i} (-1)^i \binom{x_1}{k_1} \cdots \binom{x_d}{k_d}$$

$$= \sum_{i=0}^{T} \sum_{|\mathbf{k}|=i} (-1)^i \binom{x_1}{k_1} \cdots \binom{x_d}{k_d} + (-1)^{T+1} \sum_{|\mathbf{k}|=T+1} \binom{x_1}{k_1} \cdots \binom{x_d}{k_d}$$

$$= (-1)^T \sum_{|\mathbf{k}|=T} \binom{x_1 - 1}{k_1} \binom{x_2}{k_2} \cdots \binom{x_d}{k_d} + (-1)^{T+1} \sum_{|\mathbf{k}|=T+1} \binom{x_1}{k_1} \cdots \binom{x_d}{k_d}$$

$$= (-1)^{T+1} \{ - \sum_{|\mathbf{k}|=T} \binom{x_1 - 1}{k_1} \binom{x_2}{k_2} \cdots \binom{x_d}{k_d} + \sum_{|\mathbf{k}|=T+1} \binom{x_1}{k_1} \cdots \binom{x_d}{k_d} \}$$

$$= (-1)^{T+1} \{ \sum_{|\mathbf{k}|=T} \binom{x_1}{k_1 + 1} \binom{x_2}{k_2} \cdots \binom{x_d}{k_d} - \sum_{|\mathbf{k}|=T} \binom{x_1 - 1}{k_1} \binom{x_2}{k_2} \cdots \binom{x_d}{k_d}$$

$$+ \sum_{|\mathbf{k}|=T+1} \binom{x_1}{0} \binom{x_2}{k_2} \cdots \binom{x_d}{k_d} \}$$

$$= (-1)^{T+1} \{ \sum_{|\mathbf{k}|=T} \binom{x_1 - 1}{k_1 + 1} \binom{x_2}{k_2} \cdots \binom{x_d}{k_d} + \sum_{|\mathbf{k}|=T+1} \binom{x_1}{0} \binom{x_2}{k_2} \cdots \binom{x_d}{k_d} \}.$$

By putting the first two summations together and using

$$\binom{x_1}{k_1+1} - \binom{x_1-1}{k_1} = \binom{x_1-1}{k_1+1},$$

we get

$$\sum_{i=0}^{T+1} \sum_{|\mathbf{k}|=i} (-1)^i \binom{x_1}{k_1} \cdots \binom{x_d}{k_d} = (-1)^{T+1} \sum_{|\mathbf{k}|=T+1} \binom{x_1-1}{k_1} \binom{x_2}{k_2} \cdots \binom{x_d}{k_d},$$

since

$$\sum_{|\mathbf{k}|=T} \binom{x_1-1}{k_1+1} \binom{x_2}{k_2} \cdots \binom{x_d}{k_d} + \sum_{|\mathbf{k}|=T+1} \binom{x_1}{0} \binom{x_2}{k_2} \cdots \binom{x_d}{k_d}$$

$$= \sum_{|\mathbf{k}|=T+1} \binom{x_1-1}{k_1} \binom{x_2}{k_2} \cdots \binom{x_d}{k_d}.$$

This completes the proof of Lemma 6.1.1.   □

Lemma 6.1.2 Let a be any non-negative integer and write

$$P(\mathbf{s}) = P(\mathbf{v} = \mathbf{s}) = P(v_1 = s_1, ..., v_d = s_d).$$

Denoting

$$B_1(a) = \sum_{|\mathbf{k}|=a} \sum_{p=|\mathbf{r}|+a+1}^{|\mathbf{n}|} \sum_{|\mathbf{s}|=p} P(\mathbf{s}) \binom{\mathbf{s}}{\mathbf{r}} \binom{s_1-r_1-1}{k_1} \binom{s_2-r_2}{k_2} \cdots \binom{s_d-r_d}{k_d},$$

we have

$$\sum_{i=0}^{a} (-1)^i \sum_{|\mathbf{k}|=i} \binom{\mathbf{k}+\mathbf{r}}{\mathbf{k}} S_{\mathbf{k}+\mathbf{r}} = P(\mathbf{r}) + (-1)^a B_1(a). \tag{6.6}$$

Proof: By (6.2), notice that $|\mathbf{k}| = i$, we have

$$\sum_{i=0}^{a} (-1)^i \sum_{|\mathbf{k}|=i} \binom{\mathbf{k}+\mathbf{r}}{\mathbf{k}} S_{\mathbf{k}+\mathbf{r}}$$

$$= \sum_{i=0}^{a} (-1)^i \sum_{|\mathbf{k}|=i} \binom{\mathbf{k}+\mathbf{r}}{\mathbf{k}} \sum_{p=i+|\mathbf{r}|}^{|\mathbf{n}|} \sum_{|\mathbf{s}|=p} \binom{\mathbf{s}}{\mathbf{k}+\mathbf{r}} P(\mathbf{s}).$$

Thus, letting $J = \min(a, |\mathbf{n}| - |\mathbf{r}|)$, we have

$$\sum_{i=0}^{a} (-1)^i \sum_{|\mathbf{k}|=i} \binom{\mathbf{k}+\mathbf{r}}{\mathbf{k}} S_{\mathbf{k}+\mathbf{r}}$$

$$= \sum_{p=|\mathbf{r}|}^{|\mathbf{n}|} \sum_{|\mathbf{s}|=p} P(\mathbf{s}) \sum_{i=0}^{J} \sum_{|\mathbf{k}|=i} (-1)^i \binom{\mathbf{k}+\mathbf{r}}{\mathbf{r}} \binom{\mathbf{s}}{\mathbf{k}+\mathbf{r}}$$

$$= \sum_{p=|\mathbf{r}|}^{|\mathbf{n}|} \sum_{|\mathbf{s}|=p} P(\mathbf{s}) \binom{\mathbf{s}}{\mathbf{r}} \sum_{i=0}^{J} \sum_{|\mathbf{k}|=i} (-1)^i \binom{\mathbf{s}-\mathbf{r}}{\mathbf{k}},$$

because

$$\binom{k+r}{r}\binom{s}{k+r} = \binom{s}{r}\binom{s-r}{k}.$$ (6.7)

By Lemma 6.1.1, we have

$$\sum_{i=0}^{a}(-1)^i \sum_{|k|=i}\binom{k+r}{k}S_{k+r}$$

$$= P(r) + \sum_{p=|r|+1}^{|n|}\sum_{|s|=p}P(s)\binom{s}{r}\sum_{i=0}^{J}\sum_{|k|=i}(-1)^i\binom{s-r}{k}$$

$$= P(r) + \sum_{p=|r|+1}^{|n|}\sum_{|s|=p}P(s)\binom{s}{r}(-1)^J\sum_{|k|=J}\binom{s_1-r_1-1}{k_1}\binom{s_2-r_2}{k_2}\cdots\binom{s_d-r_d}{k_d}.$$

Now

$$\sum_{i=0}^{a}(-1)^i \sum_{|k|=i}\binom{k+r}{k}S_{k+r}$$

$$= P(r) + \sum_{p=|r|+1}^{|n|}\sum_{|s|=p}P(s)\binom{s}{r}(-1)^a\sum_{|k|=a}\binom{s_1-r_1-1}{k_1}\binom{s_2-r_2}{k_2}\cdots\binom{s_d-r_d}{k_d}$$

for $a \le |n| - |r|$;

$$= P(r) + (-1)^a\sum_{|k|=a}\sum_{p=|r|+1}^{|n|}\sum_{|s|=p}P(s)\binom{s}{r}\binom{s_1-r_1-1}{k_1}\binom{s_2-r_2}{k_2}\cdots\binom{s_d-r_d}{k_d}.$$

Therefore

$$\sum_{i=0}^{a}(-1)^i \sum_{|k|=i}\binom{k+r}{k}S_{k+r} = P(r) + (-1)^a B_1(a),$$

where

$$B_1(a) = \sum_{|k|=a}\sum_{p=|r|+a+1}^{|n|}\sum_{|s|=p}P(s)\binom{s}{r}\binom{s_1-r_1-1}{k_1}\binom{s_2-r_2}{k_2}\cdots\binom{s_d-r_d}{k_d},$$

because only when $p \ge |r| + a + 1$, we have

$$\binom{s_1-r_1-1}{k_1}\binom{s_2-r_2}{k_2}\cdots\binom{s_d-r_d}{k_d} \ne 0.$$

This completes the proof of Lemma 6.1.2.   □

With lemmas 6.1.1 and 6.1.2, we can prove Theorem 6.1.1 as follows.

Proof of Theorem 6.1.1: To show the upper bounds of Theorem 6.1.1, we set $a = 2t$, which makes (6.6) become

$$\sum_{i=0}^{2t}(-1)^i \sum_{|k|=i}\binom{k+r}{k}S_{k+r} = P(r) + B_1(2t).$$

Recall

$$
\begin{aligned}
A_1(2t) &= \sum_{|\mathbf{k}|=2t} S_{k_1+r_1+1,k_2+r_2,\ldots,k_d+r_d} \binom{k_1+r_1+1}{r_1}\binom{k_2+r_2}{r_2}\cdots\binom{k_d+r_d}{r_d}\frac{k_1+1}{n_1-r_1} \\
&= \sum_{|\mathbf{k}|=2t}\sum_{p=|\mathbf{k}|+|\mathbf{r}|+1}^{|\mathbf{n}|}\sum_{|\mathbf{s}|=p} P(\mathbf{s})\binom{s_1}{k_1+r_1+1}\binom{s_2}{k_2+r_2}\cdots\binom{s_d}{k_d+r_d} \\
&\quad \binom{k_1+r_1+1}{r_1}\binom{k_2+r_2}{r_2}\cdots\binom{k_d+r_d}{r_d}\frac{k_1+1}{n_1-r_1}.
\end{aligned}
$$

We have $A_1(2t) \leq B_1(2t)$ because, by (6.2),

$$
S_{k_1+r_1+1,k_2+r_2,\ldots,k_d+r_d} = \sum_{p=|\mathbf{k}|+|\mathbf{r}|+1}^{|\mathbf{n}|}\sum_{|\mathbf{s}|=p} P(\mathbf{s})\binom{s_1}{k_1+r_1+1}\binom{s_2}{k_2+r_2}\cdots\binom{s_d}{k_d+r_d}.
$$

Notice that the term $B_1(2t)$ can be decomposed in the following way:

$$
B_1(2t) = \sum_{|\mathbf{k}|=2t}\sum_{p=|\mathbf{r}|+2t+1}^{|\mathbf{n}|}\sum_{|\mathbf{s}|=p} P(\mathbf{s})\binom{s_1}{r_1}\cdots\binom{s_d}{r_d}\binom{s_1-r_1-1}{k_1}\binom{s_2-r_2}{k_2}\cdots\binom{s_d-r_d}{k_d}.
$$

Comparing term by term the two expressions of $A_1(2t)$ and $B_1(2t)$, is essentially comparing

$$
\binom{s_1}{k_1+r_1+1}\binom{k_1+r_1+1}{r_1} \quad\text{vs}\quad \binom{s_1}{r_1}\binom{s_1-r_1-1}{k_1}\frac{n_1-r_1}{k_1+1}.
$$

Now, consider the inequality

$$
\binom{s_1}{k_1+r_1+1}\binom{k_1+r_1+1}{r_1} \leq \frac{n_1-r_1}{k_1+1}\binom{s_1}{r_1}\binom{s_1-r_1-1}{k_1}, \tag{6.8}
$$

and the identity (6.7), we have

$$
\binom{s_i}{k_i+r_i}\binom{k_i+r_i}{r_i} = \binom{s_i}{r_i}\binom{s_i-r_i}{k_i} \quad\text{for}\quad i=2,\ldots,d. \tag{6.9}
$$

Thus, for each joint probability $P(\mathbf{s})$ in $A_1(2t)$ and $B_1(2t)$, we can compare pairs of associated coefficients in $A_1(2t)$ and $B_1(2t)$ individually. By (6.8) and (6.9), we have $A_1(2t) \leq B_1(2t)$.

Notice that x_1 can be replaced by any one of $x_2, \ldots, x_d$ establishes the upper bounds in Theorem 6.1.1 when $2t \leq |\mathbf{n}| - |\mathbf{r}|$. Taking $a = 2t$ in (6.6) yields

$$
\begin{aligned}
\sum_{i=0}^{2t}(-1)^i \sum_{|\mathbf{k}|=i}\binom{\mathbf{k}+\mathbf{r}}{\mathbf{k}}S_{\mathbf{k}+\mathbf{r}} \\
= \quad P(\mathbf{r}) + B_1(2t) \\
\geq \quad P(\mathbf{r}) + A_1(2t)
\end{aligned}
$$

Rearranging the inequality and taking maximum value among all permissible integer i in the equation yields Theorem 6.1.1.   □

So far, we have two d dimension inequalities (6.3) and (6.4) for the joint probability $P(\mathbf{v} = \mathbf{r})$. To compare these two inequalities, we have the following theorem.

Theorem 6.1.2: Inequality (6.3) can be written as

$$
P(\mathbf{v} = \mathbf{r})
$$
$$
\leq \sum_{i=0}^{2t}(-1)^i \sum_{|\mathbf{k}|=i} \binom{\mathbf{k}+\mathbf{r}}{\mathbf{r}} S_{\mathbf{k}+\mathbf{r}} - \sum_{i=1}^{d}\frac{n_i - r_i}{|\mathbf{n}| - |\mathbf{r}|} A_i(2t),
$$

and (6.4) is sharper than (6.3).

We need the following lemmas to establish Theorem 6.1.2.

Lemma 6.1.3. For positive numbers a_1, ..., a_n with positive weights c_1, ..., c_n, respectively, such that $\sum_{i=1}^{n} c_i = 1$,

$$
\max_{1 \leq i \leq n} a_i \geq c_1 a_1 + ... + c_n a_n,
$$

and the equality can be attained only when a_i=constant for all $i = 1, ..., n$.
Proof: For the inequality, consider that

$$
c_1 a_1 + ... + c_n a_n
$$
$$
\leq (c_1 + ... + c_n) \max_{1 \leq i \leq n} a_i
$$
$$
= \max_{1 \leq i \leq n} a_i.
$$

Now, if there exists a set of values a_1, ..., a_n such that

$$
\min_{1 \leq i \leq n} a_i < \max_{1 \leq i \leq n} a_i,
$$

it follows that

$$
\sum_{i=1}^{n} c_i a_i < (c_1 + ... + c_n) \max_{1 \leq i \leq n} a_i,
$$

thus the equality is attained only when $a_i = a_j$ for all $i = 1, ..., n$ and $j = 1, ..., n$.   □

Lemma 6.1.4 For any two sets of events, denote

$$
A_1(2t) = \sum_{|\mathbf{k}|=2t} S_{k_1+r_1+1, k_2+r_2, ..., k_d+r_d} \binom{k_1+r_1+1}{r_1}\binom{k_2+r_2}{r_2}...\binom{k_d+r_d}{r_d}\frac{k_1+1}{n_1-r_1}.
$$

Then

$$
A_1(2t)\frac{n_1 - r_1}{|\mathbf{n}| - |\mathbf{r}|} + ... + A_d(2t)\frac{n_d - r_d}{|\mathbf{n}| - |\mathbf{r}|} = \frac{2t+1}{|\mathbf{n}| - |\mathbf{r}|}\sum_{|\mathbf{k}|=2t+1}\binom{\mathbf{k}+\mathbf{r}}{\mathbf{r}}S_{\mathbf{k}+\mathbf{r}}. \quad (6.10)
$$

Proof: For notational convenience, we prove Lemma 6.1.4 in a bivariate setting. The proof for the multivariate setting follows analogously. Notice that in the bivariate setting, $d = 2$, and

$$A_1(2t)\frac{n_1 - r_1}{|\mathbf{n}| - |\mathbf{r}|} + A_2(2t)\frac{n_2 - r_2}{|\mathbf{n}| - |\mathbf{r}|}$$

$$= \sum_{k_1+k_2=2t} S_{k_1+r_1+1,k_2+r_2}\binom{k_1+r_1+1}{r_1}\binom{k_2+r_2}{r_2}\frac{k_1+1}{|\mathbf{n}| - |\mathbf{r}|} +$$

$$\sum_{k_1+k_2=2t} S_{k_1+r_1,k_2+r_2+1}\binom{k_1+r_1}{r_1}\binom{k_2+r_2+1}{r_2}\frac{k_2+1}{|\mathbf{n}| - |\mathbf{r}|}$$

$$= \sum_{k_1=0}^{2t-1} S_{k_1+r_1+1,2t-k_1+r_2}\binom{k_1+r_1+1}{r_1}\binom{2t-k_1+r_2}{r_2}\frac{k_1+1}{|\mathbf{n}| - |\mathbf{r}|} +$$

$$S_{r_1+2t+1,r_2}\binom{2t+r_1+1}{r_1}\frac{2t+1}{|\mathbf{n}| - |\mathbf{r}|} + S_{r_1,2t+r_2+1}\binom{2t+r_2+1}{r_2}\frac{2t+1}{|\mathbf{n}| - |\mathbf{r}|} +$$

$$\sum_{k_1=1}^{2t} S_{k_1+r_1,2t-k_1+r_2+1}\binom{k_1+r_1}{r_1}\binom{2t-k_1+r_2+1}{r_2}\frac{2t-k_1+1}{|\mathbf{n}| - |\mathbf{r}|}.$$

Now, making a transformation, $k_1 + 1 = j$, in the first summation yields

$$A_1(2t)\frac{n_1 - r_1}{|\mathbf{n}| - |\mathbf{r}|} + A_2(2t)\frac{n_2 - r_2}{|\mathbf{n}| - |\mathbf{r}|}$$

$$= \sum_{j=1}^{2t} S_{j+r_1,2t-j+r_2+1}\binom{j+r_1}{r_1}\binom{2t-j+r_2+1}{r_2}\frac{j}{|\mathbf{n}| - |\mathbf{r}|} +$$

$$S_{r_1+2t+1,r_2}\binom{2t+r_1+1}{r_1}\frac{2t+1}{|\mathbf{n}| - |\mathbf{r}|} + S_{r_1,2t+r_2+1}\binom{2t+r_2+1}{r_2}\frac{2t+1}{|\mathbf{n}| - |\mathbf{r}|} +$$

$$\sum_{k_1=1}^{2t} S_{k_1+r_1,2t-k_1+r_2+1}\binom{k_1+r_1}{r_1}\binom{2t-k_1+r_2+1}{r_2}\frac{2t-k_1+1}{|\mathbf{n}| - |\mathbf{r}|}$$

$$= S_{r_1+2t+1,r_2}\binom{2t+r_1+1}{r_1}\frac{2t+1}{|\mathbf{n}| - |\mathbf{r}|} + S_{r_1,2t+r_2+1}\binom{2t+r_2+1}{r_2}\frac{2t+1}{|\mathbf{n}| - |\mathbf{r}|} +$$

$$\sum_{k_1=1}^{2t} S_{k_1+r_1,2t-k_1+r_2+1}\binom{k_1+r_1}{r_1}\binom{2t-k_1+r_2+1}{r_2}\frac{2t+1}{|\mathbf{n}| - |\mathbf{r}|}$$

$$= \frac{2t+1}{|\mathbf{n}| - |\mathbf{r}|}\sum_{|\mathbf{k}|=2t+1}\binom{\mathbf{k}+\mathbf{r}}{\mathbf{r}}S_{\mathbf{k}+\mathbf{r}}.$$

This completes the proof of Lemma 6.1.4. $\square$

With lemmas 6.1.3 and 6.1.4, we are ready to prove Theorem 6.1.2 as follows.
Proof of Theorem 6.1.2: By Lemma 6.1.4, consider various forms of the last term in

Theorem 6.1.1,

$$P(\mathbf{v} = \mathbf{r}) \leq \sum_{i=0}^{2t} (-1)^i \sum_{k_1+k_2=i} \binom{\mathbf{k}+\mathbf{r}}{\mathbf{r}} S_{\mathbf{k}+\mathbf{r}} - A_1(2t) \tag{6.11}$$

$$\vdots$$

$$P(\mathbf{v} = \mathbf{r}) \leq \sum_{i=0}^{2t} (-1)^i \sum_{k_1+k_2=i} \binom{\mathbf{k}+\mathbf{r}}{\mathbf{r}} S_{\mathbf{k}+\mathbf{r}} - A_d(2t) \tag{6.12}$$

where, as usual,

$$A_1(2t) = \sum_{|\mathbf{k}|=2t} S_{k_1+r_1+1,k_2+r_2,\ldots,k_d+r_d} \binom{k_1+r_1+1}{r_1}\binom{k_2+r_2}{r_2}\ldots\binom{k_d+r_d}{r_d}\frac{k_1+1}{n_1-r_1},$$

$$\vdots$$

$$A_d(2t) = \sum_{|\mathbf{k}|=2t} S_{k_1+r_1+1,k_2+r_2,\ldots,k_d+r_d} \binom{k_1+r_1}{r_1}\binom{k_2+r_2}{r_2}\ldots\binom{k_d+r_d+1}{r_d}\frac{k_d+1}{n_1-r_1},$$

Multiplying (6.11) by $\frac{n_1-r_1}{|\mathbf{n}|-|\mathbf{r}|}$, ..., (6.12) by $\frac{n_d-r_d}{|\mathbf{n}|-|\mathbf{r}|}$, respectively, then adding all the products together, yields

$$P(\mathbf{v} = \mathbf{r}) \leq \sum_{i=0}^{2t} (-1)^i \sum_{k_1+k_2=i} \binom{\mathbf{k}+\mathbf{r}}{\mathbf{r}} S_{\mathbf{k}+\mathbf{r}} - A_1(2t)\frac{n_1-r_1}{|\mathbf{n}|-|\mathbf{r}|} - \ldots - A_d(2t)\frac{n_d-r_d}{|\mathbf{n}|-|\mathbf{r}|}. \tag{6.13}$$

Now, using Lemma 6.1.4, we get

$$P(\mathbf{v} = \mathbf{r})$$

$$\leq \sum_{i=0}^{2t} (-1)^i \sum_{|\mathbf{k}|=i} \binom{\mathbf{k}+\mathbf{r}}{\mathbf{r}} S_{\mathbf{k}+\mathbf{r}} - \frac{2t+1}{|\mathbf{n}|-|\mathbf{r}|} \sum_{|\mathbf{k}|=2t+1} \binom{\mathbf{k}+\mathbf{r}}{\mathbf{r}} S_{\mathbf{k}+\mathbf{r}}, \tag{6.14}$$

which is the upper bound stated in Theorem 6.1.2.

For the comparison between (6.4) and (6.3), since the last term in (6.3) is actually a weighted average of d upper bounds in Theorem 6.1.1, applying Lemma 6.1.3 to (6.14) completes the proof of Theorem 6.1.2. □

The improvement of (6.4) over (6.3) is shown in Example 6.1.1. In what follows, we conclude this section with an example illustrating the upper bounds discussed in this section, and comparing the multivariate bound (6.4) with the bivariate upper bounds discussed in Examples 5.2.1 and 5.2.2.

Example 6.1.2: First, consider the two sets of events described in Example 6.1.1, in which the numerical value of the upper bound from (6.4) is 0.34. Although this value is much better than the value of the upper bound of (6.3), the value of the upper bound (5.9) is 0.3, which reaches the exact bivariate probability for evaluation. □

Table 6.2 *Numerical example comparing vectorized bounds*

Non-zero probability	value
$P(A_1A_2A_3A_4^cB_1B_2B_3^cB_4^c)$	0.1
$P(A_1^cA_2A_3A_4B_1^cB_2^cB_3B_4)$	0.1
$P(A_1A_2^cA_3A_4B_1^cB_2B_3B_4^c)$	0.1
$P(A_1A_2A_3^cA_4B_1B_2B_3^cB_4^c)$	0.1
$P(A_1A_2A_3^cA_4B_1B_2^cB_3^cB_4^c)$	0.2
$P(A_1^cA_2^cA_3^cA_4^cB_1^cB_2^cB_3^cB_4^c)$	0.1
$P(A_1A_2A_3A_4^cB_1^cB_2^cB_3^cB_4)$	0.3

In the above example, the upper bound (5.9) is sharper than the multivariate upper bound (6.4). However, this is not always the case, as shown in the following example. **Example 6.1.3**: Consider A_1, A_2, A_3, A_4 and B_1, B_2,B_3, B_4 with non-zero probabilities of elementary conjunctions specified in Table 6.2.

Such an assignment gives the corresponding non-zero probabilities of the number of exact occurrences of A and B as

$$P(v_1 = 3, v_2 = 2) = 0.4 \quad P(v_1 = 3, v_2 = 1) = 0.5 \quad P(v_1 = 0, v_2 = 0) = 0.1,$$

with corresponding Bonferroni summations:

$$S_{31} = 1.3 \quad S_{32} = 0.4 \quad S_{33} = 0 \quad S_{34} = 0$$

$$S_{41} = 0 \quad S_{42} = 0 \quad S_{43} = 0 \quad S_{44} = 0.$$

According to Example 5.2.2, the value of the upper bound of (5.9) for $P(v_1 = 3, v_2 = 1)$ is 1.03, which is a trivial upper bound. However, (6.4) provides an upper bound

$$P(v_1 = 3, v_2 = 1) \leq 1.3 - (2 \times 0.4 + 4 \times 0) = 0.5,$$

which is the exact value to be evaluated in this example. $\square$

We now move to another direction of extension of bivariate upper bounds, the hybrid upper bounds.

6.2 Hybrid Upper Bounds

Consider two sets of events A_1, ..., A_n, and B_1, ..., B_m in a probability space. Let v_1 and v_2 be the number of occurrences of events A and B, respectively. To find upper bounds for quantities $P(v_1 = t, v_2 = k)$ and $P(v_1 \geq t, v_2 \geq k)$, we discuss a new bound approach which is not within the Bonferroni framework. Related references in this regard include Hoppe (1985, 1993), Hoppe and Seneta (1989, 1990), Mi and Sampson (1993), and Worsley (1985).

We start with two identities for the two terms of interest: $P(v_1 = t, v_2 = k)$ and $P(v_1 \geq t, v_2 \geq k)$.

Theorem 6.2.1: For any two integers t and k: $0 \le t \le n$, and $0 \le k \le m$

$$P(v_1 = t, v_2 = k) = S_{tk} - \sum_{\substack{1 \le i_1 < \ldots < i_t \le n \\ 1 \le j_1 < \ldots < j_k \le m}} P((\bigcup_{k_1} A_{k_1} D^*) \bigcup (\bigcup_{k_2} B_{k_2} D^*)) \tag{6.15}$$

where $D^* = A_{i_1} \ldots A_{i_t} B_{j_1} \ldots B_{j_k}$, and

$$k_1 \in \{1, \ldots, n\} \setminus \{i_1, \ldots, i_t\}, \qquad k_2 \in \{1, \ldots, m\} \setminus \{j_1, \ldots, j_k\}.$$

Proof: To see the validity of (6.15), notice that the event $\{v_1 = t, v_2 = k\}$ refers to the exact occurrences of t, k out of n, m events from $\{A_i\}, \{B_j\}$, respectively. We can decompose such event as a union of several events as follows.

$$
\begin{aligned}
&\{v_1 = t, v_2 = k\} \\
= &\bigcup_{\substack{1 \le i_1 < \ldots < i_t \le n \\ 1 \le j_1 < \ldots < j_k \le m}} \{A_{i_1} \ldots A_{i_t} A^c_{k_1} \ldots A^c_{k_{t_1}} B_{j_1} \ldots B_{j_k} B^c_{l_1} \ldots B^c_{l_{t_2}}\}
\end{aligned} \tag{6.16}
$$

where the indexes satisfy the conditions

$$1 \le k_1 < \ldots < k_{t_1} \le n, \quad k_j \in \{1, \ldots, n\} \setminus \{i_1, \ldots, i_t\},$$

$$1 \le l_1 < \ldots < l_{t_2} \le m, \quad l_j \in \{1, \ldots, m\} \setminus \{j_1, \ldots, j_k\},$$

and

$$t_2 = m - j_k \quad t_1 = n - i_t.$$

Notice that the index

$$(i_1, \ldots, i_t, k_1, \ldots, k_{t_1}, j_1, \ldots, j_k, l_1, \ldots, l_{t_2})$$

is assigned uniquely by $(i_1, \ldots, i_t, j_1, \ldots, j_k)$ from the conditions on

$$(k_1, \ldots, k_{t_1}, l_1, \ldots, l_{t_2}).$$

Thus the events in the union are disjoint, and the probability can be written as the summation of associated events as follows.

$$
\begin{aligned}
&P(v_1 = t, v_2 = k) \\
= &\sum_{\substack{1 \le i_1 < \ldots < i_t \le n \\ 1 \le j_1 < \ldots < j_k \le m}} P(A_{i_1} \ldots A_{i_t} A^c_{k_1} \ldots A^c_{k_{t_1}} B_{j_1} \ldots B_{j_k} B^c_{l_1} \ldots B^c_{l_{t_2}}).
\end{aligned} \tag{6.17}
$$

For any integer $q \ge 1$ and events $D, D_1, \ldots D_q$, we have

$$P(DD_1^c \ldots D_q^c) = P(D) - P(\bigcup_{t=1}^{q} DD_t). \tag{6.18}$$

Letting $D = A_{i_1} \ldots A_{i_t} B_{j_1} \ldots B_{j_k}$ and using (6.18) in (6.17) yields

$$P(v_1 = t, v_2 = k)$$

$$= \sum_{\substack{1 \le i_1 < \ldots < i_t \le n \\ 1 \le j_1 < \ldots < j_k \le m}} (P(A_{i_1} \ldots A_{i_t} B_{j_1} \ldots B_{j_k}) - P(\bigcup_{i=1}^{t_1} A_{k_i} D \bigcup \bigcup_{j=1}^{t_2} B_{l_j} D)),$$

by denoting

$$\{D_t, \quad t = 1, \ldots, q\} = \{A_{k_1} \ldots A_{k_{t_1}} B_{l_1} \ldots B_{l_{t_2}}\}.$$

Now, recall

$$S_{tk} = \sum_{\substack{1 \le i_1 < \ldots < i_t \le n \\ 1 \le j_1 < \ldots < j_k \le m}} P(A_{i_1} \ldots A_{i_t} B_{j_1} \ldots B_{j_k})$$

and changing subscript for the union, we get (6.15). □

We now derive the corresponding result for the quantity $P(v_1 \ge t, v_2 \ge k)$.

Theorem 6.2.2. For any two integers t and k: $0 \le t \le n$, and $0 \le k \le m$

$$P(v_1 \ge t, v_2 \ge k) = S_{tk} - \sum_{\substack{1 \le i_1 < \ldots < i_t \le n \\ 1 \le j_1 < \ldots < j_k \le m}} P((\bigcup_{k_1} A_{k_1} D^*) \bigcup (\bigcup_{k_2} B_{k_2} D^*)), \tag{6.19}$$

where $k_1 \in \{1, \ldots, i_t\} \backslash \{i_1, \ldots, i_t\}$, $k_2 \in \{1, \ldots, j_k\} \backslash \{j_1, \ldots, j_k\}$.

Proof: $\forall \omega \in \{v_1 \ge t, v_2 \ge k\}$ there are at least t events from $\{A_i\}$ and k events from $\{B_j\}$ to which ω belongs. Denote the *first* such t events by $A_{i_1}, \ldots, A_{i_t}, 1 \le i_1 < \ldots < i_t \le n$ and the *first* such k events by $B_{j_1}, \ldots, B_{j_k}, 1 \le j_1 < \ldots < j_k \le m$.

Let $t_1 = i_t - t$, $t_2 = j_k - k$, and let $\{k_1, \ldots, k_{t_1}\} = \{1, \ldots, i_t\} \backslash \{i_1, \ldots, i_t\}, \{l_1, \ldots, l_k\} = \{1, \ldots, j_k\} \backslash \{j_1, \ldots, j_k\}$, thus the indexes satisfy $1 \le k_1 < \ldots < k_{t_1}$ and $1 \le l_1 < \ldots < l_{t_2}$. This implies that

$$\omega \in A_{i_1} \ldots A_{i_t} A_{k_1}^c \ldots A_{k_{t_1}}^c B_{j_1} \ldots B_{j_k} B_{l_1}^c \ldots B_{l_{t_2}}^c. \tag{6.20}$$

Since the events in the union are designed to be disjoint, the event $\{v_1 \ge t, v_2 \ge k\}$ is the union of them. Furthermore, let T be any event of the form

$$A_{i_1} \ldots A_{i_t} A_{k_1}^c \ldots A_{k_{t_1}}^c B_{j_1} \ldots B_{j_k} B_{l_1}^c \ldots B_{l_{t_2}}^c,$$

with indexes satisfying the following conditions

- $1 \le i_1 < \ldots < i_t \le n$, $1 \le j_1 < \ldots < j_k \le m$,
- $t_1 = i_t - t$, $t_2 = j_k - k$,
- $\{k_1, \ldots, k_{t_1}\} = \{1, \ldots, i_t\} \backslash \{i_1, \ldots, i_t\}$,
- $\{l_1, \ldots, l_{t_2}\} = \{1, \ldots, j_k\} \backslash \{j_1, \ldots, j_k\}$,

T is of form (6.20). Therefore, we can decompose the joint event of the occurrences as follows.

$$
\{v_1 \geq t, v_2 \geq k\}
$$

$$
= \bigcup_{\substack{1 \leq i_1 < \ldots < i_t \leq n \\ 1 \leq j_1 < \ldots < j_k \leq m}} (A_{i_1} \ldots A_{i_t} A_{k_1}^c \ldots A_{k_{t_1}}^c B_{j_1} \ldots B_{j_k} B_{l_1}^c \ldots B_{l_{t_2}}^c).
$$

and

$$
P(v_1 \geq t, v_2 \geq k) = \sum_{\substack{1 \leq i_1 < \ldots < i_t \leq n \\ 1 \leq j_1 < \ldots < j_k \leq m}} P(A_{i_1} \ldots A_{i_t} A_{k_1}^c \ldots A_{k_{t_1}}^c B_{j_1} \ldots B_{j_k} B_{l_1}^c \ldots B_{l_{t_2}}^c).
$$

By (6.18), we get (6.19) similarly to the proof of (6.15). This completes the proof of Theorem 6.2.2.   □

With the two identities in Theorems 6.2.1 and 6.2.2, we have the following hybrid upper bounds.

Theorem 6.2.3: For any two integers $0 \leq t \leq n$, and $0 \leq k \leq m$, we have

$$
P(v_1 = t, v_2 = k) \leq S_{tk} - \sum_{\substack{1 \leq i_1 < \ldots < i_t \leq n \\ 1 \leq j_1 < \ldots < j_k \leq m}} \max_{\substack{k_1 \geq 1 \\ k_2 \geq 1}} P(A_{k_1} D^* \cup B_{k_2} D^*), \tag{6.21}
$$

where k_1 and k_2 range in $\{1, \ldots, n\} \backslash \{i_1, \ldots, i_t\}$ and $\{1, \ldots, m\} \backslash \{j_1, \ldots, j_k\}$, respectively.

Proof: Since $P(\bigcup_{i=1}^{n} A_i) \geq \max_i P(A_i)$, we have

$$
P(\bigcup_i A_i \bigcup \bigcup_j B_j) \geq \max_{i,j} P(A_i \bigcup B_j).
$$

Applying the above inequality to the second term in (6.15) gets the theorem.   □

Theorem 6.2.4. For any two integers $0 \leq t \leq n$, and $0 \leq k \leq m$,

$$
P(v_1 \geq t, v_2 \geq k) \leq S_{tk} - \max_{\pi_1, \pi_2} \sum_{\substack{1 \leq i_1 < \ldots < i_t \leq n \\ 1 \leq j_1 < \ldots < j_k \leq m}} \max_{\substack{k_1 \geq 1 \\ k_2 \geq 1}} P(A_{k_1} D^* \bigcup B_{k_2} D^*), \tag{6.22}
$$

where k_1 and k_2 range in $\{1, \ldots, i_t\} \backslash \{i_1, \ldots, i_t\}$, $\{1, \ldots, j_k\} \backslash \{j_1, \ldots, j_k\}$ respectively, π_1 and π_2 range over the permutations of the subscripts of $\{A_1, \ldots, A_n\}$ and $\{B_1, \ldots, B_m\}$, respectively, $D^* = A_{i_1} \ldots A_{i_t} B_{j_1} \ldots B_{j_k}$ as usual.

Proof: The proof of (6.22) is similar to that of (6.21) except that the summation is dependent on the permutation of the events $A_1, \ldots, A_n$ and $B_1, \ldots, B_m$ while $P(v_1 \geq t, v_2 \geq k)$ and S_{tk}'s are not dependent on this. Optimizing over different permutations yields (6.22).   □

We note that when $t = 0$ or $k = 0$, the above results reduce to results in the univariate case. The following theorem compares upper bound (6.22) with the degree-two linear optimal upper bound discussed in Chapter 5. It shows that the hybrid bound is at least as sharp as the optimal bound.

Theorem 6.2.5. For any two positive integers $0 \leq t \leq n$, and $0 \leq k \leq m$, the hybrid bound

$$P(v_1 \geq t, v_2 \geq k) \leq S_{tk} - \max_{\pi_1, \pi_2} \sum_{\substack{1 \leq i_1 < \ldots < i_t \leq n \\ 1 \leq j_1 < \ldots < j_k \leq m}} \max_{\substack{k_1 \geq 1 \\ k_2 \geq 1}} P(A_{k_1} D^* \cup B_{k_2} D^*)$$

is not weaker than the optimal bound

$$P(v_1 \geq t, v_2 \geq k)$$

$$\leq S_{t,k} - \left(\frac{t+1}{n-t} - \binom{n}{t+1}^{-1}\right) S_{t+1,k} - \left(\frac{k+1}{m-k} - \binom{m}{k+1}^{-1}\right) S_{t,k+1} +$$

$$+ \left(\frac{t+1}{n-t} - \binom{n}{t+1}^{-1}\right)\left(\frac{k+1}{m-k} - \binom{m}{k+1}^{-1}\right) S_{t+1,k+1}. \qquad (6.23)$$

Proof: From (6.21), we have

$$P(v_1 \geq t, v_2 \geq k) \leq S_{t,k} - \sum_{\substack{1 \leq i_1 < \ldots < i_t \leq n \\ 1 \leq j_1 < \ldots < j_k \leq m}} P(A_{k_1} D^* \cup B_{k_2} D^*)$$

where $D^* = A_{i_1} \ldots A_{i_t} B_{j_1} \ldots B_{j_k}$ holds for any choices of $k_1 \in \{1, \ldots, i_t\} \backslash \{i_1, \ldots, i_t\}$ and $k_2 \in \{1, \ldots, j_k\} \backslash \{j_1, \ldots, j_k\}$.

Since $P(A \cup B) = P(A) + P(B) - P(A \cap B)$, we have

$$P(v_1 \geq t, v_2 \geq k)$$

$$= S_{t,k} - \sum_{\substack{1 \leq i_1 < \ldots < i_t \leq n \\ 1 \leq j_1 < \ldots < j_k \leq m}} [P(A_{k_1} D^*) + P(B_{k_2} D^*) - P(A_{k_1} B_{k_2} D^*)]$$

$$= S_{t,k} - \sum_{i_t=t+1}^{n} \sum_{\substack{1 \leq i_1 < \ldots < i_{t-1} \\ 1 \leq j_1 < \ldots < j_k \leq m}} P(A_{k_1} D^*) -$$

$$- \sum_{j_k=k+1}^{m} \sum_{\substack{1 \leq i_1 < \ldots < i_t \leq n \\ 1 \leq j_1 < \ldots < j_{k-1}}} P(B_{k_2} D^*) +$$

$$+ \sum_{j_k=k+1}^{m} \sum_{i_t=t+1}^{n} \sum_{\substack{1 \leq i_1 < \ldots < i_{t-1} \\ 1 \leq j_1 < \ldots < j_{k-1}}} P(B_{k_2} A_{k_1} D^*).$$

Summing the above inequality over all allowable k_1, k_2 and noting that for fixed integers i_t and j_k, there are $i_t - 1 - (t-1) = i_t - t$ possible values of k_1 and $j_k - k$ possible values of k_2, we have

$$P(v_1 \geq t, v_2 \geq k)$$

$$\leq \quad S_{t,k} - \sum_{i_t=t+1}^{n} \frac{1}{i_t - t} \sum_{\substack{k_1=1 \\ k_1 \neq i_1,\ldots,i_{t-1}}}^{i_t-1} \sum_{\substack{1 \leq i_1 < \ldots < i_{t-1} \\ 1 \leq j_1 < \ldots < j_k \leq m}} P(A_{k_1} D^*) -$$

$$- \sum_{j_k=k+1}^{m} \frac{1}{j_k - k} \sum_{\substack{k_2=1 \\ k_2 \neq j_1,\ldots,j_{k-1}}}^{j_k-1} \sum_{\substack{1 \leq i_1 < \ldots < i_t \leq n \\ 1 \leq j_1 < \ldots < j_{k-1}}} P(B_{k_2} D^*) +$$

$$+ \sum_{\substack{j_k=k+1 \\ i_t=t+1}}^{m,n} \frac{1}{(i_t-t)(j_k-k)} \sum_{\substack{k_1=1 \\ k_1 \neq i_1,\ldots,i_{t-1}}}^{i_t-1} \sum_{\substack{k_2=1 \\ k_2 \neq j_1,\ldots,j_{k-1}}}^{j_k-1} \sum_{\Pi} P(B_{k_2} A_{k_1} D^*).$$

where $\Pi = \{1 \leq i_1 < \ldots < i_{t-1}, \quad 1 \leq j_1 < \ldots < j_{k-1}\}$.
Now, averaging each summation over all permutations of the subscripts of $A_1, \ldots, A_n$ (i.e. over π_1), and over all permutations of the subscripts of $B_1, \ldots, B_n$ (i.e. over π_2), yields

$$P(v_1 \geq t, v_2 \geq k)$$

$$\leq \quad S_{t,k} - \sum_{i_t=t+1}^{n} \frac{1}{i_t - t} \sum_{\substack{k_1=1 \\ k_1 \neq i_1,\ldots,i_{t-1}}}^{i_t-1} \sum_{\substack{1 \leq i_1 < \ldots < i_{t-1} \\ 1 \leq j_1 < \ldots < j_k \leq m}} \frac{1}{n!} \sum_{\pi_1} P(A_{k_1} D^*) -$$

$$- \sum_{j_k=k+1}^{m} \frac{1}{j_k - k} \sum_{\substack{k_2=1 \\ k_2 \neq j_1,\ldots,j_{k-1}}}^{j_k-1} \sum_{\substack{1 \leq i_1 < \ldots < i_t \leq n \\ 1 \leq j_1 < \ldots < j_{k-1}}} \frac{1}{m!} \sum_{\pi_2} P(B_{k_2} D^*) +$$

$$+ \sum_{\substack{j_k=k+1 \\ i_t=t+1}}^{m,n} \frac{1}{(i_t-t)(j_k-k)} \sum_{\substack{k_1=1 \\ k_1 \neq i_1,\ldots,i_{t-1}}}^{i_t-1} \sum_{\substack{k_2=1 \\ k_2 \neq j_1,\ldots,j_{k-1}}}^{j_k-1} \sum_{\Pi} \frac{1}{n!} \frac{1}{m!} \sum_{\pi_1,\pi_2} P(B_{k_2} A_{k_1} D^*).$$

$$(6.24)$$

Consider

$$\binom{n}{t+1} \sum_{\substack{\pi_1 \\ 1 \leq j_1 < \ldots < j_k \leq m}} P(A_{k_1} A_{i_1} \ldots A_{i_t} B_{j_1} \ldots B_{j_k})$$

$$= \quad n! \sum_{\substack{1 \leq j_1 < \ldots < j_k \leq m \\ 1 \leq i_1 < \ldots < i_{t+1}}} P(A_{i_1} \ldots A_{i_{t+1}} B_{j_1} \ldots B_{j_k})$$

$$= \quad n! S_{t+1,k},$$

$$\sum_{1 \le i_1 < ... < i_{t-1}} 1 = \sum_{1 \le i_1 < ... < i_{t-1} \le i_t - 1} 1 = \binom{i_t - 1}{t - 1},$$

$$\sum_{\substack{k_1 = 1 \\ k_1 \ne i_1 ... i_t - 1}}^{i_t - 1} 1 = i_t - 1 - (t - 1) = i_t - t,$$

and

$$\sum_{i_t = t+1}^{n} \binom{i_t - 1}{t - 1} = \sum_{j=1}^{n-t} \binom{j - 1 + t}{j}$$

$$= \binom{n}{t} - 1,$$

since

$$\sum_{j=0}^{m} \binom{j + a}{j} = \binom{m + a + 1}{a + 1}.$$

Similarly, we have

$$\binom{m}{k+1} \sum_{\substack{\pi_2 \\ 1 \le i_1 < ... < i_t \le n}} P(A_{i_1} ... A_{i_t} B_{k_2} B_{j_1} ... B_{j_k}) = m! S_{t,k+1},$$

$$\binom{n}{t+1}\binom{m}{k+1} \sum_{\substack{\pi_1 \\ \pi_2}} P(A_{k_1} A_{i_1} ... A_{i_t} B_{k_2} B_{j_1} ... B_{j_k}) = m! n! S_{t+1,k+1},$$

$$\sum_{1 \le j_1 < ... < j_{k-1}} 1 = \binom{j_k - 1}{k - 1},$$

$$\sum_{\substack{k_2 = 1 \\ k_2 \ne j_1 ... j_k - 1}}^{j_k - 1} 1 = j_k - k,$$

and

$$\sum_{j_k = k+1}^{m} \binom{j_k - 1}{k - 1} = \binom{m}{k} - 1.$$

Therefore from (6.24), it follows that

$$P(v_1 \ge t, v_2 \ge k)$$

$$\le S_{t,k} - \sum_{i_t = t+1}^{n} \binom{i_t - 1}{t - 1}\binom{n}{t + 1}^{-1} S_{t+1,k} -$$

$$- \sum_{j_k = k+1}^{m} \binom{j_k - 1}{k - 1}\binom{m}{k + 1}^{-1} S_{t,k+1} +$$

$$+ \sum_{i_t=t+1}^{n} \sum_{j_k=k+1}^{m} \binom{i_t-1}{t-1} \binom{j_k-1}{k-1} \binom{n}{t+1}^{-1} \binom{m}{k+1}^{-1} S_{t+1,k+1}$$

$$= S_{t,k} - \left(\frac{t+1}{n-t} - \binom{n}{t+1}^{-1}\right) S_{t+1,k} -$$

$$- \left(\frac{k+1}{m-k} - \binom{m}{k+1}^{-1}\right) S_{t,k+1} +$$

$$+ \left(\frac{t+1}{n-t} - \binom{n}{t+1}^{-1}\right) \left(\frac{k+1}{m-k} - \binom{m}{k+1}^{-1}\right) S_{t+1,k+1}.$$

Since we have shown that (6.23) can be derived from (6.22), so (6.22) is not worse than (6.23). □

The next example contains numerical evaluations in which the associated values of (6.23) are lower than those of (6.22).

Example 6.2.1 Table 6.3 contains non-zero probabilities of 2^6 elementary conjunctions involving the events $C_1, ..., C_6$ and their complements.

Let $A_1 = C_1, A_2 = C_2, A_3 = C_3$; $B_1 = C_4, B_2 = C_5, B_3 = C_6$. We have two sets of events $\{A_1, A_2, A_3\}$, and $\{B_1, B_2, B_3\}$.

The corresponding Bonferroni summations become

$$S_{11} = 1.259, S_{21} = 0.225, S_{12} = 0.37, S_{22} = 0.055,$$

$$S_{13} = S_{23} = S_{31} = S_{32} = S_{33} = 0.$$

Based on the Bonferroni summations, the values of the upper bounds are obtained in Table 6.4. Note that we denote $P(v_1 = t, v_2 = k) = y_{tk}$ and $P(v_1 \geq t, v_2 \geq k) = z_{tk}$ in the table. This example confirms that (6.22) is better than (6.23).

6.3 Applications in Successive Comparisons

To detect whether the treatment effect of a drug increases as the dose level increases, Lee and Spurrier (1995), and Chen and Hoppe (2004) discussed a procedure for a special type of multiple comparisons, successive comparisons of treatment means. The method can be outlined as follows.

For dose levels $i = 1, ..., k$, let the response of the jth patient, $j = 1, ..., n_i$, from the ith treatment be

$$Y_{ij} = \mu_i + \varepsilon_{ij},$$

where μ_i is the unknown treatment effect of dose level i, $i = 1, ..., k$. The question of interest is to find the simultaneous confidence intervals for $\mu_i - \mu_{i+1}$ with $i = 1, ..., k-1$.

Table 6.3 *Numerical example of hybrid upper bounds*

Event C	P(C)
$C_1^c C_2^c C_3 C_4^c C_5^c C_6^c$	0.03
$C_1 C_2^c C_3^c C_4 C_5 C_6^c$	0.03
$C_1 C_2^c C_3^c C_4 C_5^c C_6$	0.09
$C_1^c C_2^c C_3^c C_4 C_5 C_6^c$	0.03
$C_1^c C_2^c C_3^c C_4 C_5^c C_6$	0.07
$C_1^c C_2^c C_3^c C_4 C_5^c C_6^c$	0.01
$C_1 C_2^c C_3^c C_4 C_5^c C_6^c$	0.07
$C_1 C_2^c C_3^c C_4^c C_5 C_6^c$	0.02
$C_1 C_2^c C_3^c C_4^c C_5^c C_6$	0.04
$C_1^c C_2^c C_3^c C_4^c C_5 C_6^c$	0.01
$C_1 C_2^c C_3^c C_4^c C_5^c C_6^c$	0.01
$C_1^c C_2 C_3 C_4 C_5 C_6^c$	0.05
$C_1^c C_2 C_3 C_4 C_5^c C_6$	0.005
$C_1^c C_2 C_3 C_4^c C_5^c C_6$	0.03
$C_1^c C_2 C_3 C_4 C_5^c C_6^c$	0.055
$C_1^c C_2 C_3 C_4^c C_5 C_6^c$	0.012
$C_1^c C_2 C_3 C_4^c C_5^c C_6$	0.048
$C_1^c C_2^c C_3^c C_4^c C_5^c C_6^c$	0.03
$C_1^c C_2 C_3 C_4^c C_5^c C_6^c$	0.021
$C_1^c C_2 C_3^c C_4 C_5 C_6^c$	0.039
$C_1^c C_2 C_3^c C_4 C_5^c C_6$	0.011
$C_1^c C_2^c C_3^c C_4^c C_5^c C_6$	0.03
$C_1^c C_2 C_3^c C_4 C_5^c C_6^c$	0.049
$C_1^c C_2 C_3^c C_4^c C_5 C_6^c$	0.03
$C_1^c C_2 C_3^c C_4^c C_5^c C_6$	0.02
$C_1^c C_2^c C_3 C_4^c C_5 C_6^c$	0.01
$C_1^c C_2 C_3^c C_4^c C_5^c C_6^c$	0.04
$C_1^c C_2^c C_3 C_4 C_5^c C_6^c$	0.08
$C_1^c C_2^c C_3 C_4 C_5^c C_6$	0.01
$C_1^c C_2^c C_3 C_4 C_5^c C_6^c$	0.02

Table 6.4 *Numerical values comparing hybrid bounds*

t	k	y_{tk}	z_{tk}	bound (6.22)	bound (6.23)
1	1	0.289	0.719	0.849	0.887
1	2	0.260	0.315	0.320	0.333
2	1	0.115	0.170	0.170	0.188
2	2	0.055	0.055	0.055	0.055

6.3.1   Equal Variances

Lee and Spurrier (1995) considered the scenario where ε_{ij}, $j = 1, ..., n_i$, $i = 1, ..., k$, are independent and identically distributed random variables with mean zero and common variance σ^2 (such as in a double-blinded experiment). Denote $\overline{Y}_i$ the sample mean of the ith treatment and S^2 the pooled sample variance. The test statistic is

$$\mathbf{t} = (T_1, ..., T_{k-1})',$$

where

$$T_i = \frac{\overline{Y}_i - \overline{Y}_{i+1} - (\mu_i - \mu_{i+1})}{\sqrt{S^2[(1/n_i) + (1/n_{i+1})]}}$$

for $i = 1, ..., k-1$.

Here, the random vector $\mathbf{t}$ follows a new type of multivariate student t distribution. The simultaneous confidence intervals for the difference of unknown treatment effects are given by

$$\mu_i - \mu_{i+1} \in (\overline{Y}_i - \overline{Y}_{i+1} - \Delta, \overline{Y}_i - \overline{Y}_{i+1} + \Delta),$$

where $\Delta = c'S\sqrt{(1/n_i) + (1/n_{i+1})}$ and c' is the constant such that

$$P(\max\{|T_1|, ..., |T_{k-1}|\} \leq c') \geq 1 - \alpha. \tag{6.25}$$

To compute the appropriate value c' that satisfies (6.25), Lee and Spurrier (1995) approximated the values of c' by a degree-two Hunter's upper bound, which is a type of hybrid upper bound. In Hunter's upper bound, the use of a degree-two Bonferroni summation is replaced by a summation of weights over a spanning tree where the vertexes are events under consideration. When two or more endpoints are of interest, we replace Hunter's upper bound with bivariate upper bounds or vector upper bounds discussed in this chapter, to correspondingly accommodate needs of various applications.

Chen and Hoppe (2004) proved the equivalence of the density function for successive comparisons and the statistic used in a two-sample ranking procedure. The assumption of equal variance is a critical condition in the derivation. When the variability across all treatments can not be assumed equal, the conventional multiple testing procedure can not be applied. This leads to the discussion of the following inequality method for unequal variance situations.

6.3.2   Unequal Variances, Behrens-Fisher Problem

When the variances can not be assumed equal, a convenient approach for the simultaneous confidence intervals of $\mu_i - \mu_{i+1}$ for $i = 1, ..., k-1$ is the Bonferroni adjustment. In what follows in this section, first we focus on the construction of the confidence interval for the difference of two population means when the variances are not equal, and then use the method of inequality to find the simultaneous confidence set for $\mu_i - \mu_{i+1}$ with $i = 1, ..., k-1$. For notational convenience, we simplify the discussion as follows.

Consider two independent random samples

$$X_1, X_2, \ldots, X_{n_1} \sim N(\mu_1, \sigma_1^2), \quad \text{and} \quad Y_1, Y_2, \ldots, Y_{n_2} \sim N(\mu_2, \sigma_2^2),$$

where all the parameters μ_1, μ_2, σ_1^2, σ_2^2 are unknown, and the variances are not necessarily equal.

The well-known Behrens-Fisher problem is to test

$$H_0 : \mu_1 = \mu_2 \quad \text{vs} \quad H_a : \mu_1 \neq \mu_2,$$

when $\sigma_1 \neq \sigma_2$.

For the comparison of any two populations with unequal variances, we consider Chapman's (1950) two-stage procedure which takes into consideration both the confidence level and estimation accuracy.

For any given confidence level $1 - \alpha$ and accuracy level d, let n_0 be the number of observations in the first stage of sampling of the two populations of interest. Denote $\overline{X}, \overline{Y}, S_1, S_2$ the corresponding mean and standard deviation of the initial sample for the two populations, respectively.

Let t_v^* be the number satisfying

$$P(-t_v^* < T_1 - T_2 < t_v^*) = 1 - \alpha,$$

as tabulated in Chapman (1950). Denote

$$\eta_0 = [d/t_{n_0-1,\alpha/2}^*]^2,$$

and

$$N_1 = \max(n_0 + 1, [\frac{S_1^2}{\eta_0}] + 1), \qquad N_2 = \max(n_0 + 1, [\frac{S_2^2}{\eta_0}] + 1), \qquad (6.26)$$

where $[t]$ is the largest integer less than or equal to the value t.

Now, draw an additional $N_1 - n_0$ and $N_2 - n_0$ samples from the two populations, respectively. Denote the following quantities,

$$p_1 = \frac{1 - (N_1 - n_0)b_1}{n_0}, \quad q_1 = \frac{1}{N_1}\left\{ 1 + \sqrt{\frac{(N_1 c - S_1^2)n_0}{(N_1 - n_0)S_1^2}} \right\}.$$

$$p_2 = \frac{1 - (N_2 - n_0)b_2}{n_0}, \quad q_2 = \frac{1}{N_2}\left\{ 1 + \sqrt{\frac{(N_2 c - S_2^2)n_0}{(N_2 - n_0)S_2^2}} \right\}.$$

and the combined sample means

$$\overline{\overline{X}} = p_1 \sum_{j=1}^{n_0} X_j + q_1 \sum_{j=n_0+1}^{N_1} X_j$$

$$\overline{\overline{Y}} = p_2 \sum_{j=1}^{n_0} Y_j + q_2 \sum_{j=n_0+1}^{N_2} Y_j,$$

then $(\overline{\overline{X}} - \overline{\overline{Y}} - d, \overline{\overline{X}} - \overline{\overline{Y}} + d)$ is a $(1 - \alpha)100\%$ confidence set for the mean difference $\mu_1 - \mu_2$:

$$P(\overline{\overline{X}} - \overline{\overline{Y}} - d \le \mu_1 - \mu_2 \le \overline{\overline{X}} - \overline{\overline{Y}} + d) \ge 1 - \alpha. \tag{6.27}$$

We use the following two propositions to prove the validity of (6.27). The proofs of the two propositions are different from that of Chapman (1950).

Proposition 6.3.1 Assume that the observations in the first sample are not identical so that $S_1 \ne 0$. If the conditional distribution of $\overline{\overline{X}}$ given S_1,

$$\frac{\overline{\overline{X}} - \mu_1}{\sqrt{\eta_0}}\bigg|S_1 \sim N(0, \frac{\sigma_1^2}{S_1^2}),$$

then

$$T_1 = \frac{\overline{\overline{X}} - \mu_1}{\sqrt{\eta_0}}$$

follows a Student's t-distribution with degree of freedom $n_0 - 1$.

Proof: Denote $W = (n_0 - 1)\frac{S_1^2}{\sigma_1^2}$, since S_1^2 is the sample variance of the population following $N(\mu_1, \sigma^2)$, we have

$$\frac{(n_0 - 1)S_1^2}{\sigma_1^2} \sim \chi_{n_0-1}^2,$$

and the density of W is a standard χ^2:

$$f_W(w) = \frac{1}{\Gamma(\frac{m}{2})2^{\frac{m}{2}}} w^{\frac{m}{2}-1} e^{-\frac{w}{2}}.$$

Since the conditional distribution of T_1 given $W/(n_0 - 1) = w/(n_0 - 1)$ is $N(0, (n_0 - 1)w^{-1})$, denoting $V_1 = \frac{S_1^2}{\sigma_1^2}$ for notational convenience, we have

$$
\begin{aligned}
P(T_1 < t) &= \int_0^\infty P(T_1 < t|V_1 = v) f_{V_1}(v) dv \\
&= \int_0^\infty \int_0^t \frac{1}{\sqrt{2\pi}v^{-\frac{1}{2}}} e^{-\frac{z^2}{2v^{-1}}} dz \frac{m}{\Gamma(\frac{m}{2})2^{\frac{m}{2}}} (mv)^{\frac{m}{2}-1} e^{-\frac{mv}{2}} dv,
\end{aligned}
$$

substituting the density of W, hence the density of V_1 gets

$$
\begin{aligned}
P(T_1 < t) &= \frac{m^{\frac{m}{2}}}{\sqrt{\pi}2^{\frac{m+1}{2}}\Gamma(\frac{m}{2})} \int_0^t \Gamma(\frac{m+1}{2})(\frac{2}{z^2+m})^{\frac{m+1}{2}} dz \\
&= \frac{m^{\frac{m}{2}}\Gamma(\frac{m+1}{2})}{\sqrt{\pi}2^{\frac{m+1}{2}}\Gamma(\frac{m}{2})} \int_0^t (\frac{2}{m})^{\frac{m+1}{2}}(1 + \frac{z^2}{m})^{-\frac{m+1}{2}} dz \\
&= \frac{\Gamma(\frac{m+1}{2})}{\Gamma(\frac{m}{2})\sqrt{m\pi}} \int_0^t (1 + \frac{z^2}{m})^{-\frac{m+1}{2}} dz.
\end{aligned}
$$

Comparing the above term with the cumulative distribution function of a Student's t variable, gets

$$T_1 \sim t_{n_0-1}.$$

This completes the proof of Proposition 6.3.1. $\square$

Proposition 6.3.1 shows the validity of using the Student-t statistic in Chapman's (1950) paper under a condition on the conditional distribution. The following theorem shows that the conditional distribution criterion is valid for Chapman's two stage method.

Proposition 6.3.2 For Chapman's two stage sampling, the conditional distribution of $\overline{\overline{X}}$ given S_1,

$$\frac{\overline{\overline{X}} - \mu_1}{\sqrt{\eta_0}} \Big| S_1 \sim N(0, \frac{\sigma_1^2}{S_1^2}).$$

Proof: The weighted mean $\overline{\overline{X}}$ with weight w_i can be written as

$$\overline{\overline{X}} = \sum_{j=1}^{N_1} w_j X_j$$

$$= p_1 \sum_{j=1}^{n_0} X_j + q_1 \sum_{j=n_0+1}^{N_1} X_j,$$

where

$$p_1 = \frac{1 - (N_1 - n_0)b_1}{n_0}, \quad q_1 = \frac{1}{N_1} \left\{ 1 + \sqrt{\frac{(N_1\eta_0 - S_1^2)n_0}{(N_1 - n_0)S_1^2}} \right\},$$

which is one of the solutions to the following set of equations.

$$\sum_{j=1}^{N_1} w_j = 1, \quad w_1 = w_2 = \ldots = w_{n_0}, \quad \sum_{j=1}^{N_1} w_j^2 = \frac{\eta_0}{S_1^2},$$

with

$$w_1 = w_2 = \ldots = w_{n_0} = p_1, \quad w_{n_0+1} = w_{n_0+2} = \ldots = w_{N_1} = q_1.$$

Notice that the weights are hidden functions of sample variance S_1^2 of the observations from the first stage, thus they are random variables. For this reason, the weighted means $\overline{\overline{X}}$ do not follow normal distributions. Once S_1 is given, the linear combination of the observation follows a normal model, and we will show that the normal conditional distribution is actually $N(\mu_1, \sigma_1^2)$.

In terms of the expected value of the weighted sample mean, notice that

$$E(\overline{\overline{X}}|S_1) = E(\sum_{j=1}^{N_1} w_j X_j | S_1)$$

$$= E(p_1 \sum_{j=1}^{n_0} X_j + q_1 \sum_{j=n_0+1}^{N_1} X_j | S_1)$$

$$= E(p_1 n_0 \overline{X}_{n_0} | S_1) + E(q_1 \sum_{j=n_0+1}^{N_1} X_j | S_1).$$

By the independence between the sample mean and sample variance for the observations in the first stage, and the independence of the observations between the two stages, we have

$$E(\overline{\overline{X}} | S_1) = E(p_1 n_0 \overline{X}_{n_0}) + E(q_1 \sum_{j=n_0+1}^{N_1} X_j)$$

$$= p_1 n_0 E(\overline{X}_{n_0}) + q_1 \sum_{j=n_0+1}^{N_1} E(X_j)$$

$$= p_1 n_0 \mu_1 + q_1 (N - n_0) \mu_1$$

$$= \mu_1,$$

which is the population mean.

In terms of the the variance of weighted mean $\overline{\overline{X}}$, consider

$$Var(\overline{\overline{X}} | S_1) = Var(p_1 \sum_{j=1}^{n_0} X_j + q_1 \sum_{j=n_0+1}^{N_1} X_j | S_1)$$

$$= Var(p_1 n_0 \overline{X}_{n_0} | S_1) + Var(q_1 \sum_{j=n_0+1}^{N_1} X_j | S_1)$$

$$= Var(p_1 n_0 \overline{X}_{n_0}) + Var(q_1 \sum_{j=n_0+1}^{N_1} X_j),$$

due to the independence between the sample mean and the sample variance, and the independence between the observations in the two stages. Thus,

$$Var(\overline{\overline{X}} | S_1) = p_1^2 n_0^2 Var(\overline{X}_{n_0}) + q_1^2 \sum_{j=n_0+1}^{N_1} Var(X_j)$$

$$= p_1^2 n_0 \sigma_1^2 + q_1^2 (N_1 - n_0) \sigma_1^2$$

$$= \sum_{j=1}^{N_1} w_j^2 \sigma_1^2$$

$$= \eta_0 \frac{\sigma_1^2}{S_1^2}.$$

Therefore the conditional distribution of $\overline{\overline{X}}$ given S_1,

$$\frac{\overline{\overline{X}} - \mu_1}{\sqrt{\eta_0}} \Big| S_1 \sim N(0, \frac{\sigma_1^2}{S_1^2}).$$

This proves Proposition 6.3.2. □

Combining propositions 6.3.1 and 6.3.2, it follows that the distribution of T_2 given S_2 is $N(0, v_2^{-1})$, where $V_2 = \frac{S_2^2}{\sigma_2^2}$, and then $T_2 \sim t_{n_0-1}$.

T_1 and T_2 are independent because X_i and Y_i are independent and η_0 is a constant. Thus we can use Chapman's (1950) table to construct pairwise confidence interval for the mean difference between any two populations. We now combine the pairwise confidence intervals and probability inequalities to construct simultaneous confidence sets for successive comparisons when the population variances can not be plausibly assumed identical.

Let A_i be the $1 - \alpha/(k - 1)$ confidence intervals for the successive differences $\mu_i - \mu_{i+1}$ as in (6.27), respectively. By the Bonferroni inequality, we have

$$
\begin{aligned}
P(\mu_i - \mu_{i+1} \in A_i \, i = 1, ..., k-1) &= 1 - P\left(\bigcup_{i=1}^{k-1} \{\mu_i - \mu_{i+1} \notin A_i\}\right) \\
&\geq 1 - \sum_{i=1}^{k-1} P(\{\mu_i - \mu_{i+1} \notin A_i\}) \\
&= 1 - \sum_{i=1}^{k-1} \frac{\alpha}{k-1} \\
&= 1 - \alpha.
\end{aligned}
$$

Certainly, the above derivation can be improved by a higher degree inequality when the associated information is available.

We address the use of probability bounds in successive comparisons in this section. Similar applications of probability bounds can be found in the work of Chen and Glaz (2004), Worsley (1979), Iyengar (1988), and Glaz and Zhang (2006).

Chapter 7

Bivariate Lower Bounds

After discussing bivariate upper bounds and their extensions in Chapters 5 and 6, we now correspondingly address bivariate lower bounds and their extensions in Chapters 7 and 8. Similar to Chapter 5, this chapter consists of three main topics focusing on three types of bivariate bounding approaches: factorization, high-degree bounds, and optimization.

In Section 7.1, we use factorization of the bivariate indicator functions to derive two types of bivariate lower bounds, which extend the univariate Dawson-Sankoff lower bounds. Using numerical examples, we compare different types of bounding techniques and discuss the improvements of the bivariate lower bounds in three different directions — high order bounds in Section 7.2, optimal factorized bounds in Section 7.3, and optimal bounds using linear programming techniques in Section 7.4.

7.1 Bivariate Factorized Lower Bounds

As mentioned in Chapter 2, the univariate Dawson-Sankoff lower bound is not uniformly linear in S_1 and S_2, but is a Fréchet optimal lower bound. For bivariate situations, published work includes Galambos and Xu (1993), Móri (1996), Chen and Seneta (1998, 2006) for the optimality of bivariate lower bounds. In this section, we synthesize these results to show the bounding approach of factorization for bivariate lower bounds.

As usual, consider two sets of events $\{A_i\}$, $i = 1,...,n$, $\{B_j\}$, $j = 1,...,m$ in a probability space, denote v_1 and v_2 the number of occurrences for the two event sets, respectively. We need the following two lemmas to facilitate the discussion on the main results in this section; more details of derivation can be found in, for example, Seneta (1988, 1992), Seneta and Chen (2002).

Lemma 7.1.1. For integers x, y satisfying $1 \leq x \leq n$, $1 \leq y \leq m$, denote

$$F_{a,b}(x,y) = \frac{x(2a+1-x)y(2b+1-y)}{a(a+1)b(b+1)},$$

then

$$F_{a,b}(v_1, v_2) \leq tI(v_1 \geq 1, v_2 \geq 1),$$

Table 7.1 *Parabola values of special points*

x	1	a	$a+1$	n
values	$\frac{2}{a+1}$	1	1	$\frac{n(2a+1-n)}{a(a+1)}$

where

$$t = \max\left(1, \frac{n(2a+1-n)m(2b+1-m)}{a(a+1)b(b+1)}\right).$$

Proof: Notice that $F(.)$ is the function of (x,y) with parameters a and b,

$$F_{a,b}(x,y) = \frac{x(2a+1-x)y(2b+1-y)}{a(a+1)b(b+1)}.$$

Actually, in the product of two terms above, $x(2a+1-x)/(a(a+1))$ is a parabola in x, with maximum at $x = a+\frac{1}{2}$. We consider integer x in the range $1 \le x \le n$. The maximum and minimum values occur among the values at $x = 1, a, a+1, n$ (which are not necessarily in order since a or $a+1$ may exceed n). These values are given in Table 7.1.

Notice that the value of the parabola at $x = 1$ always satisfies $0 < 2/(a+1) \le 1$. If $2a+1-n \ge 0$, the value at n is non-negative and is at most as large as the value of the parabola at $x = a, a+1$, so it is less than or equal to 1. Clearly the value of the parabola at n can be negative.

Similar considerations hold for y in the second term in the product of the function $F(.)$, thus for integer x, y, $1 \le x \le n$, $1 \le y \le m$, we have

$$F_{a,b}(x,y) \le \max\left(1, \frac{n(2a+1-n)m(2b+1-m)}{a(a+1)b(b+1)}\right).$$

Letting $t = \max\left(1, \frac{n(2a+1-n)m(2b+1-m)}{a(a+1)b(b+1)}\right)$ yields

$$L_{a,b}(v_1, v_2) \le tI(v_1 \ge 1, v_2 \ge 1).$$

This completes the proof of Lemma 7.1.1.   □

Since the product-type bivariate lower bound stems from the univariate Dawson-Sankoff bound, the next lemma clarifies the optimal point of the Dawson-Sankoff lower bound regarding the constraints on the parameters for the bivariate Dawson-Sankoff type lower bound.

Lemma 7.1.2 For any sequence of events $A_1, ..., A_n$, the highest value of the form

$$\frac{2}{k+1}S_1(A) - \frac{2}{k(k+1)}S_2(A),$$

over $k \ge 1$, is unique and can be reached at $k^0 \in \{[\frac{2S_2}{S_1}] + 1, [\frac{2S_2}{S_1}]\}$, where $k^0 - 1 \le k^0 \le n$.

Proof: Let

$$f(k) = \frac{2}{k+1}S_1 - \frac{2}{k(k+1)}S_2, \quad \text{for any integer} \quad k \geq 1.$$

We have shown that k^0 gives the maximum of $f(k)$ in Examples 2.2.2 and 2.2.3. Here, we essentially just need to verify its uniqueness.

$$f(k+1) - f(k) = -\frac{2S_1}{k(k+1)(k+2)}\left(k - \frac{2S_2}{S_1}\right).$$

When $\frac{2S_2}{S_1}$ is an integer,

$$f(k^0) - f(k^0 - 1) = 0$$

and

$$f(k+1) - f(k) \begin{cases} > 0, & \text{if } k < \frac{2S_2}{S_1}, \\ < 0, & \text{if } k > \frac{2S_2}{S_1}. \end{cases}$$

When $\frac{2S_2}{S_1}$ is not an integer,

$$f(k+1) - f(k) \begin{cases} > 0, & \text{if } k \leq \lceil\frac{2S_2}{S_1}\rceil, \\ < 0, & \text{if } k > \lceil\frac{2S_2}{S_1}\rceil. \end{cases}$$

and

$$f(k^0) - f(k^0 - 1) > 0.$$

Thus, we can conclude that $f(k)$ has its *only* maximal value either at k^0 or at $k^0 - 1$, and from the discussion in Example 2.2.3,

$$k^0 - 1 \leq k^0 \leq n.$$

This completes the proof of Lemma 7.1.2. □

With the above two lemmas, we start the main results on factorizing lower bounds as follows.

Theorem 7.1.1 Consider two sets of events $A_1, ..., A_n$ and $B_1, ..., B_m$. For any two positive integers a and b, if $n - 2a - 1 \leq 0$ or $m - 2b - 1 \leq 0$, then

$$P(v_1 \geq 1, v_2 \geq 1) \geq \tag{7.1}$$

$$\frac{4S_{11}}{(a+1)(b+1)} - \frac{4S_{21}}{a(a+1)(b+1)} - \frac{4S_{12}}{b(a+1)(b+1)} + \frac{4S_{22}}{ab(a+1)(b+1)}$$

Note: The conditions of Theorem 7.1.1 clearly hold (and so $t = 1$) if either $a \geq n/2$ or $b \geq m/2$. The proof of Theorem 7.1.1 can be found in Galambos and Xu (1993); alternatively, it can be viewed as a special case of Theorem 7.1.2.

Example 7.1.1 Choosing each B_j, $j = 1, ..., m$ as the sample space Ω and $b = m - 1$, $m \geq 2$ then

$$S_{11} = \sum_{\substack{1 \leq i \leq n \\ 1 \leq j \leq m}} P(A_i B_j) = m S_1(A),$$

$$S_{12} = \sum_{\substack{1 \leq i \leq n \\ 1 \leq j_1 < j_2 \leq m}} P(A_i B_{j_1} B_{j_2}) = m \frac{m-1}{2} S_1(A),$$

$$S_{21} = \sum_{\substack{1 \leq i_1 < i_2 \leq n \\ 1 \leq j \leq m}} P(A_{i_1} A_{i_2} B_j) = m S_2(A),$$

and

$$S_{22} = \sum_{\substack{1 \leq i_1 < i_2 \leq n \\ 1 \leq j_1 < j_2 \leq m}} P(A_{i_1} A_{i_2} B_{j_1} B_{j_2}) = m \frac{(m-1)}{2} S_2(A).$$

The bound in Theorem 7.1.1 reads, since $b = m - 1 \geq m/2$,

$$
\begin{aligned}
& P(\nu_1 \geq 1) \\
= \ & P(\nu_1 \geq 1, \nu_2 \geq 1) \quad (\text{ for } \{B_j\} = \{\Omega, ..., \Omega\}) \\
\geq \ & \frac{4S_{1,1}}{(a+1)(b+1)} - \frac{4S_{2,1}}{a(a+1)(b+1)} - \frac{4S_{1,2}}{b(a+1)(b+1)} + \frac{4S_{2,2}}{ab(a+1)(b+1)} \\
= \ & \frac{4mS_1(A)}{(a+1)m} - \frac{4mS_2(A)}{a(a+1)m} - \frac{4m(m-1)S_1(A)}{2(a+1)m(m-1)} + \frac{4m(m-1)S_2(A)}{2a(a+1)m(m-1)} \\
= \ & \frac{2S_1(A)}{a+1} - \frac{2S_2(A)}{a(a+1)}.
\end{aligned}
$$

This is the Dawson-Sankoff bound for the set of events $A_1, ..., A_n$. $\quad\square$

Theorem 7.1.1 has the constraint that either the value a or b needs to satisfy. However, Lemma 7.1.2 shows that k^0 and $k^0 - 1$ are the only possible optimal points. Therefore, it suggests that the constraint on the associated parameters a, b satisfy $1 \leq a \leq n$, $1 \leq b \leq m$. So, there is a discrepancy on the optimal domains for the two parameters. Now, since the optimal point may not always be larger than $\frac{n-1}{2}$, when both parameters are less than $\frac{n-1}{2}$ and $\frac{m-1}{2}$, respectively, the bivariate lower bound stated in Theorem 7.1.1 can be improved. Namely, the information on the Bonferroni summations is not fully used. Toward this end, the following theorem removes the constraint on the two parameters a and b in the lower bound stated in Theorem 7.1.1.

Theorem 7.1.2 Consider two sets of events $A_1, ..., A_n$ and $B_1, ..., B_m$. For any two positive integers $a \geq 1$ and $b \geq 1$,

$$P(\nu_1 \geq 1, \nu_2 \geq 1) \geq \tag{7.2}$$
$$\frac{1}{t} \left(\frac{4S_{11}}{(a+1)(b+1)} - \frac{4S_{21}}{a(a+1)(b+1)} - \frac{4S_{12}}{b(a+1)(b+1)} + \frac{4S_{22}}{ab(a+1)(b+1)} \right),$$

where
$$t = \max\left(1, \frac{nm(n-2a-1)(m-2b-1)}{ab(a+1)(b+1)}\right).$$

Proof: Let
$$I = I(v_1 \geq 1, v_2 \geq 1) = \begin{cases} 1, & \text{if } v_1 \geq 1, v_2 \geq 1 \\ 0, & \text{otherwise,} \end{cases}$$

so that $E(I) = P(v_1 \geq 1, v_2 \geq 1)$. Denote

$$L_{a,b}(x,y) = \frac{4xy}{(a+1)(b+1)} - \frac{2x(x-1)y}{a(a+1)(b+1)} - \frac{2xy(y-1)}{b(a+1)(b+1)} + \frac{xy(x-1)(y-1)}{ab(a+1)(b+1)},$$

for integers $x \geq 1, y \geq 1$. Notice that

$$S_{k,t} = E\left[\binom{v_1}{k}\binom{v_2}{t}\right]$$

and taking account of $L_{ab}(v_1, v_2)$ being a function of v_1, v_2, we have

$$
\begin{aligned}
&E[L_{a,b}(v_1,v_2)] \\
&= \frac{4S_{11}}{(a+1)(b+1)} - \frac{4S_{21}}{a(a+1)(b+1)} - \frac{4S_{12}}{b(a+1)(b+1)} + \frac{4S_{22}}{ab(a+1)(b+1)}.
\end{aligned}
$$

By Lemma 7.1.1,
$$EL_{a,b}(v_1,v_2) \leq tE(I(v_1 \geq 1, v_2 \geq 1)),$$

which implies that

$$
\begin{aligned}
&P(v_1 \geq 1, v_2 \geq 1) \\
&\geq \frac{1}{t}\left[\frac{4S_{1,1}}{(a+1)(b+1)} - \frac{4S_{2,1}}{a(a+1)(b+1)} - \frac{4S_{12}}{b(a+1)(b+1)} + \frac{4S_{22}}{ab(a+1)(b+1)}\right].
\end{aligned}
$$

This completes the proof of Theorem 7.1.2. □

We now use examples to illustrate properties of the bivariate factorized lower bound established in Theorem 7.1.2. Example 7.1.2 shows that Theorem 7.1.2 is a natural extension of the Dawson-Sankoff bound in a product space.

Example 7.1.2. Consider any events A_1, ..., A_n with $n \geq 2$ in a probability space. According to the (univariate) Dawson-Sankoff bound (Example 2.2.2)

$$P(v_1 \geq 1) \geq \frac{2}{a+1}S_1(A) - \frac{2}{a(a+1)}S_2(A), \tag{7.3}$$

for each integer $a \geq 1$. Now consider a product probability space with corresponding events in the copy of the original space denoted by $B_1, ..., B_m$ (where $m = n$). Then

$$
\begin{aligned}
&P(v_1 \geq 1, v_2 \geq 1) \\
&= P(v_1 \geq 1)P(v_2 \geq 1) \\
&\geq \left(\frac{2}{a+1}S_1(A) - \frac{2}{a(a+1)}S_2(A)\right)\left(\frac{2}{b+1}S_1(B) - \frac{2}{b(b+1)}S_2(B)\right)
\end{aligned}
$$

for any integers $a \geq (n-1)/2$, $b \geq (n-1)/2$ such that each univariate bound is non negative; the right-hand side of above inequality is then equal to

$$\frac{4S_{1,1}}{(a+1)(b+1)} - \frac{4S_{2,1}}{a(a+1)(b+1)} - \frac{4S_{1,2}}{b(a+1)(b+1)} + \frac{4S_{2,2}}{ab(a+1)(b+1)},$$

since by independence $S_{i,j} = S_i(A)S_j(B)$. $\qquad\square$

It should be mentioned that the condition of $a \geq (n-1)/2$ and $b \geq (n-1)/2$ is sufficient but not necessary for the validity of the inequality in Theorem 7.1.1. The following example shows that for the numerical example discussed in Example 2.4.7, the product-type lower bound discussed in Example 7.1.2 reaches the optimal points when the parameters a and b are both less than $\frac{n+1}{2}$ and $\frac{m+1}{2}$, respectively.

Example 7.1.3 Consider a set of events A_1, ..., A_6 with probabilities of elementary conjunctions given in Table 2.1 and Table 2.2 in Example 2.4.7. In this case, $n = 6$ and $S_1(A) = 2.364$, $S_2(A) = 2.252$, thus the univariate optimal point is $a = 2$. The optimized value of the bound in (7.3) is $(2/3)S_1(A) - (1/3)S_2(A) = 0.8253$. By Lemma 7.1.2, for all $a \geq 1$, the optimal value is attained only at $a = 2$.

Now, from the numerical specification of the events under consideration, the actual value of $P(v_1(A) \geq 1) = 0.9505$. Hence, following the product space formulation in Example 7.1.3, the bivariate lower bound in Theorem 7.1.2 reads:

$$(0.9505)^2 > (0.8253)^2.$$

This shows that even though the conditions of Theorem 7.1.1 are broken (in particular neither $a \geq n/2$ nor $b \geq m/2$ here, since $6/2 = 3$ and $a = b = 2$), (7.1) nevertheless holds. At the end of this section, after examining another type of extension of the Dawson-Sankoff lower bound and discussing the numerical property of the factorized lower bound, Example 7.1.6 shed a different light with another numerical example to show that the term t in the bound (7.2) is necessary. $\qquad\square$

We now present a lower bound in terms of $S_{1,1}$, $S_{1,2}$, $S_{2,1}$ only, originating from Chen and Seneta (1995a). This is another extension of the univariate Dawson-Sankoff lower bound into the bivariate setting in another factorized form.

Theorem 7.1.3 Consider two sets of events A_1, ..., A_n and B_1, ..., B_m. For any two positive integers $a \geq 1$ and $b \geq 1$, denote $A = \min(a+1,n)$ and $B = \min(b+1,m)$, we have

$$P(v_1 \geq 1, v_2 \geq 1) \geq \frac{1}{f}\left(S_{1,1} - S_{2,1}/a - S_{1,2}/b\right), \qquad (7.4)$$

where the value f is defined as

$$f = \begin{cases} 2, & \text{if } a = b = 2 \\ AB(1 - \frac{A-1}{2a})(1 - \frac{B-1}{2b}) & \text{otherwise.} \end{cases}$$

Bound (7.4) reverts to a univariate Dawson-Sankoff bound as shown in the following example.

Example 7.1.4. Consider a set of events $A_1, ..., A_n$ and another event-set chosen with $m = 1$ and $B_1 = \Omega$. For these two sets of events,

$$S_{11} = S_1(A), \qquad S_{12} = 0, \qquad S_{21} = S_2(A), \qquad S_{22} = 0, \qquad (7.5)$$

for any $a + 1 \leq n$ integer. We choose $b = 1$, so that $B = 1$ and $A = a + 1$ with the value $f = \frac{a+1}{2}$, under this setting, Bound (7.4) reads

$$
\begin{aligned}
P(v_1 \geq 1) &= P(v_1 \geq 1, v_2 \geq 1) \quad (\text{since} \quad \{B_j\} = \{\Omega\}) \\
&\geq \frac{2}{(a+1)} S_1(A) - \frac{2}{(a+1)} (S_2(A)/a) \quad (\text{Combining (7.4) with (7.5)}) \\
&= \frac{2S_1(A)}{a+1} - \frac{2S_2(A)}{a(a+1)}.
\end{aligned}
$$

This is the univariate Dawson-Sankoff bound for $k + 1 \leq n$ since $a + 1 \leq n$.

Obviously the Bound (7.4) is another extension of the Dawson-Sankoff lower bound with information on S_{11}, S_{12}, and S_{21} only. It should be mentioned that the factorized bound (7.2) uses all degree-two Bonferroni summations S_{11}, S_{12}, S_{21} and S_{22}; however, Bound (7.4) uses only S_{11}, S_{12} and S_{21}. We now check the validity of Bound (7.4).

Proof of Theorem 7.1.3: Denote $z_{ij} = P(v_1 \geq i, v_2 \geq j)$. When $a = b = 2$, using the binomial property of the Bonferroni summations, we have

$$
\begin{aligned}
S_{11} - S_{21}/2 - S_{12}/2 &= \sum_{i=1}^{n} \sum_{j=1}^{m} (1 - (i-1)/2 - (j-1)/2) z_{ij} \\
&\leq z_{11} + z_{21}/2 + z_{12}/2,
\end{aligned}
$$

since the other terms on the right are non-positive. Therefore

$$S_{11} - S_{21}/2 - S_{12}/2 \leq 2z_{11}, \qquad (7.6)$$

because $z_{12} \leq z_{11}, z_{21} \leq z_{11}$.

Equation (7.6) implies that

$$P(v_1 \geq 1, v_2 \geq 1) \geq \frac{1}{2}(S_{1,1} - S_{1,2}/2 - S_{2,1}/2).$$

Now, for any integers a and b such that $a \geq 1$ and $b \geq 1$ except $(a = b = 2)$, we have

$$
\begin{aligned}
S_{11} - S_{21}/a - S_{12}/b &= \sum_{i=1}^{n} \sum_{j=1}^{m} (1 - (i-1)/a - (j-1)/b) z_{ij} \\
&\leq \sum_{i=1}^{A} \sum_{j=1}^{B} (1 - (i-1)/a - (j-1)/b) z_{ij}
\end{aligned}
$$

$$\leq \sum_{i=1}^{A}\sum_{j=1}^{B}\left(1-(i-1)/a\right)\left(1-(j-1)/b\right)z_{ij}$$

$$\leq z_{11}\sum_{i=1}^{A}\sum_{j=1}^{B}\left(1-(i-1)/a\right)\left(1-(j-1)/b\right)$$

$$= z_{11}\{A-((A-1)A)/(2a)\}\{B-((B-1)B)/(2b)\}$$

which is tantamount to (7.4).

Note that the above argument holds for $a=b=2$ as well. However, the value of f in this case becomes $9/4 > 2$, which makes the bound weaker than what is stated in the theorem for the case of $a=b=2$:

$$P(v_1 \geq 1, v_2 \geq 1) \geq 1/2\left(S_{1,1}-S_{2,1}/2-S_{1,2}/2\right). \tag{7.7}$$

This completes the proof of Theorem 7.1.3. $\square$

The above three theorems (Theorems 7.1.1, 7.1.2, and 7.1.3) extend the univariate Dawson-Sankoff lower bound in three different versions of factorized bounds. The bound (7.1) is a special case of (7.2). In fact, in an appropriate setting, the bound (7.4) also reduces to the form of (7.2). To see this point, consider $1 \leq a \leq n-1$ and $1 \leq b \leq m-1$ and $(a,b) \neq (2,2)$ in Theorem 7.1.3, we have (7.4) reduces to

$$P(v_1 \geq 1, v_2 \geq 1) \geq \frac{4}{(a+1)(b+1)}\left(S_{1,1}-S_{2,1}/a-S_{1,2}/b\right), \tag{7.8}$$

which is of the form of (7.2) with $t=1$ and $S_{2,2}=0$.

In the literature, the first version of a bivariate lower bound with incomplete Bonferroni summations is due to Lee (1992), who proved that for integer $a, b, 2 \leq a \leq n,$ $2 \leq b \leq m,$

$$P(v_1 \geq 1, v_2 \geq 1) \geq \frac{2}{ab}\{S_{11}-S_{21}/a-S_{12}/b\}. \tag{7.9}$$

When $a=b=2$, (7.9) becomes (7.7). When $(a,b) \neq (2,2)$, since $0 \leq (b-1)(a-1)-2$, we have

$$ab \geq a+b+1,$$

which is equivalent to

$$2/(ab) \leq 4/((a+1)(b+1)),$$

so (7.8) gives a sharper bound than (7.9) when $S_{1,1}-S_{2,1}/a-S_{1,2}/b > 0$.

It should be mentioned that Theorem 7.1.3 is of a form similar to the result of Galambos and Xu (1993), which reads

$$P(v_1 \geq 1, v_2 \geq 1) \geq 1/(ab)\{(3-(1/a)-(1/b))S_{1,1}-2S_{2,1}/a-2S_{1,2}/b\}. \tag{7.10}$$

Clearly (7.10) is an improvement on (7.9), since for $a, b \geq 2, 3-1/a-1/b \geq 2$.

Notice that when $a=b=2$, all three bounds (7.7) (7.9), and (7.10) coincide. This raises the question of the differences of the three bounds. We have shown that Bounds (7.7) and (7.8) are sharper than Bound (7.9), thus, we just need to compare Bounds (7.7) and (7.8) with Bound (7.10).

The following example shows that the bounds (7.7) and (7.8) cannot be improved by any member of the family (7.10).

Example 7.1.5. Consider a case with 6 events. The construction of the numerical example amounts to specifying the corresponding probability space and probabilities corresponding to the 2^6 elementary conjunctions of intersections taken 6 at a time of 6 events $C_1, C_2, ..., C_6$ and their complements. The non-zero probability assignments are given as follows.

The only elementary conjunctions with non-zero probability are

- The $\binom{6}{1}=6$ events with precisely one C_i and five C_i^c's each with probability 0.01;

- The $\binom{3}{1}\binom{3}{1}=9$ events with just one of the first 3 entries and just one of the last 3 being a C_i and the others being C_i^c's (for example, $C_1^c C_2 C_3^c C_4 C_5^c C_6^c$), are each given probability 0.096;

- The $2\binom{3}{1}\binom{3}{2}=18$ events where just one (respectively 2) of the first 3 entries is a C_i and 2 (respectively one) of the second 3 (for example, $C_1 C_2^c C_3^c C_4 C_5 C_6^c$), are each given probabilities 0.001;

- $P(C_1^c C_2^c C_3^c C_4^c C_5^c C_6^c) = 0.058$.

Now, taking $A_1 = C_1, A_2 = C_2, A_3 = C_3, B_1 = C_4, B_2 = C_5, B_3 = C_6$, we have

$$P(A_iB_j) = P(C_iC_{3+j}) = 0.1 \quad \text{for} \quad i=1,2,3, j=1,2,3.$$

We provide details for the calculation of $P(C_1C_4)$; calculations for the rest are then self-evident.

By the specification of the probabilities in this example,

$$
\begin{aligned}
P(C_1C_4) &= P(C_1C_2^cC_3^cC_4C_5^cC_6^c) + P(C_1C_2^cC_3^cC_4C_5C_6^c) + \\
&\quad P(C_1C_2^cC_3^cC_4C_5^cC_6) + P(C_1C_2C_3^cC_4C_5^cC_6^c) + P(C_1C_2^cC_3C_4C_5^cC_6^c) \\
&= 0.096 + 2 \times 0.001 + 2 \times 0.001 \\
&= 0.1.
\end{aligned}
$$

$$P(A_iB_{j_1}B_{j_2}) = P(C_iC_{3+j_1}C_{3+j_2}) = 0.001 \quad \text{for} \quad i=1,2,3, \quad 1 \le j_1 \le j_2 \le 3,$$

for example,

$$P(C_1C_4C_5) = P(C_1C_2^cC_3^cC_4C_5C_6^c) + 0 = 0.001.$$

Similarly

$$P(A_{i_1}A_{i_2}B_j) = 0.001 \quad \text{for} \quad 1 \le i_1 < i_2 \le n \quad j=1,2,3.$$

Thus,

$$S_{1,1} = \sum_{\substack{1 \le i \le 3 \\ 1 \le j \le 3}} P(A_iB_j) = 9 \times 0.1 = 0.9$$

$$S_{1,2} = \sum_{\substack{1 \le i \le 3 \\ 1 \le j_1 < j_2 \le 3}} P(A_iB_{j_1}B_{j_2}) = 9 \times 0.001 = 0.009$$

and

$$S_{2,1} = 0.009, \qquad S_{2,2} = 0,$$

$$\begin{aligned}
P(v_1 \geq 1, v_2 \geq 1) &= 9 \times 0.096 + 18 \times 0.001 \\
&= 0.864 + 0.018 \\
&= 0.882.
\end{aligned}$$

Therefore in this example, we have, $n = m = 3$, $S_{1,1} = 0.9$, $S_{1,2} = S_{2,1} = 0.009$ and $S_{2,2} = 0$. At $a = b = 1$, (7.7) becomes $S_{11} - S_{21} - S_{12} = 0.882$, which is exactly $P(v_1 \geq 1, v_2 \geq 1)$.

However, the highest value of the expression (7.10) occurs at $a = 1$ and $b = 1$ with the value of the bound as 0.864. □

Notice that Theorem 7.1.1 (without the restriction on the parameters) also gives a lower bound 0.882, at $a = b = 1$ in the above example. This leads to the question regarding the necessity of including the quantity t in Theorem 7.1.2 (see also Example 7.1.3). We finish this section with an example showing that for some a, b and $S_{i,j}$, $i = 1, 2$, $j = 1, 2$,

$$L_{a,b} = \frac{4S_{1,1}}{(a+1)(b+1)} - \frac{4S_{2,1}}{a(a+1)(b+1)} - \frac{4S_{1,2}}{b(a+1)(b+1)} + \frac{4S_{2,2}}{ab(a+1)(b+1)}$$

may be greater than 1, so we do need $t > 1$ in (7.2) to keep $\frac{1}{t}L_{a,b}$ as a lower bound. In this example, $n = 4$, $m = 4$, and the value of $L_{a,b} > 1$ at $a = b = 1$, the conditions on a and b in Theorem 7.1,1 are broken.

Example 7.1.6. Consider the numerical example specified in Example 4.2.2. For $a = b = 1$,

$$L_{1,1} = S_{1,1} - S_{1,2} - S_{2,1} + S_{2,2} = 3.203.$$

Here, $L_{1,1} > 1$. So we do need $t > 1$ to make $t^{-1}L_{1,1} \leq 1$. In fact, in this case $t = 4$ and we get

$$t^{-1}L_{1,1} = \frac{1}{4} \times 3.203 = 0.80075,$$

while the actual value to be evaluated is

$$\begin{aligned}
P(v_1 \geq 1, v_2 \geq 1) &= 1 - [P(v_1 = 0, v_2 = 0) + P(v_1 \neq 0, v_2 = 0) + P(v_1 = 0, v_2 \neq 0)] \\
&= 1 - P(C_1^c ... C_8^c) - (2^4 - 1)\frac{0.01}{254} \times 2 \\
&= 0.80882,
\end{aligned}$$

which is larger than the lower bound

$$0.80075 = t^{-1}L_{11}.$$

Thus, it is necessary to have the coefficient t to validate the lower bound, although such condition affects the optimality of the bound. □

7.2 Bivariate High-degree Lower Bounds

Information on the probabilities of higher degree intersections usually leads to improved probability inequalities. This principle remains valid for bivariate lower bounds. In this section, we select two high degree lower bounds as examples to enhance this bounding principle. Technical treatments and derivations of these lower bounds were thoroughly discussed in Chapter 3 where the method of indicator functions was introduced. For concreteness and completeness of the material, we do not get into details of the proofs for high degree lower bounds mentioned in this section.

Consider two sets of events in a probability space. Assume that the first set of events has n and the second has m events. For integers r, u, a, b satisfying $0 \le r \le n$, $0 \le u \le m$, $r + a \le n$, $b + u \le m$, we have

Theorem 7.2.1 When both integers a, b are odd, the joint probability of at least r occurrences in the first and u occurrences in the second, is

$$P(v_1 \ge r, \quad v_2 \ge u)$$

$$\ge \sum_{k=0}^{a} \sum_{t=0}^{b} (-1)^{k+t} \binom{k+r-1}{r-1} \binom{t+u-1}{u-1} S_{k+r,t+u}$$

$$- \binom{b+u}{u} \binom{a+r}{r} \frac{r}{a+1} \frac{u}{b+1} S_{r+a+1,u+b+1}. \tag{7.11}$$

Theorem 7.2.2 When both integers a, b are odd, the joint probability of exactly r occurrences in the first and u occurrences in the second, is

$$P(v_1 = r, \quad v_2 = u)$$

$$\ge \sum_{k=0}^{a} \sum_{h=0}^{b} (-1)^{k+h} \binom{k+r}{r} \binom{t+u}{u} S_{k+r,h+u}$$

$$- \binom{b+u+1}{b} \binom{a+r+1}{a} \frac{r+1}{a+1} \frac{u+1}{b+1} S_{r+a+1,u+b+1}. \tag{7.12}$$

In the literature, efforts have been devoted to seek the extension of the high degree classical Bonferroni inequalities in a bivariate setting. In this regard, Galambos and Lee (1994) proposed bounds extending the classical Bonferroni bounds to a bivariate setting ($d = 2$), which were then discussed by Chen and Seneta (1996) as follows.

Theorem 7.2.3. For non-negative integers u_1, u_2, r_1, and r_2, the lower bound for the joint probability of exactly r_1 occurrences from the first, and r_2 from the second, is

$$P(v_1 = r_1, v_2 = r_2)$$

$$\ge \sum_{k_1=0}^{2u_1+1} \sum_{k_2=0}^{2u_2+1} (-1)^{k_1+k_2} \binom{k_1+r_1}{r_1} \binom{k_2+r_2}{r_2} S_{k_1+r_1,k_2+r_2} + \Delta_1 + \Delta_2 - \Delta_3$$

where

$$\Delta_1 = \begin{cases} 0, & \text{for } n_1 = r_1, \\ \binom{r_1+2u_1+1}{r_1} \frac{r_1+2u_1+2}{n_1-r_1} \sum_{y=0}^{n_2-r_2} (-1)^y S_{r_1+2u_1+2,r_2+y} \binom{r_2+y}{r_2}, & \text{for } n_1 > r_1. \end{cases}$$

$$f(k+1) - f(k) \begin{cases} > 0, & \text{if } k < \frac{2S_2}{S_1}, \\ < 0, & \text{if } k > \frac{2S_2}{S_1}. \end{cases}$$

$$\Delta_2 = \begin{cases} 0, & \text{for } n_2 = r_2, \\ \binom{r_2+2u_2+1}{r_2} \frac{r_2+2u_2+2}{n_2-r_2} \sum_{x=0}^{n_1-r_1} (-1)^x S_{r_1+x,r_2+2u_2+2} \binom{r_1+x}{r_1}, & \text{for } n_2 > r_2. \end{cases}$$

$$\Delta_3 = \begin{cases} 0, & \text{for } n_1 = r_1 \text{ or } n_2 = r_2, \\ D, & \text{for } n_1 > r_1 \text{ and } n_2 > r_2, \end{cases}$$

with

$$D = \binom{r_1+2u_1+1}{r_1} \binom{r_2+2u_2+1}{r_2} \frac{r_1+2u_1+2}{2u_1+2} \frac{r_2+2u_2+2}{2u_2+2} S_{r_1+2u_1+2,r_2+2u_2+2}.$$

As delineated in Section 3.2, the nature of this extension is to use the quantity

$$\sum_{i=0}^{a} \sum_{j=0}^{b} (-1)^{i+j} \binom{i+r_1}{r_1} \binom{j+r_2}{r_2} S_{i+r_1,j+r_2}$$

as the basic term for inclusion and exclusion, with lower bounds obtained by setting a and b both odd, to bound the difference between $P(v_1 = r_1, v_2 = r_2)$ and the associated linear combinations of Bonferroni summations. The main idea relies on the expression of the binomial moment:

$$\begin{aligned} S_{kt} &= E(\binom{v_1}{k} \binom{v_2}{t}) \\ &= \sum_{i=k}^{n_1} \sum_{j=t}^{n_2} \binom{v_1}{k} \binom{v_2}{t} P(v_1 = i, v_2 = j), \end{aligned}$$

in conjunction with lemmas on combinatorial coefficients via the method of indicators. This direction of generalization in theory separately involves all the degrees from 0 to $a + r_1$ and 0 to $b + r_2$, respectively, for event sets $\{A_{11}, ..., A_{n_1 1}\}$ and $\{A_{12}, ..., A_{n_2 2}\}$.

7.3   Bivariate Optimal Factorized Bounds

Section 7.1 describes a situation where the product of two univariate Fréchet optimal lower bounds does not form a bivariate lower bound, and the need of bivariate adjustment. Section 7.2 outlines high order bivariate lower bounds. Although bounds in these sections improve the approximation of the probability of the union of two sets of events, we have not discussed the potential space of improvement for those bounds. For an optimal bound, there is no space for improvement. Thus the discussion of inequality improvement necessitates the concept of bivariate optimal lower bound introduced in Chapter 2.

We use two parts to discuss bivariate optimal lower bounds. This section addresses conditions facilitating the transition of optimality from univariate optimal bounds to bivariate lower bounds, the bivariate optimal factorized bounds. The next section focuses on an algorithm that implements the theoretical results of bivariate lower bounds addressed in Chapter 4.

Theorem 7.1.2 shows that in general, the product of two factorized univariate lower bounds needs a bivariate adjustment to form a bivariate lower bound; the bivariate adjustment ruins the optimality of the bivariate Dawson-Sankoff-type lower bound (Seneta and Chen, 2002). The following theorem specifies conditions that optimize the bivariate lower bound in Theorem 7.1.2. We start with an illustrative example.

Example 7.3.1. Let $A_1, A_2, \cdots, A_n$ be a set of events in a probability space. Consider the product probability space with corresponding events in the copy of the original space denoted by $B_1, \cdots, B_m$ (where $n = m$). Then $S_{ij} = S_i(A)S_j(B)$. Denote the corresponding univariate Bonferroni summations $v_i = S_i(A)$ for $i = 1, 2$ and $w_j = S_j(B)$ for $j = 1, 2$. We have the decomposition $S_{ij} = v_i w_j$ for $i, j = 1, 2$.

Now using the Fréchet optimality of the Dawson-Sankoff lower bound, we know that there exists a set of events $\{A_i^*, i = 1, ..., n\}$ and $\{B_j^*, j = 1, ..., m\}$ such that

$$P(v(A) \geq 1) = \left(\frac{2S_1(A)}{a^* + 1} - \frac{2S_2(A)}{a^*(a^* - 1)} \right)$$

and

$$P(v(B) \geq 1) = \left(\frac{2S_1(B)}{b^* + 1} - \frac{2S_2(B)}{b^*(b^* - 1)} \right)$$

where $v(A)$ and $v(B)$ are the numbers of occurrences of the two sets of events, respectively. $a^* = b^* = [2S_2(A)/S_1(A)] + 1$. Thus two sets of events in the product space can be constructed as

$$A_i^* \times \Omega, i = 1, ..., n \qquad \Omega \times B_j^*, j = 1, ..., m$$

such that

$$
\begin{aligned}
& P(v(A^* \times \Omega) \geq 1, v(\Omega \times B^*) \geq 1) \\
= \ & P(v(A^*) \geq 1)P(v(B^*) \geq 1) \\
= \ & [\frac{2}{a^* + 1}S_1(A) - \frac{2}{(a^* + 1)a^*}S_2(A)][\frac{2}{b^* + 1}S_1(B) - \frac{2}{(b^* + 1)b^*}S_2(B)] \\
= \ & \frac{4S_{11}}{(a^* + 1)(b^* + 1)} - \frac{4S_{21}}{a^*(a^* + 1)(b^* + 1)} - \frac{4S_{12}}{b^*(a^* + 1)(b^* + 1)} + \frac{4S_{22}}{a^*b^*(a^* + 1)(b^* + 1)} \\
= \ & L(a^*, b^*).
\end{aligned}
$$

Notice that when

$$t = \max\left(1, \frac{nm(n - 2a^* - 1)(m - 2b^* - 1)}{a^*b^*(a^* + 1)(b^* + 1)}\right) = 1,$$

the term $L(a^*, b^*)$ is a bivariate lower bound. Thus it is an Fréchet optimal lower bound due to the existence of two sets of events that equal both sides of the bivariate inequality. $\square$

Example 7.3.1 can be extended into the following theorem regarding the optimality of factorized lower bounds without the limitation on the setting of product space.

Theorem 7.3.1. For a set of consistent Bonferroni summations $\{S_{11}, S_{12}, S_{21}, S_{22}\}$, if each summation can be decomposed into a product form, $S_{ij} = v_i w_j$, $i, j = 1, 2$ where $v_i > 0$, $w_j > 0$, $i, j = 1, 2$, and

$$t = \max\left(1, \frac{nm(n - 2a - 1)(m - 2b - 1)}{ab(a+1)(b+1)}\right) = 1$$

at $a = [2v_2/v_1] + 1$ and $b = [2w_2/w_1] + 1$, then the following lower bound is Fréchet optimal.

$$P(v_1 \geq 1, v_2 \geq 1) \geq$$
$$\frac{4S_{11}}{(a+1)(b+1)} - \frac{4S_{21}}{a(a+1)(b+1)} - \frac{4S_{12}}{b(a+1)(b+1)} + \frac{4S_{22}}{ab(a+1)(b+1)}.$$

Proof: By Theorem 7.1.2, it suffices to show that there is a probability space and two sets of events $\{A_i^*\}, \{B_j^*\}$ in it such that

$$P(v_1(A^*) \geq 1, v_2(B^*) \geq 1) =$$
$$\frac{4S_{11}}{(a+1)(b+1)} - \frac{4S_{21}}{a(a+1)(b+1)} - \frac{4S_{12}}{b(a+1)(b+1)} + \frac{4S_{22}}{ab(a+1)(b+1)}.$$

Notice that for the **B**-matrix,

$$\mathbf{B} = \begin{pmatrix} 1 & 1 & 1 & 1 & 1 \\ 0 & ab & (a+1)b & a(b+1) & (a+1)(b+1) \\ 0 & \frac{ab(b-1)}{2} & \frac{(a+1)b(b-1)}{2} & \frac{ab(b+1)}{2} & \frac{(a+1)(b+1)b}{2} \\ 0 & \frac{ab(a-1)}{2} & \frac{(a+1)ab}{2} & \frac{a(a-1)(b+1)}{2} & \frac{(a+1)(b+1)a}{2} \\ 0 & \frac{ab(b-1)(a-1)}{4} & \frac{(a+1)ab(b-1)}{4} & \frac{ab(a-1)(b+1)}{4} & \frac{ab(a+1)(b+1)b}{4} \end{pmatrix}$$

The associated inverse matrix $\mathbf{B_1}^{-1}$ reads

$$\mathbf{B}^{-1} = \begin{pmatrix} 1 & \frac{-4}{(a+1)(b+1)} & \frac{4}{b(a+1)(b+1)} & \frac{4}{a(a+1)(b+1)} & \frac{-4}{ab(a+1)(b+1)} \\ 0 & 1 & \frac{-2}{b} & \frac{-2}{a} & \frac{4}{ab} \\ 0 & \frac{-(a-1)}{a+1} & \frac{2(a-1)}{(a+1)b} & \frac{2}{a+1} & \frac{-4}{b(a+1)} \\ 0 & \frac{-(b-1)}{b+1} & \frac{2}{b+1} & \frac{2(b-1)}{a(b+1)} & \frac{-4}{a(b+1)} \\ 0 & \frac{(a-1)(b-1)}{(a+1)(b+1)} & \frac{-2(a-1)}{(a+1)(b+1)} & \frac{-2(b-1)}{(a+1)(b+1)} & \frac{4}{(a+1)(b+1)} \end{pmatrix}.$$

Denote $\mathbf{b} = (1, S_{11}, S_{12}, S_{21}, S_{22})'$. Since $S_{ij} = v_i w_j$ for $i, j = 1, 2$, we have

$$a = a^* = [2v_2/v_1] + 1 \leq n - 1,$$

$$b = b^* = [2w_2/w_1] + 1 \leq m - 1,$$

thus $\mathbf{B}^{-1}\mathbf{b}$ can be expressed as

$$\mathbf{B}^{-1}\mathbf{b} = \begin{pmatrix} 1 - L(a^*, b^*) \\ \left(v_1 - \frac{2v_2}{a}\right)\left(w_1 - \frac{2}{b}w_2\right) \\ \frac{1}{a+1}(2v_2 - (a-1)v_1)\left(w_1 - \frac{2}{b}w_2\right) \\ \frac{1}{b+1}(2w_2 - (b-1)w_1)\left(v_1 - \frac{2}{a}v_2\right) \\ \frac{1}{(a+1)(b+1)}(2v_2 - (a-1)v_1)(2w_2 - (b-1)w_1) \end{pmatrix}.$$

With the choice of a as a^*, and b as b^*, each of the multiplicative factors is non-negative. Hence

$$0 \leq \frac{2w_2 - (b^* - 1)w_1}{b^* + 1} + w_1 - \frac{2}{b^*}w_2$$

$$= \frac{2w_1}{b^* + 1} - \frac{2w_2}{b^*(b^* + 1)} \leq 1$$

Similarly

$$0 \leq \frac{2w_1}{a^* + 1} - \frac{2w_2}{a^*(a^* + 1)} \leq 1.$$

Notice that when $t = 1$, the first entry of $\mathbf{B}^{-1}\mathbf{b}$ is $1 - L(a^*, b^*) \geq 0$. Under the conditions of Theorem 7.3.1, we have $\mathbf{x}_B = \mathbf{B}^{-1}\mathbf{b} \geq \mathbf{0}$. By Lemma 4.2.2, the bivariate lower bound is Fréchet optimal. □

Theorem 7.3.1 provides sufficient conditions for a factorized lower bound to be Fréchet optimal. However, when the bivariate Bonferroni summations can not be factorized in the specific forms stated in Theorem 7.3.1, the connection between a factorized bound and Fréchet optimal lower bound is unknown. Toward this end, we provide the following result.

Theorem 7.3.2. For consistent bivariate Bonferroni summations $\mathbf{b} = (1, S_{11}, S_{12}, S_{21}, S_{22})'$, denote

$$\mathbf{B} = \begin{pmatrix} 1 & 1 & 1 & 1 & 1 \\ 0 & ab & (a+1)b & a(b+1) & (a+1)(b+1) \\ 0 & \frac{ab(b-1)}{2} & \frac{(a+1)b(b-1)}{2} & \frac{ab(b+1)}{2} & \frac{(a+1)(b+1)b}{2} \\ 0 & \frac{ab(a-1)}{2} & \frac{(a+1)ab}{2} & \frac{a(a-1)(b+1)}{2} & \frac{(a+1)(b+1)a}{2} \\ 0 & \frac{ab(b-1)(a-1)}{4} & \frac{(a+1)ab(b-1)}{4} & \frac{ab(a-1)(b+1)}{4} & \frac{ab(a+1)(b+1)b}{4} \end{pmatrix}.$$

If $\mathbf{B}^{-1}\mathbf{b} \geq \mathbf{0}$ for integers $a = a_0$, $b = b_0$ where $1 \leq a_0 \leq n - 1$, $1 \leq b_0 \leq m - 1$, and

$$t = \max\left(1, \frac{nm(n - 2a_0 - 1)(m - 2b_0 - 1)}{a_0 b_0 (a_0 + 1)(b_0 + 1)}\right) = 1,$$

then there are two sets of events $\{A_i^*\}$, $\{B_j^*\}$ in a probability space such that $L(a_0, b_0)$ is a Fréchet-optimal lower bound.

Proof: The condition of $\mathbf{B}^{-1}\mathbf{b} \geq \mathbf{0}$ implies that there exist two sets of events in a probability space where

$$P(v(A^*) \geq 1, v(B^*) \geq 1) = L(a_0, b_0).$$

Now that $t = 1$, $L(a_0, b_0)$ is a lower bound for $P(v(A) \geq 1, v(B) \geq 1)$ for *any* events $\{A_i\}, i = 1, \cdots, n$, $\{B_j\}, j = 1, \cdots, m$ in *any* probability space with the given $\{S_{ij}\}$. Hence $L(a_0, b_0)$ is Fréchet-optimal. $\square$

Remark: When the conditions of Theorem 7.3.1 are satisfied, (as with the S_{ij}'s initially being determined by a product space, with $S_{ij} = S_i(A) S_j(B)$), if $t = 1$ at $a = a_0 = a^*$, $b = b_0 = b^2$, $L(a^*, b^*)$ is Fréchet-optimal for such S_{ij}'s, which actually is the product of the two Fréchet-optimal Dawson-Sankoff bounds.

Example 7.3.2 Consider a probability space and events where $n = m = 3$, $S_{11} = 0.9$, $S_{12} = S_{21} = 0.009$, $S_{22} = 0$, for which the lower bound $L(1, 1) = 0.882$ as shown in Example 7.1.5. Notice that the lower bound 0.882 is in fact the exact value of the probability $P(v(A) \geq 1, v(B) \geq 1)$ for the probability space generating the example.

Taking $a = b = 1$ results in

$$\mathbf{B}^{-1}\mathbf{b} = (0.118, 0.864, 0.009, 0.009, 0)'$$

so that Theorem 7.3.2 holds and $1 - 0.118 = 0.882$. This also shows that the Fréchet-optimal bound is achieved for any probability space for which the non-zero probabilities read,

$$P(v(A) = 0 \quad \text{or} \quad v(B) = 0) = 0.118, \quad Pr(v(A) = 1, v(B) = 1) = 0.864,$$
$$P(v(A) = 1, v(B) = 2) = P(v(A) = 2, v(B) = 1) = 0.009,$$
$$P(v(A) = 2, v(B) = 2) = 0. \quad \square$$

Conditions in Theorems 7.3.1 and 7.3.2 require $\mathbf{B}^{-1}\mathbf{b} \geq \mathbf{0}$ in conjunction with a product-type probability space. However, these conditions are not always true. In the next section, we introduce an approach to finding the optimal lower bound for any set of consistent Bonferroni summations S_{ij} for $i, j = 1, 2$.

7.4   Bivariate Optimal Algorithm Bounds

Chapter 4 shows the existence of a linear programming lower bound for the bivariate probability. Such lower bounds are in fact Fréchet optimal bivariate lower bounds.

As discussed in Chapter 4, the iteration process successfully leads to an optimal solution that results in the value of a Fréchet optimal lower bound. However, the use of the perturbation term $c_i(\varepsilon)$ makes the procedure difficult to apply. In this section, we provide a practical algorithm seeking such optimal bounds without the involvement of the perturbation term.

Notice that we may set the perturbation term $c_i(\varepsilon)$

$$c_i(\varepsilon) = \varepsilon^i, i \neq 1, \quad c_1(\varepsilon) = 1, \tag{7.13}$$

in every run of the iteration process.

For this specific function of the perturbation term, we construct an iteration algorithm without the involvement of the perturbation term ε as follows.

Setting of the algorithm:

- With given numbers of events m and n, construct the matrix $\mathbf{A}$:

$$\begin{pmatrix} 1 & \cdots & 1 & \cdots & 1 & 1 & \cdots & 1 & \cdots & 1 \\ 0 & \cdots & j & \cdots & m & 2 & \cdots & 2m & \cdots & nm \\ 0 & \cdots & \frac{j(j-1)}{2} & \cdots & \frac{m(m-1)}{2} & 0 & \cdots & 2\frac{m(m-1)}{2} & \cdots & n\frac{m(m-1)}{2} \\ 0 & \cdots & 0 & \cdots & 0 & 1 & \cdots & m & \cdots & \frac{n(n-1)}{2}m \\ 0 & \cdots & 0 & \cdots & 0 & 0 & \cdots & \frac{m(m-1)}{2} & \cdots & \frac{nm(n-1)(m-1)}{4} \end{pmatrix}.$$

- Choose $a = [n/2] + 1, b = [m/2] + 1$.
- Denote vector $\mathbf{b} = (1, S_{11}, S_{12}, S_{21}, S_{22})'$.
- For the integers a and b chosen above, set up $\mathbf{B_1}$.

$$\mathbf{B_1} = \begin{pmatrix} 1 & 1 & 1 & 1 & 1 \\ 0 & ab & (a+1)b & a(b+1) & (a+1)(b+1) \\ 0 & \frac{ab(b-1)}{2} & \frac{(a+1)b(b-1)}{2} & \frac{ab(b+1)}{2} & \frac{(a+1)(b+1)b}{2} \\ 0 & \frac{ab(a-1)}{2} & \frac{(a+1)ab}{2} & \frac{a(a-1)(b+1)}{2} & \frac{(a+1)(b+1)a}{2} \\ 0 & \frac{ab(b-1)(a-1)}{4} & \frac{(a+1)ab(b-1)}{4} & \frac{ab(a-1)(b+1)}{4} & \frac{ab(a+1)(b+1)b}{4} \end{pmatrix}$$

The ith of the iteration algorithm when the B-matrix is $\mathbf{B_i}$

Step 1 Calculate $\mathbf{B_i}^{-1}$.

Step 2 Calculate $\mathbf{x}(i) = \mathbf{B_i}^{-1}\mathbf{b}$.

Step 3 If $\mathbf{x}(i) \geq \mathbf{0}$, stop the process; $\mathbf{x}(i)$ is used to construct the optimal lower bound. If, on the other hand, $\mathbf{x}(i) \ngeq \mathbf{0}$, there exist elements in $\mathbf{x}(i)$, which are negative. Denote by r the location of the smallest element among all negative elements of $\mathbf{x}(i)$.

Step 4 Calculate $y_{rj} = \mathbf{s_r}'\mathbf{a_j}$ where $\mathbf{s_r}'$ is the rth row in $\mathbf{B_i}^{-1}$ and $\mathbf{a_j}$ is the jth column in the matrix $\mathbf{A}$.

Step 5 Select a set of all the indexes of columns $\mathbf{a_j}$ that have negative values: $H = \{j_1, ..., j_t\}$ where $y_{rj_1} < 0, ... y_{rj_t} < 0$.

Step 6 Use the sub-scheme A to choose an appropriate vector $\mathbf{a_{p*}}$ to replace $\mathbf{b_r}$ and form a new B-matrix, $\mathbf{B_{i+1}}$.

Step 7 With this newly formed B-matrix, $\mathbf{B_{i+1}}$, repeat Steps **1** to **6** until the optimal probability space is reached.

Figure (7.1) provides a flow-chart for the main iteration algorithm.

Note: Theorem 4.2.5 ensures that the above iteration process eventually stops after a finite number of iterations.

Sub-scheme A for **Step 6** in the iteration process:

- Calculate $\max\{y_{1j_1}/y_{rj_1}, ..., y_{1j_t}/y_{rj_t}\}$.
- If the maximum point is unique, say at $\mathbf{a}_{j_k}$, we take $p^* = j_k$ and replace the rth column of $\mathbf{B}$ with $\mathbf{a}_{p^*}$ (equivalently $\mathbf{a}_{j_k}$) and thereby gain a new $\mathbf{B}$-matrix, $\mathbf{B}_{i+1}$;
- If, on the other hand, a tie occurs, there exist integers p_i, $i \geq 2$, (with $p_i \neq r$, by Lemma 4.2.4.2), such that

$$\frac{y_{1p_1}}{y_{rp_1}} = \frac{y_{1p_2}}{y_{rp_2}} = \frac{y_{1p_3}}{y_{rp_3}} = \cdots .$$

Use the next step to appropriately choose a column from the matrix $\mathbf{A}$ to replace $\mathbf{b}_r$ when a tie occurs.

- Consider $\{p_i, i \geq 2\} = \{p_1, p_2, p_3\}$ for convenience. The approach for the case where three or more vectors are involved in a tie will be self-evident. Under this scenario,

$$\frac{y_{1p_1}}{y_{rp_1}} = \frac{y_{1p_2}}{y_{rp_2}} = \frac{y_{1p_3}}{y_{rp_3}}.$$

By the definition of $\theta(\varepsilon)$ in Chapter 4 and the selection of the specific function of $c_i(\varepsilon)$ in (7.13), we have

$$
\begin{aligned}
\theta(\varepsilon) &= \max_{p_1, p_2, p_3} \left(\frac{y_{1p}}{y_{rp}} + \sum_{i=2}^{5} c_{t_i}(\varepsilon) \frac{y_{ip}}{y_{rp}} - \frac{c_p(\varepsilon)}{y_{rp}} \right) \\
&= \max_{p_1, p_2, p_3} \left(\frac{y_{1p}}{y_{rp}} + \sum_{i=2}^{5} \varepsilon^{t_i} \frac{y_{ip}}{y_{rp}} - \frac{\varepsilon^p}{y_{rp}} \right) \\
&= \max_{p_1, p_2, p_3} \left(\frac{y_{1p}}{y_{rp}} + \sum_{i=2, i\neq r}^{5} \varepsilon^{t_i} \frac{y_{ip}}{y_{rp}} + \varepsilon^{t_r} - \frac{\varepsilon^p}{y_{rp}} \right), \qquad (7.14)
\end{aligned}
$$

where $y_{rp} < 0$ and the corresponding location of the column from $\mathbf{A}$ acting as $\mathbf{b}_r$ in $\mathbf{B}$, t_r, is one of the integers in the set $\{t_2, ..., t_5\}$.

- Check whether $t_i > \min(p_1, p_2, p_3)$ for each $t_i \in \{t_2, ...t_5\} \backslash \{t_r\}$. If so, by (7.14), $p^* = \min(p_1, p_2, p_3)$; if not, also according to (7.14), select the smallest value among $\{t_2...t_5\} \backslash \{t_r\}$, say t_{i*}, so $t_{i*} < \min(p_1, p_2, p_3)$, (by Lemma 4.2.4.2, equality is not possible).
- Find

$$\max_{p_1, p_2, p_3} \frac{y_{i*p}}{y_{rp}}. \qquad (7.15)$$

If the maximizing value of p (out of p_1, p_2, p_3) in (7.15) is unique, select this maximizing value as p^*. If non-unique, discard any non-maximizing value (justified by considering all ε small enough in (7.14), and denote the remaining p values (which consist of either two out of p_1, p_2, p_3; or the whole three of p_1, p_2, p_3) by K^*.

- Suppose t_{i**} is the second smallest of $\{t_2, ..., t_5\} \backslash \{t_r\}$. If $t_{i**} > \min\{p \in K^*\}$, by (7.14), take $p^* = \min\{p \in K^*\}$.

- If $t_{i**} < \min\{p \in K^*\}$, consider

$$\max_{p \in K^*} \frac{y_{i**p}}{y_{rp}}.$$

If the maximum is achieved for a unique value p, then take this p as p^*; if non-unique, discard any non-maximizing $p \in K^*$, and form a new set K^{**}, $K^{**} \subseteq K^*$ of the remaining p's.

Note 1: The selecting process in sub-scheme A must end, at worst, with a unique maximum at the largest t_{i***} out of $\{t_2,...,t_5\}\backslash\{t_r\}$, for otherwise, we will have co-incidence of $\frac{y_{ip}}{y_{rp}}$ for each t_i of the set $\{t_1,...,t_5\}$ and two distinct p's, which is a contradiction to Theorem 4.2.3.

Note 2: The selected index p^* clearly is the one producing $\theta(\varepsilon)$ for all ε small enough; however, we use the index as the proxy of ε in $\theta(\varepsilon)$ to select the appropriate column from $\mathbf{A}$ toward the optimal solution in the iteration process.

For convenience, denote t_2, ..., t_5 locations of columns in matrix $\mathbf{A}$ forming a current $\mathbf{B}$-matrix, denote p_1, p_2, ... locations of columns in $\mathbf{A}$ involving in the tie of $\max \frac{y_{1j}}{y_{rj}}$ with current $\mathbf{B}$ used. Figure 7.2 provides a flow-chart for the implementation of sub-scheme A.

Besides finding the optimal lower bound, we will use an example to show that the iteration procedure can also be used to examine the consistency of a given set of bivariate Bonferroni summations.

Now, we use examples to illustrate the application of the optimizing algorithm described above.

Example 7.4.1 (Avoiding a dead cycle caused by degeneracy): Consider the case where $n = m = 6$ and the given S_{ij}'s are $S_{11} = 2.001$, $S_{12} = 3.006$, $S_{21} = 3.001$, $S_{22} = 4.421$. We want to find a Fréchet optimal lower bound for $P(v_1 \geq 1, v_2 \geq 1)$, and examine whether the given Bonferroni summations S_{ij}'s are consistent. Once the algorithm identifies a probability space associated with a set of given Bonferroni summations, the consistency of the Bonferroni summations follows. In practice, for various reasons, the observed values of the Bonferroni summations may drift away from the true values. Since the input for a linear programming bound is essentially just the values of the S_{ij}'s, theoretically it is desirable to check the consistency to ensure the validity of the output of the algorithm.

First we will follow the main-scheme (and the sub-scheme when needed) to demonstrate how it works. In this example, since $m = n = 6$, the corresponding $\mathbf{A}$ matrix reads

$$\mathbf{A} = \begin{pmatrix} 1 & 1 & 1 & ... & 1 & 1 & ... & 1 & ... & 1 & 1 & ... & ... & 1 \\ 0 & 1 & 2 & ... & 6 & 2 & ... & 12 & ... & 18 & 14 & ... & ... & 36 \\ 0 & 0 & 1 & ... & 15 & 0 & ... & 30 & ... & 45 & 0 & ... & ... & 90 \\ 0 & 0 & 0 & ... & 0 & 1 & ... & 6 & ... & 18 & 6 & ... & ... & 90 \\ 0 & 0 & 0 & ... & 0 & 0 & ... & 15 & ... & 45 & 0 & ... & ... & 225 \end{pmatrix}$$

Main-scheme of iteration process

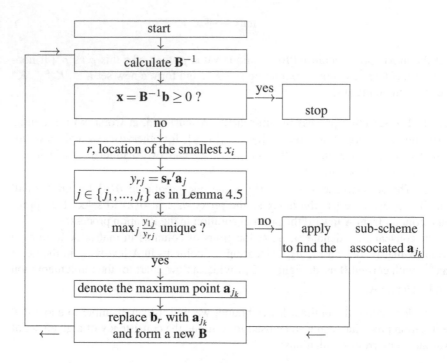

Figure 7.1 *Iteration procedure*

and $a = b = 6/2 + 1 = 4$, so the initial B matrix reads

$$\mathbf{B_1} = \begin{pmatrix} 1 & 1 & 1 & 1 & 1 \\ 0 & 16 & 20 & 20 & 25 \\ 0 & 24 & 30 & 40 & 50 \\ 0 & 24 & 40 & 30 & 50 \\ 0 & 36 & 60 & 60 & 100 \end{pmatrix}$$

which is formed by the first, 23th, 29th, 24th and 30th columns of $\mathbf{A}$. Denote it by

$$\mathbf{B_1} = (\mathbf{a}_1, \mathbf{a}_{23}, \mathbf{a}_{29}, \mathbf{a}_{24}, \mathbf{a}_{30}).$$

Then

$$\mathbf{x} = \mathbf{B_1}^{-1}\mathbf{b} = (0.876, 0.103, 0.107, 0.018, -0.014)',$$

so $r = 5$ (r is the index corresponding to the smallest elements of $\mathbf{x}$). The associated index set of indexes that have negative values of y_{rj}, $H(\mathbf{B_1})$, reads

$$\{j_1, ..., j_t\} = \{13, 33, 7, 19, 32, 34, 12, 27, 6, 18, 26, 28\}$$

Sub-scheme A —- Selection of Replacing Columns

*One way implementation of selecting a column in the matrix **A** to update the **B**-matrix in the iteration process in Figure 7.1*

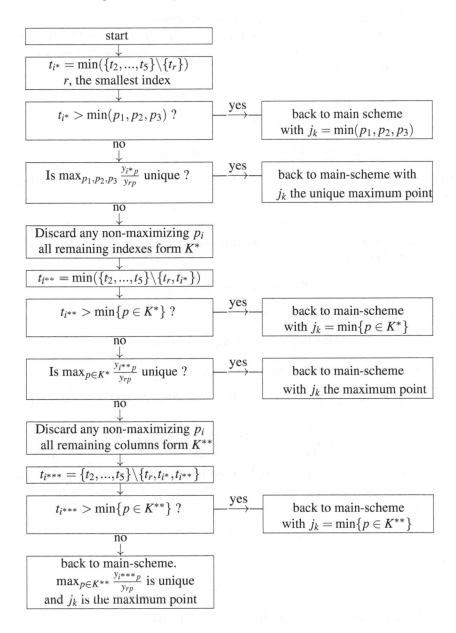

Figure 7.2 *Supplementary algorithm*

Now a tie for maximum occurs between the 19th and the 34th columns of $\mathbf{A}$, since

$$\frac{y_{1,19}}{y_{5,19}} = \frac{y_{1,34}}{y_{5,34}} = -0.132,$$

so we need to apply the selecting procedure in sub-scheme A.

Let $W = \{t_2, ..., t_5\} \setminus \{t_r\} = \{23, 29, 24\}$; for any $t_i \in W$, $t_i > \min(19, 34)$, so by the sub-scheme, we select column 19 in $\mathbf{A}$ to replace the fifth column in $\mathbf{B}_1$, and the associated $\mathbf{B}$ matrix then becomes

$$\mathbf{B}_2 = (\mathbf{a}_1, \mathbf{a}_{23}, \mathbf{a}_{29}, \mathbf{a}_{24}, \mathbf{a}_{19}).$$

Here $\mathbf{x} = \mathbf{B}_2^{-1}\mathbf{b}$ in the second iteration is

$$(0.874, 0.125, 0.009, -0.017, 0.010)',$$

thus $r = 4$, and $\{j_1, ..., j_t\} = \{33, 32, 34, 27, 26, 28, 7, 13, 21, 20, 22, 6, 12\}$ for $H(\mathbf{B}_2)$, notice that

$$\max_{p \in H(\mathbf{B}_2)} \frac{y_{1p}}{y_{rp}} = \frac{y_{1,34}}{y_{4,34}} = 0,$$

so column 34 in $\mathbf{A}$, which uniquely gives the maximum, can be used to replace column 4 in $\mathbf{B}_2$, to get $\mathbf{B}_3$:

$$\mathbf{B}_3 = (\mathbf{a}_1, \mathbf{a}_{23}, \mathbf{a}_{29}, \mathbf{a}_{34}, \mathbf{a}_{19}).$$

Here, $\mathbf{x} = \mathbf{B}_3^{-1}\mathbf{b}$ in the third run, reads

$$(0.874, 0.125, -0.008, 0.006, 0.003)',$$

so $r = 3$, and the corresponding index of $H(\mathbf{B}_3)$, $\{j_1, ..., j_t\}$ equals to

$$\{7, 13, 32, 33, 12, 27, 6, 26, 11, 21, 16, 5, 10, 15, 17, 20, 22, 9, 4, 14, 18, 28, 3, 8, 2\},$$

$$\max_{p \in H(\mathbf{B}_3)} \frac{y_{1p}}{y_{3p}} = \frac{y_{1,18}}{y_{3,18}} = \frac{y_{1,28}}{y_{3,28}} = -0.056.$$

Thus, a tie between columns 18 and 28 occurs. By the sub-scheme, note that $\min(18, 28) = 18$, $\{t_2, ..., t_5\} \setminus \{t_r\} = \{23, 34, 19\}$, we can use column 18 to replace the third column of $\mathbf{B}_3$, thus

$$\mathbf{B}_4 = (\mathbf{a}_1, \mathbf{a}_{23}, \mathbf{a}_{18}, \mathbf{a}_{34}, \mathbf{a}_{19}).$$

Now, $\mathbf{x} = \mathbf{B}_4^{-1}\mathbf{b}$ in the 4th run, reads

$$(0.874, 0.104, 0.022, 0.006, -0.006)'$$

$r = 5$, the associated index set of $H(\mathbf{B}_4)$ for $\{j_1, ..., j_t\}$ reads

$$\{32, 33, 27, 21, 26, 15, 16, 10, 11, 20, 22, 7, 6, 9, 12, 5, 14, 17, 4, 28, 8, 13, 3, 2\},$$

$$\max_{p \in H(\mathbf{B_4})} \frac{y_{1p}}{y_{rp}} = \frac{y_{1,28}}{y_{5,28}} = 0,$$

uniquely, so column 28 can be used to form $\mathbf{B_5}$:

$$\mathbf{B_5} = (\mathbf{a}_1, \mathbf{a}_{23}, \mathbf{a}_{18}, \mathbf{a}_{34}, \mathbf{a}_{28}).$$

With $\mathbf{B_5}$ being considered,

$$\mathbf{x}_0 = \mathbf{B_5}^{-1}\mathbf{b} = (0.874, 0.104, 0.012, 0.000, 0.011)' \geq \mathbf{0}.$$

By Lemma 4.2.2, we get an appropriate probability space $\mathbf{x}_0$ (above) associated with the given $\mathbf{b}$ and the Fréchet optimal lower bound $(1 - 0.874 = 0.126)$.

Example 7.4.2 (When a dead cycle is caused by degeneracy): From the second run and the 4th run in the preceding example, degeneracy occurs where

$$\max_{p \in H(\mathbf{B})} \frac{y_{1p}}{y_{rp}} = 0,$$

caused by $y_{1,34} = 0$ or $y_{1,28} = 0$. If we select a column from $\mathbf{A}$ to replace the associated column in $\mathbf{B}$ (and hence form a new $\mathbf{B}$ for next iteration) by simply choosing any p satisfying

$$\max_{p \in Z} \frac{y_{1p}}{y_{rp}},$$

arbitrarily, where $Z = \{p : \frac{y_{1p}}{y_{rp}} \leq 0, \mathbf{a}_p \notin \mathbf{B}\}$, a dead cycle may emerge as demonstrated below.

Starting with

$$\mathbf{B}(1) = (\mathbf{a}_1, \mathbf{a}_{23}, \mathbf{a}_{29}, \mathbf{a}_{24}, \mathbf{a}_{30}).$$

$$\mathbf{B}(1)^{-1}\mathbf{b} = (0.876, 0.103, 0.107, 0.018, -0.014)',$$

the associated maximum value is

$$\frac{y_{1,19}}{y_{5,19}} = \frac{y_{1,34}}{y_{5,34}} = -0.132,$$

according to the new criterion of arbitrary selection, either cloumn 19 or column 34 in $\mathbf{A}$ can be selected. If we choose $\mathbf{a}_{34}$,

$$\mathbf{B}(2) = (\mathbf{a}_1, \mathbf{a}_{23}, \mathbf{a}_{29}, \mathbf{a}_{24}, \mathbf{a}_{34}).$$

Then

$$\mathbf{B}(2)^{-1}\mathbf{b} = (0.874, 0.125, -0.017, 0.009, 0.010)'$$

$r = 3$ and

$$\max_{p \in Z} \frac{y_{1p}}{y_{rp}} = \frac{y_{1,19}}{y_{3,19}} = 0,$$

uniquely, so column 19 is selected, and

$$\mathbf{B}(3) = (\mathbf{a}_1, \mathbf{a}_{23}, \mathbf{a}_{19}, \mathbf{a}_{24}, \mathbf{a}_{34}).$$

Now

$$\mathbf{B}(3)^{-1}\mathbf{b} = (0.874, 0.125, 0.006, -0.008, 0.003)'$$

$r = 4$ and

$$\max_{p \in Z} \frac{y_{1p}}{y_{rp}} = \frac{y_{1,29}}{y_{4,29}} = 0,$$

uniquely, so $\mathbf{a}_{29}$ is selected to replace $\mathbf{a}_{24}$.

$$\mathbf{B}(4) = (\mathbf{a}_1, \mathbf{a}_{23}, \mathbf{a}_{19}, \mathbf{a}_{29}, \mathbf{a}_{34}).$$

Therefore

$$\mathbf{B}(4)^{-1}\mathbf{b} = (0.874, 0.125, 0.003, -0.008, 0.006)',$$

$r = 4$ and

$$\max_{p \in Z} \frac{y_{1p}}{y_{rp}} = \frac{y_{1,24}}{y_{4,24}} = 0,$$

uniquely, $\mathbf{a}_{24}$ is selected to replace $\mathbf{a}_{29}$ by the new criterion.

$$\mathbf{B}(5) = (\mathbf{a}_1, \mathbf{a}_{23}, \mathbf{a}_{19}, \mathbf{a}_{24}, \mathbf{a}_{34}).$$

$\mathbf{B}(5)$ is exactly the same as $\mathbf{B}(3)$ and a dead cycle would then run infinitely. The pattern of the procedure would be described as

$$\mathbf{B}(1) \to \mathbf{B}(2) \to \mathbf{B}(3) \to \mathbf{B}(4) \to \mathbf{B}(3) \to \dots.$$

The problem is caused by the changing of the criterion selecting $\mathbf{a}_j$ to form the new $\mathbf{B}$. In fact, with our algorithm as exposited in the last section, at the 4th iteration, column 29 can not be selected because $y_{4,29} = 1 > 0$ which means that $\mathbf{a}_{29}$ does not belong to our $H(\mathbf{B})$ in Lemma 4.5, so it should not be chosen. In the preceding, we see that the dead cycle can be efficiently avoided by following our algorithm. □

Example 7.4.3 (Optimal factorized bounds) This example will confirms that the value of the bound in Example 7.1.5 agrees with a Fréchet optimal lower bound for events in an associated probability space (which is different from the probability space in Example 7.1.5).

We start with the Bonferroni summations S_{ij}'s ($i = 1, 2; j = 1, 2$) in Example 7.1.5. For these S_{ij}'s the value of the Fréchet optimal bound (0.882) will be seen to coincide with the value produced by Theorem 7.1.3. However, this does not mean that the bound in Theorem 7.1.3 is Fréchet optimal for arbitrary consistent Bonferroni summations S_{ij}'s. Moreover, even in the case with the special values of S_{ij}'s of Example 7.1.5, the probability space produced by the algorithm is different from the probability space producing the S_{ij}'s in Example 7.1.5.

For $n = m = 3$ and $\mathbf{b} = (1, 0.9, 0.009, 0.009, 0)'$ (see Example 7.1.5), to find the Fréchet optimal lower bound, we consider the following.

$n = m = 3$, the associated $\mathbf{A}$ matrix reads

$$\mathbf{A} = \begin{pmatrix} 1 & 1 & 1 & 1 & 1 & 1 & 1 & 1 & 1 & 1 \\ 0 & 1 & 2 & 3 & 2 & 4 & 6 & 3 & 6 & 9 \\ 0 & 0 & 1 & 3 & 0 & 2 & 6 & 0 & 3 & 9 \\ 0 & 0 & 0 & 0 & 1 & 2 & 3 & 3 & 6 & 9 \\ 0 & 0 & 0 & 0 & 0 & 1 & 3 & 0 & 3 & 9 \end{pmatrix},$$

and $a = b = [3/2] + 1 = 2$, the $\mathbf{B}$ matrix correspondingly reads

$$\mathbf{B_1} = \begin{pmatrix} 1 & 1 & 1 & 1 & 1 \\ 0 & 4 & 6 & 6 & 9 \\ 0 & 2 & 6 & 3 & 9 \\ 0 & 2 & 3 & 6 & 9 \\ 0 & 1 & 3 & 3 & 9 \end{pmatrix}$$

or

$$\mathbf{B_1} = (\mathbf{a}_1, \mathbf{a}_6, \mathbf{a}_7, \mathbf{a}_9, \mathbf{a}_{10}).$$

Now

$$\mathbf{x} = \mathbf{B_1}^{-1}\mathbf{b} = (0.604, 0.882, -0.291, -0.291, 0.096)', \text{ so } r - 3,$$

$$\max_{p \in H(\mathbf{B}_1)} \frac{y_{1p}}{y_{3p}} = \frac{y_{1,5}}{y_{3,5}} = -1 > -1.667,$$

uniquely. By the algorithm, we can choose $\mathbf{a}_5$ to replace the third column in $\mathbf{B}_1$ and form $\mathbf{B}_2$,

$$\mathbf{B_2} = (\mathbf{a}_1, \mathbf{a}_6, \mathbf{a}_5, \mathbf{a}_9, \mathbf{a}_{10}).$$

The second run of the procedure :

$$\mathbf{x} = \mathbf{B_2}^{-1}\mathbf{b} = (0.313, 0.009, 0.873, -0.291, 0.096)', \text{ so } r = 4, \text{ since}$$

$$\max_{p \in H(\mathbf{B}_2)} \frac{y_{1p}}{y_{4p}} = \frac{y_{1,2}}{y_{4,2}} = -0.667 > -1,$$

uniquely. By the algorithm, we can choose $\mathbf{a}_2$ to replace the 4th column in $\mathbf{B}_2$ and form $\mathbf{B}_3$,

$$\mathbf{B_3} = (\mathbf{a}_1, \mathbf{a}_6, \mathbf{a}_5, \mathbf{a}_2, \mathbf{a}_{10}).$$

The third run of the procedure : $\mathbf{x} = \mathbf{B_3}^{-1}\mathbf{b} = (0.119, 0.009, 0, 0.873, -0.001)'$, so $r = 5$, since

$$\max_{p \in H(\mathbf{B}_2)} \frac{y_{1p}}{y_{3p}} = \frac{y_{1,3}}{y_{5,3}} = -1 > -4,$$

uniquely. By the algorithm, we can choose $\mathbf{a}_3$ to replace the fifth column in $\mathbf{B}_3$ and form $\mathbf{B}_4$,

$$\mathbf{B_4} = (\mathbf{a}_1, \mathbf{a}_6, \mathbf{a}_5, \mathbf{a}_9, \mathbf{a}_3).$$

Now $\mathbf{x_0} = \mathbf{B_4}^{-1}\mathbf{b} = (0.118, 0.000, 0.009, 0.864, 0.009)$, the corresponding optimal bound is $1 - 0.118 = 0.882$, which is the same as the value of the bound computed by Theorem 7.1.3. Now, by Lemma 4.2.2, the associated probability space constructed from $\mathbf{x_0}$ reads,

$$(0.118, 0, 0.009, 0, 0.009, 0, 0, 0, 0.864, 0)$$

is different from the one given in Example 7.1.5:

$$(0.118, 0.864, 0.009, 0, 0.009, 0, 0, 0, 0, 0),$$

which is calculated directly from Example 7.1.5, for instance $p_{11} = P(v_1 = 1, v_2 = 1) = 9 \times 0.096 = 0.864$. $\square$

Example 7.4.4 (A factorized bound that is not optimal for any integers): This example addresses the question about the bivariate optimality of the Dawson-Sankoff-type bound under the constraint of $1 \le a \le n$, $1 \le b \le m$; we go through all permissible pairs of integers (a,b)'s and calculate all the corresponding values of the bivariate Dawson-Sankoff-type bounds, then compare these values with the bound derived from our algorithm in Figure 7.1. The result shows that the bivariate Dawson-Sankoff-type lower bound is not a Fréchet optimal lower bound.

Specifically, the focus of this example is that there exist $S_{11}, S_{12}, S_{21}, S_{22}$, coming from a probability space in which Fréchet optimal lower bound is reached, but this optimal value is strictly greater than $T(a,b)$ for all $1 \le a \le n$, and $1 \le b \le m$, where

$$T(a,b) = \frac{4S_{1,1}}{(a+1)(b+1)} - \frac{4S_{2,1}}{a(a+1)(b+1)} - \frac{4S_{12}}{b(a+1)(b+1)} + \frac{4S_{22}}{ab(a+1)(b+1)}.$$

Consider the optimal lower bound for events satisfying $n = m = 3$, and

$$\mathbf{b} = (1, S_{11}, S_{12}, S_{21}, S_{22})' = (1, 0.738, 0.522, 0.669, 0.492)'.$$

The associated $\mathbf{A}$ matrix reads

$$\mathbf{A} = \begin{pmatrix} 1 & 1 & 1 & 1 & 1 & 1 & 1 & 1 & 1 & 1 \\ 0 & 1 & 2 & 3 & 2 & 4 & 6 & 3 & 6 & 9 \\ 0 & 0 & 1 & 3 & 0 & 2 & 6 & 0 & 3 & 9 \\ 0 & 0 & 0 & 0 & 1 & 2 & 3 & 3 & 6 & 9 \\ 0 & 0 & 0 & 0 & 0 & 1 & 3 & 0 & 3 & 9 \end{pmatrix}$$

and with $a = b = [3/2] + 1 = 2$, the $\mathbf{B}$ matrix reads

$$\mathbf{B_1} = \begin{pmatrix} 1 & 1 & 1 & 1 & 1 \\ 0 & 4 & 6 & 6 & 9 \\ 0 & 2 & 6 & 3 & 9 \\ 0 & 2 & 3 & 6 & 9 \\ 0 & 1 & 3 & 3 & 9 \end{pmatrix}$$

or

$$\mathbf{B}_1 = (\mathbf{a}_1, \mathbf{a}_6, \mathbf{a}_7, \mathbf{a}_9, \mathbf{a}_{10}).$$

Thus,

$$\mathbf{x} = \mathbf{B}_1^{-1}\mathbf{b} = (0.882, 0.039, -0.003, 0.046, 0.036)'$$

so $r = 3$,

$$\max_{p \in H(\mathbf{B}_1)} \frac{y_{1p}}{y_{rp}} = \frac{y_{15}}{y_{r5}} = -1$$

uniquely, so we select $\mathbf{a}_5$ to replace the third column of $\mathbf{B}_1$, and get

$$\mathbf{B}_2 = (\mathbf{a}_1, \mathbf{a}_6, \mathbf{a}_5, \mathbf{a}_9, \mathbf{a}_{10}).$$

then

$$\mathbf{x}_0 = \mathbf{B}_2^{-1}\mathbf{b} = (0.879, 0.030, 0.009, 0.046, 0.036)' \geq \mathbf{0}$$

so the given S_{ij}'s has an optimal lower bound $1 - 0.879 = 0.121$, with the probabilities of associated events assigned from $\mathbf{x}_0$.

Note that $\mathbf{B}_2$ has its five entries with associated pairs of integers as $(0,0)$, $(2,2)$, $(2,1)$, $(2,3)$, $(3,3)$ which are not of the form $(0,0)$, (a,b), $(a+1,b)$, $(a,b+1)$, $(a+1,b+1)$ for any integer a,b.

Since $n = m = 3$ here, we have $3 \times 3 = 9$ pairs of permissible (a,b)'s. We list the corresponding $T(a,b)$ in Table 7.2.

Table 7.2 *Values of a bivariate lower bound with different parameters*

a	b	$T(a,b)$
1	1	0.039
1	2	0.036
1	3	0.029
2	1	0.085
2	2	0.118
2	3	0.104
3	1	0.078
3	2	0.112
3	3	0.099

We can see from here that for all permissible a,b used, $T(a,b) < 0.121$. Therefore, form $T(a,b)$ does not coincide with an optimal bound. □

Example 7.4.5 (Comparing product space, factorized bounds, and optimal lower bounds): In this example, we set $n = m = 4$, and the probability space generating the example is a product space. Consider first a single set of 4 events $\{A_1, A_2, A_3, A_4\}$ (The construction of a probability space amounts to assigning probabilities for the 2^4

Table 7.3 *Numerical example on optimal bivariate bounds*

$P(A_1 A_2 A_3^c A_4^c) = 0.05$	$P(A_1 A_2^c A_3 A_4^c) = 0.10$	$P(A_1 A_2^c A_3^c A_4^c) = 0.15$
$P(A_1^c A_2 A_3^c A_4^c) = 0.15$	$P(A_1^c A_2^c A_3 A_4) = 0.05$	$P(A_1^c A_2^c A_3 A_4^c) = 0.15$
$P(A_1^c A_2^c A_3^c A_4^c) = 0.2$	$P(A_1^c A_2^c A_3^c A_4) = 0.15$	

elementary conjunctions of events $A_1, \cdots, A_4$ and their complements). For convenience of illustration, all of the elementary conjunctions with non-zero probabilities are specified in Table 7.3.

Such an assignment results in $S_1(A) = 1$ and $S_2(A) = 0.2$.

Now we consider corresponding events in the copy of the original space denoted by $B_1, \cdots, B_4$. Thus $S_1(B) = 1, S_2(B) = 0.2$. When considering $A_1, \cdots, A_4$ and $B_1, \cdots, B_4$ in the product probability space, $S_{11} = S_1(A) S_1(B) = 1$, $S_{12} = S_{21} = 0.2$, $S_{22} = 0.04$. The associated bounds are tabulated as t, $T(a,b)$ and $t^{-1}T(a,b)$ in Table 7.4.

Table 7.4 *Values of a bivariate optimal lower bound*

a	b	t	$T(a,b)$	$t^{-1}T(a,b)$
1	1	4	0.64	0.16
1	2	1	0.48	0.48
1	3	1	0.37	0.37
1	4	1	0.30	0.30
2	2	1	0.36	0.36
2	3	1	0.28	0.28
2	4	1	0.23	0.23
3	3	1	0.22	0.22
3	4	1	0.18	0.18
4	4	1	0.14	0.14

Thus, for these bivariate Bonferroni summations $\{S_{ij}\}$, $i, j = 1, 2$, the highest value of the explicit bound is 0.48. Now, from the construction of the example itself, considering the product of Fréchet optimal Dawson-Sankoff "one dimensional" bounds, the product of two univariate Fréchet optimal lower bound is 0.64. However, the product $T(a,b)$ is not a lower bound since $t > 1$. In fact, applying the algorithm in Figure 7.1, we find that with the corresponding vector $\mathbf{b}' = (1,1,0.2,0.2,0.04)$ the Fréchet optimal bound is lower: $1 - 0.363333 = 0.636667$, and corresponds to the 5×5 sub-matrix

$$\mathbf{B} = \begin{bmatrix} 1 & 1 & 1 & 1 & 1 \\ 0 & 1 & 2 & 2 & 16 \\ 0 & 0 & 0 & 1 & 24 \\ 0 & 0 & 1 & 0 & 24 \\ 0 & 0 & 0 & 0 & 36 \end{bmatrix}$$

of **A**. Consequently

$$\mathbf{B}^{-1}\mathbf{b} = \left(\frac{109}{300}, \frac{13}{45}, \frac{13}{75}, \frac{13}{75}, \frac{1}{900} \right)'.$$

In fact, the value of components of the vector **p** reads

$$\begin{pmatrix} (0,0) & (1,1) & (2,1) & (3,1) & (4,1) & (1,2) & (2,2) & (3,2) & (4,2) \\ \frac{109}{300}, & \frac{13}{45}, & \frac{13}{75}, & 0, & 0, & \frac{13}{75}, & 0, & 0, & 0, \\ \\ (1,3) & (2,3) & (3,3) & (4,3) & (1,4) & (2,4) & (3,4) & (4,4) \\ 0, & 0, & 0, & 0, & 0, & 0, & 0 & \frac{1}{900} \end{pmatrix}$$

which cannot correspond to a product space, even though the original S_{ij}'s were generated from a product space example.

7.5 Applications in Seasonal Trend Analysis

Applications of Bonferroni lower bounds include the recognition of seasonal trends in a time series and identification of an annual peak period of a disease in epidemiology as discussed in David and Newell (1965), Glaz, Guerriero, and Sen (2010), and Chen (2003). The detection of seasonal effects on the occurrences of events over disjoint discrete time intervals works as follows.

For the period of time under investigation, assume that it is plausible to divide the time interval of interest into $2n$ sub-periods such that the number of occurrences in each sub-period is observable. Denote N_i the number of occurrences over n sub-periods starting from the sub-period i, $i = 1, ..., n$. Under this scenario, David and Newell considered the following statistic

$$Z = \max_{1 \leq i \leq n} \left| \frac{N_i - (N - N_i)}{\sqrt{N}} \right|,$$

where N is the total number of occurrences through the time intervals of interest. $N_i - (N - N_i)$ measures the difference between occurrences and non-occurrences of the events in the sub-period i. The sub-period corresponding to the maximum difference is the sample peak period.

Now, for each fixed value N, assume that the number of occurrences $N_1, ..., N_n$ are binomial random variables with parameters N and $p = 1/2$ in each sub-period, David and Newell (1965) used the statistics $Z_i = (2N_i - N)/\sqrt{N}$ to make inference on the annual peak period.

Notice that $Z_1, ..., Z_n$ are asymptotically standard normal distributed, and the correlation coefficient between Z_i and Z_{i+h} is $\rho_{i,i+h} = 1 - (2h)/n$. Denote $A_i = \{|Z_i| > z\}$. The distribution of $Z = \max_{1 \leq i \leq n} |Z_i|$ can be evaluated using upper and lower bound as follows.

$$\sum_{i=1}^{n} P(A_i) - \sum_{1 \le i < j \le n} P(A_i A_j)$$

$$\le \quad P(Z > z)$$

$$\le \quad \sum_{i=1}^{n} P(A_i) - \max_{\tau^*} \sum_{\tau^*} P(A_i A_j),$$

where the upper bound is the Worsley/Hunter's bound that we discussed in the previous chapter. And τ^* is the set of spanning trees with events vertexes $v_1, \ldots, v_n$ representing the events A_i for $i = 1, \ldots, n$, respectively. Notice that the lower bound here is the traditional Bonferroni degree-two lower bound. For the use of inequalities in U-statistic, see Weber (2011); in scan statistic, see Wallenstein et al. (1994), Glaz, Guerriero, and Sen (2010), Guerriero at al. (2010).

When more than one type of disease is under consideration, the above discussion is correspondingly extended to a bivariate or vector-variate setting where the evaluation of the probability of the annual peak period necessitates bivariate lower bounds discussed in this chapter.

Chapter 8

Multivariate and Hybrid Lower Bounds

This chapter goes beyond bounding results on Bonferroni-type bivariate lower bounds. It extends the previous chapter in two directions. The first one is in the direction of multivariate lower bounds that focus on high dimension settings with any p sets of events. Literature in this regard includes the work of Chen and Seneta (1996, 1988, 2000), and Galambos and Lee (1992, 1994), among others. The second direction of extension improves bivariate lower bounds by using information beyond the Bonferroni summations. In particular, we concentrate on product-type lower bounds and lower bounds using information on Hamilton circuits in Sections 8.2 and 8.3.

8.1 High Dimension Lower Bounds

In multivariate setting, consider d sets of events, A_{ij}, $i = 1, ..., n_j$ for $j = 1, ..., d$ respectively. Let $\mathbf{v} = (v_1, ..., v_d)'$, where v_j is the number of events $\{A_{1j}, ..., A_{n_jj}\}$ that occur for $j = 1, ..., d$. For a vector of integers $\mathbf{r} = (r_1, ..., r_d)'$, this section focuses on lower bounds for $P(\mathbf{v} = \mathbf{r})$ using

$$S_{\mathbf{k}} = E(\prod_{i=1}^{d} \binom{v_i}{k_i}) = E(\binom{\mathbf{v}}{\mathbf{k}}),$$

where $\mathbf{k} = (k_1, ..., k_d)'$ is a vector of non-negative integers.

Consider the high dimension lower bound discussed in Meyer (1969),

$$P(v_1 = r_1 ... v_d = r_d) \geq \sum_{t=\Sigma r_j}^{\Sigma r_j + 2k + 1} \sum_{\Sigma i_j = t} (-1)^{t - \Sigma r_j} \prod_{j=1}^{d} \binom{i_j}{r_j} S_{i_1, ..., i_j}$$

where $r_j \leq i_j \leq n_j$, $1 \leq j \leq d$.

Similar to the discussion of high dimension upper bounds in Chapter 6, in this section, we discuss multivariate inequalities which use degree p Bonferroni summations; the quantity involves the computation of p dimensional joint probabilities.

$$D_p = \sum_{|\mathbf{k}| = p} \binom{\mathbf{k}}{\mathbf{r}} S_{\mathbf{k}},$$

with $|\mathbf{k}| = \sum_{i=1}^{d} k_i$, $\mathbf{k} = (k_1,...,k_d)'$ and

$$\binom{\mathbf{k}}{\mathbf{r}} = \prod_{i=1}^{d}\binom{k_i}{r_i}, \qquad S_{\mathbf{k}} = S_{k_1,...,k_d}.$$

One of the vector-form inequalities is from Galambos and Xu (1996), aiming at extension of univariate Sobel-Uppuluri-Galambos bounds.

$$P(\mathbf{v} = \mathbf{r}) \geq \sum_{i=0}^{2t+1}(-1)^i \sum_{|\mathbf{k}|=i}\binom{\mathbf{k}+\mathbf{r}}{\mathbf{r}}S_{\mathbf{k}+\mathbf{r}} + \frac{2t+2}{|\mathbf{n}|-|\mathbf{r}|}\sum_{|\mathbf{k}|=2t+2}\binom{\mathbf{k}+\mathbf{r}}{\mathbf{r}}S_{\mathbf{k}+\mathbf{r}}. \quad (8.1)$$

(8.1) has the following extensions from Seneta and Chen (2001).

Theorem 8.1.1: Consider d sets of events and denote $\mathbf{v}$ a d dimension vector of exact numbers of occurrences of the events. For any integer $t \geq 0$ with $2t+1 \leq |\mathbf{n}| - |\mathbf{r}|$, the probability that the random vector $\mathbf{v}$ equals a numerical vector $\mathbf{r}$ can be bounded as follows.

$$P(\mathbf{v} = \mathbf{r}) \geq \sum_{i=0}^{2t+1}(-1)^i \sum_{|\mathbf{k}|=i}\binom{\mathbf{k}+\mathbf{r}}{\mathbf{r}}S_{\mathbf{k}+\mathbf{r}} + \max_i A_i(2t+1), \quad (8.2)$$

where

$$A_i(x) = \sum_{|\mathbf{k}|=x} S_{k_1+r_1,...,k_i+r_i+1...k_d+r_d}\binom{k_1+r_1}{r_1}...\binom{k_i+r_i+1}{r_i}...\binom{k_d+r_d}{r_d}\frac{k_i+1}{n_i-r_i},$$

$i = 1,...,d$. Note that $A_i(x)$ has degree $|\mathbf{r}|+x+1$.   □

Proof: We use (6.6) to prove Theorem 8.1.1. For the lower bounds, we set $a = 2t+1$, which makes (6.6) become

$$\sum_{i=0}^{2t+1}(-1)^i \sum_{|\mathbf{k}|=i}\binom{\mathbf{k}+\mathbf{r}}{\mathbf{k}}S_{\mathbf{k}+\mathbf{r}} = P(\mathbf{r}) - B_1(2t+1), \quad (8.3)$$

where the term $B_1(2t+1)$ equals

$$\sum_{|\mathbf{k}|=2t+1}\sum_{p=|\mathbf{r}|+2t+1}^{|\mathbf{n}|}\sum_{|\mathbf{s}|=p}P(\mathbf{s})\binom{s_1}{r_1}...\binom{s_d}{r_d}\binom{s_1-r_1-1}{k_1}\binom{s_2-r_2}{k_2}...\binom{s_d-r_d}{k_d}.$$

Due to (6.2), we can decompose the Bonferroni summation as follows.

$$S_{k_1+r_1+1,k_2+r_2,...,k_d+r_d} = \sum_{p=|\mathbf{k}|+|\mathbf{r}|+1}^{|\mathbf{n}|}\sum_{|\mathbf{s}|=p}P(\mathbf{s})\binom{s_1}{k_1+r_1+1}\binom{s_2}{k_2+r_2}...\binom{s_d}{k_d+r_d},$$

therefore,

$$A_1(2t+1)$$

$$= \sum_{|\mathbf{k}|=2t+1} S_{k_1+r_1+1,k_2+r_2,\ldots,k_d+r_d} \binom{k_1+r_1+1}{r_1}\binom{k_2+r_2}{r_2}\cdots\binom{k_d+r_d}{r_d}\frac{k_1+1}{n_1-r_1}$$

$$= \sum_{|\mathbf{k}|=2t+1} \sum_{p=|\mathbf{k}|+|\mathbf{r}|+1}^{|\mathbf{n}|} \sum_{|\mathbf{s}|=p} P(\mathbf{s})\binom{s_1}{k_1+r_1+1}\binom{s_2}{k_2+r_2}\cdots\binom{s_d}{k_d+r_d}$$

$$\binom{k_1+r_1+1}{r_1}\binom{k_2+r_2}{r_2}\cdots\binom{k_d+r_d}{r_d}\frac{k_1+1}{n_1-r_1}.$$

Noticing that the difference in the corresponding terms for summation essentially stems from

$$\binom{s_1}{k_1+r_1+1}\binom{k_1+r_1+1}{r_1} \quad \text{vs} \quad \binom{s_1}{r_1}\binom{s_1-r_1-1}{k_1}\frac{n_1-r_1}{k_1+1},$$

by expanding the left-hand expression, we have

$$\binom{s_1}{k_1+r_1+1}\binom{k_1+r_1+1}{r_1} \leq \frac{n_1-r_1}{k_1+1}\binom{s_1}{r_1}\binom{s_1-r_1-1}{k_1}. \qquad (8.4)$$

Now, applying a relation in (6.7) yields

$$\binom{s_i}{k_i+r_i}\binom{k_i+r_i}{r_i} = \binom{s_i}{r_i}\binom{s_i-r_i}{k_i} \quad \text{for} \quad i=2,\ldots,d. \qquad (8.5)$$

Thus, for each $P(\mathbf{s})$ in $A_1(2t+1)$ and $B_1(2t+1)$, we compare pairs of associated coefficients in $A_1(2t)$ and $B_1(2t)$ individually. Using (8.4) and (8.5), we have

$$A_1(2t) \leq B_1(2t). \qquad (8.6)$$

Now, for $2t+1 \leq |\mathbf{n}|-|\mathbf{r}|$, taking $a=2t+1$ in (6.6) yields

$$\sum_{i=0}^{2t+1}(-1)^i \sum_{|\mathbf{k}|=i}\binom{\mathbf{k}+\mathbf{r}}{\mathbf{k}}S_{\mathbf{k}+\mathbf{r}} = P(\mathbf{r})-B_1(2t+1).$$

Using (8.6), the above equation implies that

$$\sum_{i=0}^{2t+1}(-1)^i \sum_{|\mathbf{k}|=i}\binom{\mathbf{k}+\mathbf{r}}{\mathbf{k}}S_{\mathbf{k}+\mathbf{r}} \leq P(\mathbf{r})-A_1(2t+1).$$

Considering the fact that x_1 can be replaced by any one of $x_2, \ldots, x_d$ completes the proof of Theorem 8.1.1. $\square$

With Theorem 8.1.1, we now compare the two high dimensional bounds (8.2) and (8.1). We need the following lemma to facilitate the proof of Theorem 8.1.2.

Lemma 8.1.1: For any d sets of events,

$$A_1(2t+1)\frac{n_1-r_1}{|\mathbf{n}|-|\mathbf{r}|}+\ldots+A_d(2t+1)\frac{n_d-r_d}{|\mathbf{n}|-|\mathbf{r}|}=\frac{2t+2}{|\mathbf{n}|-|\mathbf{r}|}\sum_{|\mathbf{k}|=2t+2}\binom{\mathbf{k}+\mathbf{r}}{\mathbf{r}}S_{\mathbf{k}+\mathbf{r}},$$

(8.7)

Proof: For notational convenience, without loss of generality, we prove (8.7) in the case where $d=2$. Notice that

$$A_1(2t+1)\frac{n_1-r_1}{|\mathbf{n}|-|\mathbf{r}|}+A_2(2t+1)\frac{n_2-r_2}{|\mathbf{n}|-|\mathbf{r}|}$$

$$=\sum_{k_1+k_2=2t+1}S_{k_1+r_1+1,k_2+r_2}\binom{k_1+r_1+1}{r_1}\binom{k_2+r_2}{r_2}\frac{k_1+1}{|\mathbf{n}|-|\mathbf{r}|}+$$

$$\sum_{k_1+k_2=2t+1}S_{k_1+r_1,k_2+r_2+1}\binom{k_1+r_1}{r_1}\binom{k_2+r_2+1}{r_2}\frac{k_2+1}{|\mathbf{n}|-|\mathbf{r}|}$$

$$=\sum_{k_1=0}^{2t}S_{k_1+r_1+1,2t-k_1+r_2}\binom{k_1+r_1+1}{r_1}\binom{2t-k_1+r_2}{r_2}\frac{k_1+1}{|\mathbf{n}|-|\mathbf{r}|}+$$

$$S_{r_1+2t+2,r_2}\binom{2t+r_1+2}{r_1}\frac{2t+2}{|\mathbf{n}|-|\mathbf{r}|}+S_{r_1,2t+r_2+2}\binom{2t+r_2+2}{r_2}\frac{2t+2}{|\mathbf{n}|-|\mathbf{r}|}+$$

$$\sum_{k_1=1}^{2t+1}S_{k_1+r_1,2t-k_1+r_2+2}\binom{k_1+r_1}{r_1}\binom{2t-k_1+r_2+2}{r_2}\frac{2t-k_1+2}{|\mathbf{n}|-|\mathbf{r}|}.$$

We can make a variable transformation $j=k_1+1$ for the first summation to get

$$A_1(2t+1)\frac{n_1-r_1}{|\mathbf{n}|-|\mathbf{r}|}+A_2(2t+1)\frac{n_2-r_2}{|\mathbf{n}|-|\mathbf{r}|}$$

$$=\sum_{j=1}^{2t+1}S_{j+r_1,2t-j+r_2+2}\binom{j+r_1}{r_1}\binom{2t-j+r_2+2}{r_2}\frac{j}{|\mathbf{n}|-|\mathbf{r}|}+$$

$$S_{r_1+2t+2,r_2}\binom{2t+r_1+2}{r_1}\frac{2t+2}{|\mathbf{n}|-|\mathbf{r}|}+S_{r_1,2t+r_2+2}\binom{2t+r_2+2}{r_2}\frac{2t+2}{|\mathbf{n}|-|\mathbf{r}|}+$$

$$\sum_{k_1=1}^{2t+1}S_{k_1+r_1,2t-k_1+r_2+2}\binom{k_1+r_1}{r_1}\binom{2t-k_1+r_2+2}{r_2}\frac{2t-k_1+2}{|\mathbf{n}|-|\mathbf{r}|}$$

$$=S_{r_1+2t+2,r_2}\binom{2t+r_1+2}{r_1}\frac{2t+2}{|\mathbf{n}|-|\mathbf{r}|}+S_{r_1,2t+r_2+2}\binom{2t+r_2+2}{r_2}\frac{2t+2}{|\mathbf{n}|-|\mathbf{r}|}+$$

$$\sum_{k_1=1}^{2t+1}S_{k_1+r_1,2t-k_1+r_2+2}\binom{k_1+r_1}{r_1}\binom{2t-k_1+r_2+2}{r_2}\frac{2t+2}{|\mathbf{n}|-|\mathbf{r}|}$$

$$=\frac{2t+2}{|\mathbf{n}|-|\mathbf{r}|}\sum_{|\mathbf{k}|=2t+2}\binom{\mathbf{k}+\mathbf{r}}{\mathbf{r}}S_{\mathbf{k}+\mathbf{r}}.\quad\square$$

Theorem 8.1.2: Multivariate lower bound (8.2) is uniformly sharper than lower bound (8.1).

Proof: For notational convenience, without loss of generality, we state the proof of Theorem 8.1.2 in a bivariate setting. The associated proof for multivariate settings follows analogously. In the bivariate case, Theorem 8.1.1 yields

$$\sum_{i=0}^{2t+1}(-1)^i \sum_{k_1+k_2=i} \binom{\mathbf{k}+\mathbf{r}}{\mathbf{r}} S_{\mathbf{k}+\mathbf{r}} + A_1(2t+1)$$

$$\leq P(\mathbf{v}=\mathbf{r}) \qquad (8.8)$$

and

$$\sum_{i=0}^{2t+1}(-1)^i \sum_{k_1+k_2=i} \binom{\mathbf{k}+\mathbf{r}}{\mathbf{r}} S_{\mathbf{k}+\mathbf{r}} + A_2(2t+1)$$

$$\leq P(\mathbf{v}=\mathbf{r}) \qquad (8.9)$$

where

$$A_1(2t+1) = \sum_{k_1+k_2=2t+1} S_{k_1+r_1+1,k_2+r_2} \binom{k_1+r_1+1}{r_1}\binom{k_2+r_2}{r_2}\frac{k_1+1}{n_1-r_1}$$

and

$$A_2(2t+1) = \sum_{k_1+k_2=2t+1} S_{k_1+r_1,k_2+r_2+1} \binom{k_1+r_1}{r_1}\binom{k_2+r_2+1}{r_2}\frac{k_2+1}{n_2-r_2}.$$

Now we Multiply (8.8) and (8.9) by

$$\frac{n_1-r_1}{|\mathbf{n}|-|\mathbf{r}|}, \quad \frac{n_2-r_2}{|\mathbf{n}|-|\mathbf{r}|},$$

respectively, and then add them to get

$$\sum_{i=0}^{2t+1}(-1)^i \sum_{k_1+k_2=i} \binom{\mathbf{k}+\mathbf{r}}{\mathbf{r}} S_{\mathbf{k}+\mathbf{r}} + A_1(2t+1)\frac{n_1-r_1}{|\mathbf{n}|-|\mathbf{r}|} + A_2(2t+1)\frac{n_2-r_2}{|\mathbf{n}|-|\mathbf{r}|}$$

$$\leq P(\mathbf{v}=\mathbf{r}). \qquad (8.10)$$

Comparing (8.10) with (8.1), by Lemma 8.1.1,

$$A_1(2t+1)\frac{n_1-r_1}{|\mathbf{n}|-|\mathbf{r}|} + A_2(2t+1)\frac{n_2-r_2}{|\mathbf{n}|-|\mathbf{r}|} = \frac{2t+2}{|\mathbf{n}|-|\mathbf{r}|}\sum_{|\mathbf{k}|=2t+2}\binom{\mathbf{k}+\mathbf{r}}{\mathbf{r}} S_{\mathbf{k}+\mathbf{r}}.$$

Therefore, the lower bound in (8.1) is a weighted average of d lower bounds in Theorem 8.1.1. and the sharpness of lower bounds in Theorem 8.1.2 follows. □

Example 8.1.1. Using the numerical values described in Example 6.1.1, $d=2$, $n_1 = n_2 = 4$ and $r_1 = 3$, $r_2 = 1$, the value of lower bounds (8.2) is 0.3 (the same value as using (8.1)), which is the exact value of the joint probability $P(v_1 = 3, v_2 - 1)$.

Related publications on this section also include Nakamura and Douke (2003), Chen (1998), David (1996), Bailey (1977), and Eaton (1982). In the next section, we discuss a new type of multivariate bounds for k sets of events that satisfy *sub-Markovian* conditions.

8.2 Hybrid Lower Bounds

Considering k sets of events in a probability space $(\Omega, \mathcal{F}, P), A_{11}, ..., A_{1n_1}, ..., A_{k1}, ..., A_{kn_k}$, multivariate probability inequalities are sought to bound the probability of the occurrences of all the groups (Block and Chen, 2001). In this section, we focus on product-type lower bounds for the probability of at least one occurrence from each event group simultaneously.

$$P(\bigcup_{i_1=1}^{n_1} A_{1i_1} \cap \cdots \cap \bigcup_{i_k=1}^{n_k} A_{ki_k}).$$

Recall that in the framework of additive-type bounds, such as the multivariate Bonferroni bounds proposed by Meyer (1969):

$$\sum_{t=k}^{k+2m+1} \sum_{\Sigma i_j=t} (-1)^{t-k} S_{i_1,...,i_k} \tag{8.11}$$

$$\leq P(\cup_{i_1=1}^{n_1} A_{1i_1} \cap \ldots \cap \cup_{i_k=1}^{n_k} A_{ki_k})$$

$$\leq \sum_{t=k}^{k+2m} \sum_{\Sigma i_j=t} (-1)^{t-k} S_{i_1,...,i_k}, \tag{8.12}$$

where m is any positive integer. For integer $1 \leq i_j \leq n_j, \quad 1 \leq j \leq k$, the multivariate Bonferroni summations read

$$S_{i_1...i_k} = \sum_{1 \leq j_1^{(1)} < ... < j_{i_1}^{(1)} \leq n_1} \cdots \sum_{1 \leq j_1^{(k)} < ... < j_{i_k}^{(k)} \leq n_k} P(A_{1j_1^{(1)}}...A_{1j_{i_1}^{(1)}}...A_{kj_1^{(k)}}...A_{kj_{i_k}^{(k)}}). \tag{8.13}$$

For example, when $k = 2$ and $i_1 = i_2 = 2$,

$$S_{22} = \sum_{1 \leq j_1^{(1)} < j_2^{(1)} \leq n_1} \sum_{1 \leq j_1^{(2)} < j_2^{(2)} \leq n_2} P(A_{1j_1^{(1)}} A_{1j_2^{(1)}} A_{2j_1^{(2)}} A_{2j_2^{(2)}}).$$

The bounding structure is a linear function of addition and subtraction, which stems from the principle of exclusion and exclusion of the original thinking of Bonferroni (Meyer, 1969). In the literature, efforts have been devoted to improved multivariate bounds. For instance, for bivariate additive-type bounds with high degrees, an improved lower bound (for example, Galambos and Lee (1992)) reads

$$P(\bigcup_{i=1}^{n_1} A_i \cap \bigcup_{j=1}^{n_2} B_j) \tag{8.14}$$

$$\geq \sum_{k=0}^{a} \sum_{t=0}^{b} (-1)^{k+t} S_{k+1,t+1} - \binom{b+1}{1} \binom{a+1}{1} \frac{1}{a+1} \frac{1}{b+1} S_{a+2,b+2},$$

for any positive odd numbers a and b.

Inequality (8.14) is further sharpened (for example, in Chen and Seneta, 1996) as follows. For any odd numbers $a, b, a \geq 1, \quad b \geq 1$ and any integer $c \geq 0, \quad d \geq 0$:

$$P(\cup_{i=1}^{n_1} A_i \cap \cup_{j=1}^{n_2} B_j)$$

$$\geq \sum_{k=0}^{a} \sum_{t=0}^{b} (-1)^{k+t} S_{k+1,t+1} + \frac{a+1}{n_1-1} \binom{n_2}{1+c}^{-1} S_{a+2,1+c} +$$

$$+\frac{1+b}{n_2-1} \binom{n_1}{1+d}^{-1} S_{d+1,b+2} - S_{a+2,b+2}. \qquad (8.15)$$

Obviously, both inequalities (8.14) and (8.15) are in summation form (Bonferroni-type inequalities) with the use of Bonferroni-summation S_{ij} for $i = 1, ..., a+2$ and $j = 1, ..., b+2$. Compared with (8.12), bounds (8.14) and (8.15) require more information on S_{ij}'s, for instance, when $a = b = 1$, S_{33} is needed in (8.14).

To go beyond the framework of Bonferroni summations, which limits the consideration of additive-type functions in bounding, we consider product-type bounds and Hamilton circuits-type bounds in this section.

8.2.1 Setting of Hybrid Lower Bounds

The product-type bound resembles the way Markov handled the joint probability. After introducing the concept of sub-Markovian events, we combine the frameworks of Bonferroni, Fréchet, and Markov to create a new type of bounding technique, and obtain multivariate lower bounds that significantly improve the optimal Bonferroni-type lower bounds. Toward this end we need the following definitions.

Galambos and Simonelli (1996) describe different types of univariate upper and lower bounds. For example, for the union of a set of events $P(\cup_{i=1}^{n} A_i)$, among different types of bounds improving the additive type bounds, an effective way using the assumption of positive dependence is the method of product-type bounds in conjunction with graph theory. Thus, the following definitions of sub-Markovian events are stimulated. See, for example, Glaz and Johnson (1984), Block and Chen (2001), or the definition of sub-Markovian random variables in Block, Costigan and Sampson (1993).

Definition 8.2.1. For any set of events $A_1, ..., A_n$ in any probability space, if for any $1 \leq j \leq n$, there exists an integer j^*, $1 \leq j^* \leq j-1$ so that

$$P(A_j | A_1, ..., A_{j-1}) \geq P(A_j | A_{j^*}).$$

This set of events is called a set of *weak sub-Markovian* events.

Definition 8.2.2. For any set of events $A_1, ..., A_n$ in any probability space, assume for any $1 \leq j \leq n$ and for any integer j^*, $1 \leq j^* \leq j-1$,

$$P(A_j | A_1, ..., A_{j-1}) \geq P(A_j | A_{j^*}).$$

This set of events $A_1, ..., A_n$ is called a set of *strong sub-Markovian* events.

Example 8.2.1: For any set of sub-Markovian random variables $X_1, ..., X_n$ defined in Block et al. (1992), for any set of constants $c_1, ..., c_n$, denote $A_1, ..., A_n$ the events $X_1 \geq c_1, ..., X_n \geq c_n$, respectively, then $A_1, ..., A_n$ are strong sub-Markovian events.

The concept of sub-Markovian events extends the idea of sub-Markovian random variables that has various applications as documented in Block, Costigan and Sampson (1993). They established the connection between sub-Markovian random variables and multivariate total positive dependent random variables with degree-2 (hereby MTP_2) random variables. In terms of probability bounding theory, the concept of the sub-Markovian random variable can be applied to form an upper bound that is sharper than the well-known Hunter's upper bound. For example, consider a set of MTP_2 random variables $X_1, ..., X_p$:

$$P(X_1 \leq c_1, ..., X_p \leq c_p) \geq \prod_{(i,j) \in T} P(X_i \leq c_i, X_j \leq c_j) / \prod_{i=1}^{p} P(X_i \leq c_i)^{d_i - 1}, \quad (8.16)$$

where T is a spanning tree of the vertex set $\{1, ..., p\}$ and d_i is the degree of vertex i within this tree.

Block Costigan and Sampson (1993) show that the bound (8.16) improves Hunter (1976)'s bound for MTP_2 random variables. The multivariate hybrid upper bound in this regard was discussed in Chapter 6. In next sub-section, we will focus on a multivariate hybrid lower bound that combines a product-type bound with Fréchet optimal lower bounds.

The above discussion describes multivariate lower bounds using spanning trees in conjunction with sub-Markovian conditions. In this section, we further extend the refinement by using lower bounds with information depending on the joint probabilities represented as weights of edges in a Hamilton-type circuit in the setting of a connected graph. The improvement of the Hamilton-type lower bound over the Dawson-Sankoff-type lower bound is similar to the way the Hunter's upper bound improves the degree-two Bonferroni-type optimal bound. We describe the new lower bound in a univariate setting although it can be applied to evaluate multivariate probabilities when combined with other bounds. For example, the hypermultitrees of Bukszár (2003), random graph of Erdos and Renyi (1960), exchangeable events of Galambos (1982), subset optimization of Hoppe (2009), Euler methods of Naiman and Wynn (1992), abstract tubes of Dohmen (2003), and the graph approach described in Galambos and Simonelli (1996), among others.

For an integer $n \geq 1$, consider a set of arbitrary events $A_1, ..., A_n$ in an arbitrary probability space $(\Omega, \mathcal{F}, P)$. Let v be the number of events which occur at a given sample point. Recall Hunter's upper bound:

$$P(v \geq 1) \leq S_1 - \max_{T} \sum_{(i,j) \in T} P(A_i A_j), \quad (8.17)$$

where T is a spanning tree with $A_1, ..., A_n$, serving as the vertexes of the connected graph, and

$$S_t = \begin{cases} \sum_{1 \leq i_1 < ... < i_t \leq n} P(A_{i_1} ... A_{i_t}) & \text{for } 1 \leq t \leq n; \\ 1 & \text{for } t = 0; \\ 0 & \text{for } t > n. \end{cases} \quad (8.18)$$

Bound (8.17) improves the optimal degree-two Bonferroni-type upper bound (see Seneta and Chen, 1997)

$$P(v \geq 1) \leq \min(1, S_1 - \frac{2}{n}S_2), \tag{8.19}$$

at the cost of more specific information on the pairwise joint probabilities of $A_1,...,$ A_n.

The following is the mechanism of improvement over an optimal upper bound. Notice that the optimal bound (8.19) is optimal for the bounds depending on the values of S_1 and S_2 only. With extra information (such as pairwise joint probabilities specified in (8.17)), we can refine the optimal upper bound (8.19) that was grounded on S_1 and S_2 only. It is in this framework that we consider the improvement of optimal lower bounds by using Hamilton-type circuits.

In terms of univariate lower bounds, the celebrated Dawson-Sankoff (1967) bound is optimal based on the information of S_1 and S_2:

$$P(v \geq 1) \geq \frac{2}{k+1}S_1 - \frac{2}{k(k+1)}S_2 \tag{8.20}$$

for $k = 1, .., n$. As shown in Chapter 2, the lower bound (8.20) is the most stringent degree-two Bonferroni-type lower inequality with the optimal value at

$$k^* = [\frac{2S_2}{S_1}] + 1.$$

Recall that the optimality of this bound is Fréchet optimality: For any consistent Bonferroni summations S_1 and S_2 (defined in (8.18)), there exists a set of events $A_1^*,$ ..., A_n^* so that

$$S_1(A^*) = S_1, \quad S_2(A^*) = S_2$$

and the inequality becomes equality:

$$P(\cup_{i=1}^n A_i^*) = \frac{2}{k^*+1}S_1 - \frac{2}{k^*(k^*+1)}S_2. \tag{8.21}$$

(8.21) implies that for any lower bound depending on S_1 and S_2, the Dawson-Sankoff lower bound is always at least as good. Notice that the optimality is for bounds consisting of S_1 and S_2 only. Beyond this bounding domain, the Dawson-Sankoff bound loses its property of optimality. We can thus seek a way for its improvement.

Consider a set of events A_1, ..., A_n in a probability space. Denote u_1, ..., u_n the events A_1, ..., A_n, respectively, in a connected graph where vertexes u_i and u_j are joined by edge e_{ij} weighted by the joint probability between A_i and A_j, $P(A_iA_j)$. Recall that a *Hamilton circuit* refers to the circuit $\{x_0, x_1, ..., x_n, x_0\}$ in a connected graph with vertexes $x_0, x_1, ..., x_n$, $x_i \neq x_j$ if $i \neq j$. Denote $V = \{u_1,...,u_n\}$ and $E = \{e_{ij} : u_i \in V, u_j \in V\}$, we define a partition of the vertex set V of a graph $G = (V,E)$ as follows.

Definition 8.2.3: Consider m sets of vertexes B_1, ..., B_m in a connected graph $G = (V, E)$, if $B_1, \ldots, B_m$ satisfy the following conditions:

$$\bigcup_{i=1}^{m} B_i = V \quad \text{and} \quad B_i \bigcap B_j = \emptyset \quad \text{for} \quad i \neq j, \tag{8.22}$$

then $B_1, \ldots, B_m$ is called a partition of the vertex set V. $\square$

The partition of a vertex set leads to the definition of a Hamilton-type circuit as follows.

Definition 8.2.4: Let $B_1, \ldots, B_m$ be a partition of V in a connected graph $G = (V, E)$ with vertex set $V = \{u_1, \ldots, u_n\}$ and edge set E, the *circuit*, $\tau = (B_1, \ldots, B_m, B_1)$ is called a Hamilton-type circuit. $\square$

Example 8.2.2: Let $V = \{u_1, u_2, u_3, u_4, u_5\}$, $B_1 = \{u_2, u_5\}$, $B_2 = \{u_1, u_3\}$, and $B_3 = \{u_4\}$. Then, B_1, B_2, B_3 constitute a partition of the vertex set V, $\tau_1 = (B_1, B_2, B_3, B_1)$ and $\tau_2 = (B_2, B_3, B_1, B_2)$ are two Hamilton-type circuits in the graph.

Note that when the number of elements in the partition equals the number of events, $m = n$, the Hamilton-type circuit becomes a regular Hamilton circuit. Also, the length of the a Hamilton circuit m can be any positive integer $m \leq n$. In Theorem 8.3.3, we show that for a set of events where S_1 and S_2 are given, when the length of a Hamilton circuit is set to $\left[\frac{2S_2}{S_1}\right] + 1$, the lower bound defined on Hamilton circuit is Fréchet optimal with the new bounding information.

With above definitions, we have the following lemmas that lead to the main results on hybrid lower bounds. The first lemma focuses on sub-Markovian events.

Lemma 8.2.1. Assuming that A_1, ..., A_n is a set of weak sub-Markovian events, there exists a spanning tree T formed by using A_1, ..., A_n as its vertexes, such that

$$P(\cap_{i=1}^{n} A_i) \geq \frac{\prod_{(i,j) \in T} P(A_i A_j)}{\prod_{i=1}^{n} P(A_i)^{d(i)-1}}. \tag{8.23}$$

Taking into account all spanning trees T satisfying (8.23) yields

$$P(\cap_{i=1}^{n} A_i) \geq \max_{T} \frac{\prod_{(i,j) \in T} P(A_i A_j)}{\prod_{i=1}^{n} P(A_i)^{d(i)-1}}.$$

Proof: For this lemma, we need to construct a spanning tree satisfying (8.23). By the conditioning approach, we have

$$P(\bigcap_{i=1}^{n} A_i) = P(A_n | A_1, \ldots, A_{n-1}) P(A_{n-1} | A_1, \ldots, A_{n-2}) \ldots$$

$$\ldots P(A_3 | A_1, A_2) P(A_2 | A_1) P(A_1). \tag{8.24}$$

For $P(A_3 | A_1, A_2)$, since the events are sub-Markovian, there exists a number $t_3 \in \{1, 2\}$ so that

$$P(A_3 | A_1, A_2) \geq P(A_3 | A_{t_3}).$$

For the term $P(A_k|A_1,...,A_{k-1})$, with $k = 4,...,n-1$ in (8.24), by the definition of weak sub-Markovian events, there exists a number $t_k \in \{1,...,k-1\}$ so that

$$P(A_k|A_1,...,A_{k-1}) \geq P(A_k|A_{t_k}).$$

Similarly, for the term $P(A_n|A_1,...,A_{n-1})$ in (8.24), there exists a number $t_n \in \{1,...,n-1\}$ so that

$$P(A_n|A_1,...,A_{n-1}) \geq P(A_n|A_{t_n}).$$

From the way we select the integers t_3, ..., t_n, we have the edges (relationship between two events) $(2,1)$, (j,t_j), for $j = 3,...,n$. Since $t_j \in \{1,2,...,j-1\}$ in the selection of t_j, we know t_j is connected (by the relationship of conditional probability) to $\{1,2,...,j-1\}$ for integers $j = 3,...,n$. Consequently $T = \{(j,t_j), j = 2,...,n\}$ forms a spanning tree because there is no loop in the way of selecting t_j. Now, denote $t_1 = 1$, so we have

$$
\begin{aligned}
& P(\bigcap_{i=1}^{n} A_i) \\
= \ & P(A_n|A_1,...,A_{n-1})P(A_{n-1}|A_1,...,A_{n-2})...P(A_3|A_1,A_2)P(A_2|A_1)P(A_1) \\
\geq \ & \prod_{i=2}^{n} P(A_i|A_{t_i})P(A_{t_1}) \\
= \ & \prod_{(i,j)\in T} \frac{P(A_iA_j)}{P(A_j)}P(A_1)
\end{aligned}
\tag{8.25}
$$

The right-hand side of (8.25) can be simplified as follows. Denote d_i the degree of the set A_i being a vertex in T in the spanning tree , $i = 1,...,n$. Then, d_i actually is the frequency in which index i repeats in the spanning tree $\{t_1,...,t_n\}$. By (8.25), we have

$$P(\bigcap_{i=1}^{n} A_i) \geq \frac{\prod_{(i,j)\in T} P(A_i \cap A_j)}{\prod_{i=1}^{n} P(A_i)^{d(i)-1}},$$

this is because for the event A_i, if $i \notin \{t_1,...,t_n\}$, $P(A_i)$ does not appear in the denominator where $d(i) - 1 = 0$. This proves (8.23).

In the proof above, there may be more than one spanning tree satisfying the condition for the construction of a lower bound. For example, with a different permutation of the order of all the events, t_1 is changed to another number. Following the argument for (8.23), another spanning tree T^{**} is formed, so is another lower bound. Therefore, maximizing over all permissible spanning trees yields

$$P(\cap A_i) \geq \max_{T^{**}} \frac{\prod_{(i,j)\in T^{**}} P(A_i \cap A_j)}{\prod_{i=1}^{n} P(A_i)^{d(i)-1}}.$$

This completes the proof of Lemma 8.2.1.   □

We now describe a lemma regarding the Hamilton-type circuits.

Lemma 8.2.2: For any connected graph, a Hamilton-type circuit always exists.

Proof: For a connected graph G with n vertexes and any integer k, $k \leq n$, consider

$n - 1 = ks + r$ with $r \leq k - 1$. For any vertex v_i, we partition the set of all n vertexes into $k + 1$ sets as follows. Let $B_1 = \{v_i\}$ and $B_2, B_3, \ldots, B_{r+1}$ be the sets consisting of $s + 1$ distinct vertexes in each set; and let $B_{r+2}, \ldots, B_{k+1}$ be the sets consisting of s distinct vertexes in each set. In this way, Condition (8.22) is satisfied for $B_1, \ldots, B_{k+1}$, and the total number of vertexes consisted in the $k + 1$ sets is

$$1 + \sum_{i=2}^{r+1} (s+1) + \sum_{i=r+2}^{k+1} s$$
$$= 1 + (s+1)r + (k-r)s$$
$$= 1 + ks + r$$
$$= n.$$

Thus $B_1, \ldots, B_{k+1}$ forms a partition of $V = \{u_1, \ldots, u_n\}$ and $\tau_i = (B_1, \ldots, B_{k+1}, B_1)$ is a Hamilton-type circuit. $\qquad \square$

8.2.2   Main Results of Hybrid Lower Bounds

Consider k sets of events, $A_{11}, \ldots, A_{1n_1}; \ldots; A_{k1}, \ldots, A_{kn_k}$ in a probability space. Let v_i be the number of occurrences of the events $\{A_{i1}, \ldots, A_{in_i}\}$, $i = 1, \ldots, k$. We need two terms $L(i,j)$ and G_t defined as follows.

For any two sets of events $\{A_{i1}, \ldots, A_{in_i}\}$ and $\{A_{j1}, \ldots, A_{jn_j}\}$, denote $L(i,j)$ the bivariate Fréchet optimal lower bound of $P(v_i \geq 1, v_j \geq 1)$. The value of $L(i,j)$ can be computed using the algorithm described in Section 7.4.

For each event set $\{A_{t1}, \ldots, A_{tn_t}\}$, denote

$$G_t = 1 - \max \frac{\prod_{(i,j) \in T_t} P(A_{ti}^c \cap A_{tj}^c)}{\prod_{i=1}^{n_t} P(A_{ti}^c)^{d_t(i)-1}} = 1 - \frac{\prod_{(i,j) \in T_t^*} P(A_{ti}^c \cap A_{tj}^c)}{\prod_{i=1}^{n_t} P(A_{ti}^c)^{d_t(i)-1}},$$

where the maximum is over all possible spanning trees T_t formed by taking each event A_{ti}^c as a vertex, and $d_t(i)$ is the degree of event A_{ti}^c as a vertex in the spanning tree T_t.

For notational convenience, denote the union of the tth event set, $t = 1, \ldots, k$ as

$$B_t = \bigcup_{i=1}^{n_t} A_{ti}.$$

Also denote T_B the spanning tree formed by using each event B_i as its vertex, and S_B the set of all spanning trees of $B_1, \ldots, B_k$. Denote T_t the spanning tree formed by using each event A_{ti}^c as its vertex, and S_t the set of spanning trees of the event set A_{t1}^c, $\ldots, A_{tn_t}^c$.

Theorem 8.2.1 Consider k sets of events, $A_{11}, \ldots, A_{1n_1}; \ldots; A_{k1}, \ldots, A_{kn_k}$ in a probability space. Assume that events $B_1, \ldots, B_k$ (as defined above) are weak sub-Markovian events. For each $i = 1, ..., k$, assume that the set of complementary events $A_{i1}^c, ..., A_{in_i}^c$ are weak sub-Markovian events; then the set S_B is not empty, and the set S_t is not empty for any t such that

$$P(v_1 \geq 1, ..., v_k \geq 1) \geq \max_{S_B} \frac{\prod_{(i,j) \in T} L(i,j)}{\prod_{t=1}^k G_t^{d(t)-1}} = \frac{\prod_{(i,j) \in T_B^*} L(i,j)}{\prod_{t=1}^k G_t^{d(t)-1}}.$$

where $d(t)$ is the degree of event B_t being a vertex in spanning tree T_B^*.

Proof: Since the events $B_1, ..., B_k$ are sub-Markovian events, by Lemma 8.2.1, we know that there exists a spanning tree T_B such that

$$P(B_1 \cap ... \cap B_k) \geq \max_{T_B} \frac{\prod_{(i,j) \in T_B} P(B_i \cap B_j)}{\prod_{i=1}^k P(B_j)^{d(i)-1}}$$

$$= \max_{T_B} \frac{\prod_{(i,j) \in T} P(\bigcup_k A_{ik} \cap \bigcup_t A_{jt})}{\prod_{i=1}^k P(B_j)^{d(i)-1}}.$$

Now, for each set of events, say B_j, notice that

$$P(B_j) = P(\bigcup_{i=1}^{n_j} A_{ji}) = 1 - P(\bigcap_{i=1}^{n_j} A_{ji}^c).$$

For $j = 1, ..., k$, we may apply Lemma 8.2.1 to each set of the joint events $\{A_{j1}^c, ..., A_{jn_j}^c\}$ to obtain

$$P(B_j) = 1 - P(A_{j1}^c \cap ... \cap A_{jn_j}^c)$$

$$\leq 1 - \max_{T_j} \frac{\prod_{(p,q) \in T_j} P(A_{jp}^c \cap A_{jq}^c)}{\prod_{k=1}^{n_j} P(A_{jk}^c)^{d_t(k)-1}}.$$

Thus, the left-hand side of Theorem 8.2.1 becomes

$$P(v_1 \geq 1, ..., v_k \geq 1)$$

$$= P(\bigcup_{i=1}^{n_1} A_{1i} \cap ... \cap \bigcup_{i=1}^{n_k} A_{ki})$$

$$\geq \max_{T_B} \frac{\prod_{(i,j) \in T_B} P(\bigcup_k A_{ik} \cap \bigcup_t A_{jt})}{\prod_{i=1}^k (1 - \max_{T_j} \frac{\prod_{(p,q) \in T_j} P(A_{jp}^c \cap A_{jq}^c)}{\prod_{k=1}^{n_j} P(A_{jk}^c)^{d_t(k)-1}})^{d(i)-1}}.$$

Now notice that in the above expression, the joint probability

$$P(\bigcup_k A_{ik} \cap \bigcup_t A_{jt})$$

can be bounded by a Fréchet algorithm lower bound as described in Chapter 7. Denoting such a lower bound as $L(i,j)$,

$$P(\bigcup_k A_{ik} \cap \bigcup_t A_{jt}) \geq L(i,j),$$

we have

$$P(v_i \geq 1, i = 1, ..., k) \geq \max_{T_B} \frac{\prod_{(i,j) \in T_B} L(i,j)}{\prod_{i=1}^{k} (1 - \max_{T_j} \frac{\prod_{(p,q) \in T_j} P(A_{jp}^c \cap A_{jq}^c)}{\prod_{k=1}^{n_j} P(A_{jk}^c)^{d_t(k)-1}}) d(i) - 1}.$$

Note that at most there are a finite number of trees for maximization in the above inequality, thus tree T_B^* and tree T_i^* in Theorem 8.2.1 always exist. This completes the proof of Theorem 8.2.1. □

The result in Theorem 8.2.1 is technically quite involved due to the consideration of k sets of events. When $k = 1$, we have the following hybrid lower bound which is relatively simple compared with the result in Theorem 8.2.1.

Theorem 8.2.2. Consider a set of events A_1, ..., A_n in a probability space, for any event A_i, $1 \leq i \leq n$, denote the first set in a partition $B_1 = \{A_i\}$, and let $B_1, B_2, ...,$ B_{k+1} (for any $1 \leq k \leq n-1$) be a partition of the vertex set $V = \{u_1, ..., u_n\}$ with u_j representing the event A_j, $1 \leq j \leq n$. Let $\tau_i = (B_1, ..., B_{k+1}, B_1)$ be a Hamilton-type circuit that starts and ends at the set B_1 (B_1 is a set that contains only one vertex $\{u_i\}$). With this setting, a lower bound for the probability of the union of the set of events A_1, ..., A_n can be expressed as follows:

$$P(v_A \geq 1) \geq \max_{i=1,...,n} \max_{\tau_i} \frac{2}{k+1} [S_1 - \frac{1}{2} \sum_{j=1}^{k+1} \sum_{u_p, u_q \in B_j \cup B_{j+1}, \ p<q} P(A_p A_q)], \qquad (8.26)$$

where $B_{k+2} = B_1$ when $j = k+1$.

Proof: By Lemma 8.2.2, there exists a Hamilton-type circuit for the graph G starting with the set $B_1 = \{A_i\}$. For any Hamilton-type circuit $\tau_i = (B_1, ..., B_{k+1}, B_1)$,

$$P(\bigcup_{i=1}^{n} A_i) = P(\bigcup_{t=1}^{k+1} \bigcup_{u_j \in B_t} A_j)$$

$$= P(\bigcup_{t=1}^{k+1} \bigcup_{u_j \in B_t \cup B_{t+1}} A_j) \qquad (8.27)$$

with $B_{k+2} = B_1$ for notational convenience.

In the connected graph $G = (V, E)$ where $V = \{u_1, ..., u_n\}$ with u_i representing A_i, $i = 1, ..., n$, E is the set of all edges connecting u_i and u_j with $i, j = 1, ..., n$, the Hamilton-type circuit τ_i satisfies

$$P(\bigcup_{u_j \in B_t \cup B_{t+1}} A_j) \geq \sum_{u_j \in B_t \cup B_{t+1}} P(A_j) - \sum_{u_p, u_q \in B_t \cup B_{t+1}, \ p<q} P(A_p A_q). \qquad (8.28)$$

There are $k+1$ such pairs of sets $((B_t, B_{t+1}), t = 1, ..., k+1$, with $B_{k+2} = B_1)$ in a Hamilton-type circuit.

Notice that in a Hamilton-type circuit $\tau_i = (B_1,...,B_{k+1},B_1)$, the sets $B_i, i = 1,...,k+1$ satisfy (8.22), and $B_{k+2} = B_1$, thus

$$\sum_{t=1}^{k+1}[\sum_{u_j \in B_t \cup B_{t+1}} P(A_j)] = 2\sum_{i=1}^{n} P(A_i).$$

Now, applying the above identity and putting Equation (8.28) into (8.27) for all $t = 1,...,k+1$ yields

$$P(\bigcup_{i=1}^{n} A_i) \geq \frac{1}{k+1}\sum_{t=1}^{k+1}[\sum_{u_j \in B_t \cup B_{t+1}} P(A_j) - \sum_{u_p, u_q \in B_t \cup B_{t+1},\, p<q} P(A_p A_q)]$$

$$= \frac{1}{k+1}\sum_{t=1}^{k+1}[2S_1 - \sum_{u_p, u_q \in B_t \cup B_{t+1},\, p<q} P(A_p A_q)]. \qquad (8.29)$$

Maximizing (8.29) over all permissible Hamilton-type circuits (τ_i), and then over all beginning vertexes $(i = 1,...,n)$ for the circuits, yields Theorem 8.2.2. $\square$

The bound in (8.26) can be found via an algorithm of optimization (traveling salesman algorithm) as follows. When n is large, the algorithm greatly reduces the amount of computation. It is also available in commonly available software such as Maple, Mathematica, SAS, R, or MiniTab.

Notice that the optimal solution in (8.26) is essentially to find a Hamilton circuit that optimizes the total weight in a weighted connected graph. Such optimization is equivalent to the solution of the traveling salesman problem, which has been well developed in graph theory. When n is large, approximation algorithms developed in graph theory can be used to reduce the burden of computation, for example, Rosen (2011). The following algorithm was tailored to suit the optimization problem of Hamilton-type circuits in (8.26).

Optimizing algorithm of Theorem 8.2.2

(1) Denote the set of index of the vertexes $T = \{1, \cdots n\}$. For any integer k such that $1 \leq k \leq n-1$, we can find a positive integers s and r so that

$$n-1 = ks+r \quad \text{with} \quad r \leq k-1. \qquad (8.30)$$

(2) For the set of events $\{A_1,...,A_n\}$, select any one event to start with, say i_0. Define $T_1 = T - \{i_0\}$.

(3) Let $p = 1; V_0 = \{A_{i_0}\}; m = 1$ and $g = 1$.

(4) If $g \leq k - r$, find s vertexes from the set T_p to get a set of vertexes V^* so that

$$\sum_{i,j \in V^* \cup V_{p-1}} P(A_i A_j) = \min_{V^*}[\sum_{i,j \in V^* \cup V_{p-1}} P(A_i A_j)].$$

(5) If $m \leq r$, find $s+1$ vertexes from the set T_p to get a set of vertexes V^{**} so that

$$\sum_{i,j \in V^{**} \cup V_{p-1}} P(A_i A_j) = \min_{V^{**}}[\sum_{i,j \in V^{**} \cup V_{p-1}} P(A_i A_j)].$$

(6) Let V_p be the set satisfying

$$\sum_{i,j\in V_p\cup V_{p-1}} P(A_iA_j) = \min\{\sum_{i,j\in V^{**}\cup V_{p-1}} P(A_iA_j), \sum_{i,j\in V^*\cup V_{p-1}} P(A_iA_j)\}.$$

(7) If $V_p = V^*$, set $g = g+1$; if $V_p = V^{**}$, set $m = m+1.8$. Let $T_{p+1} = T_p\backslash V_p$; if $T_{p+1} \neq \emptyset$, do Step 4; if $T_{p+1} = \emptyset$, end.

It should be noted that the above algorithm takes r runs with $s+1$ and $k-r$ runs with s vertexes, after k iterations, the set of vertexes T_p contains no vertex when the algorithm ends because in each run of iteration, either s or $s+1$ indexes are selected from T_p. In this way, since the algorithm takes k runs to get the Hamilton-type circuit τ_i that gives a local maximum lower bound for each A_{i_0}. Since there are n events involved in the union, the algorithm can then be run n times with each event as the starting event (i_0 in the algorithm) to screen for the optimal lower bound in (8.26).

Another way to simplify Theorem 8.2.1 focuses on the existence of the spanning trees, which complicates the maximization process. The following result uses the condition of strong sub-Markovian events and obtains a lower bound for *any* spanning tree.

Theorem 8.2.3 Assume that events $B_1, ..., B_k$ are strong sub-Markovian events. For each $i = 1, ..., k$, assume that the set of complementary events $A_{i1}^c, ..., A_{in_i}^c$ are strong sub-Markovian events. Let S be any set of spanning trees of $B_1, ..., B_k$,

$$P(v_1 \geq 1, ..., v_k \geq 1) \geq \max_S \frac{\prod_{(i,j)\in T} L(i,j)}{\prod_{t=1}^k G_t^{d(t)-1}} = \frac{\prod_{(i,j)\in T_B^*} L(i,j)}{\prod_{t=1}^k G_t^{d(t)-1}}.$$

where $d(t)$ is the degree of event B_t being a vertex in spanning tree T_B^*.

Proof: For any tree $T_B \in S$, assume that the edges are $(1,t_1), ..., (k,t_k)$. Selecting the $k-1$ edges of the spanning tree as $(k,t_k), ..., (2,t_2)$, we have

$$P(v_1 \geq 1, ..., v_k \geq 1)$$
$$= P(B_1\cap...\cap B_k)$$
$$= P(B_k|B_1, ..., B_{k-1})P(B_{k-1}|B_1, ..., B_{k-2})...P(B_3|B_1, B_2)P(B_2|B_{t_2})P(B_{t_2})$$

Now, since $B_1, ..., B_n$ are strong sub-Markovian events, for any $i, i = 3, ..., k$, we have

$$P(B_i|B_1, ..., B_{i-1}) \geq P(B_i|B_{t_i}).$$

Thus,

$$P(v_1 \geq 1, ..., v_k \geq 1)$$
$$\geq P(B_k|B_{t_k})...P(B_2|B_{t_2})P(B_{t_2})$$
$$= \frac{\prod_{(i,j)\in T} P(B_i\cap B_j)}{\prod_{t=1}^k G_t^{d(t)-1}}$$
$$= \frac{\prod_{(i,j)\in T} L(i,j)}{\prod_{t=1}^k G_t^{d(t)-1}},$$

where $L(i, j)$ is the optimal bivariate algorithm lower bound for

$$P(B_i \cap B_j) = P(\bigcup_{t=1}^{n_i} A_{it} \cap \bigcup_{s=1}^{n_j} A_{js}).$$

The maximization over all possible spanning trees in the set of candidate graphs, S, gets the theorem. □

Although both Theorem 8.2.1 and Theorem 8.2.3 use the condition of sub-Markovian events to construct product-type lower bounds in conjunction with Fréchet optimal lower bounds, the conditions in the two theorems are distinguishable. In fact, Theorem 8.2.1 is weaker than Theorem 8.2.3, but the conditions in Theorem 8.2.1 are easier to verify than those in Theorem 8.2.3. With the strong sub-Markovian condition, bounds obtained from Theorem 8.2.3 are at least as sharp as bounds derived from Theorem 8.2.1, because the set of trees for maximization in Theorem 8.2.3 includes all of the trees for maximization in Theorem 8.2.1.

Toward the goal of simplifying the conditions in Theorems 8.2.1, we consider the following theorem which provides product-type lower bounds for k sets of random variables satisfying the general condition of MTP_2.

Denote $C_1 = \bigcup_{i=1}^{n_1} \{X_{1i} \leq c_{1i}\}$, ..., $C_k = \bigcup_{i=1}^{n_k} \{X_{ki} \leq c_{ki}\}$. And

$$G_t = 1 - \max_{T_t} \frac{\prod_{(i,j) \in T_t} P(\{X_{ti} > c_{ti}\} \cap \{X_{tj} > c_{tj}\})}{\prod_{i=1}^{n_t} P(\{X_{ti} > c_{ti}\})^{d_t(i)-1}},$$

where the maximum is over all spanning trees T_t among the events $\{X_{t1} > c_{t1}\}$, ..., $\{X_{tn_t} > c_{tn_t}\}$ for $t = 1, ..., k$. $d_t(i)$ is the degree of event $\{X_{ti} > c_{ti}\}$ being a vertex in the tree T_t.

Theorem 8.2.4. For k sets of MTP_2 random variables X_{11}, ..., X_{1n_1}, ..., X_{k1}, ..., X_{kn_k},

$$P(\bigcup_{i=1}^{n_1} \{X_{1i} \leq c_{1i}\} \cap \cdots \cap \bigcup_{i=1}^{n_k} \{X_{ki} \leq c_{ki}\}) \geq \max_T \frac{\prod_{(i,j) \in T} L(i,j)}{\prod_{t=1}^{k} G_t^{d(t)-1}},$$

where the maximum is over all spanning trees T among events C_1, ..., C_k, and $L(i, j)$ is the bivariate Fréchet optimal lower bound of $P(\bigcup_p \{X_{ip} \leq c_{ip}\} \cap \bigcup_q \{X_{jq} \leq c_{jq}\})$, $d(t)$ is the degree of event $\bigcup_{i=1}^{n_t} \{X_{ti} \leq c_{ti}\}$ being a vertex in tree T.

Proof: By Theorem 8.2.3, it suffices to show that the events associated with the MTP_2 random variables C_1, ..., C_k are strong sub-Markovian events and $\{X_{t1} > c_{t1}\}$, ..., $\{X_{tn_t} > c_{tn_t}\}$ are also strong sub-Markovian events .

First, for $C_j = \bigcup_{i=1}^{n_j} \{X_{ji} \leq c_{ji}\}$, $j = 1, ..., k$, denote

$$D = C_1 \cap \cdots \cap C_{j^*-1} \cap C_{j^*+1} \cap \cdots \cap C_{j-1}$$

for any $1 \leq j \leq n$ and $2 \leq j^* \leq j - 2$.

Now, denote $f(.)$ the joint probability density of MTP_2 random variables X_{11}, ..., X_{1n_1}, ..., X_{k1}, ..., X_{kn_k}. Put

$$g(\mathbf{x}) = \begin{cases} 1; & \mathbf{x} \in C_{j^*} \\ 0; & \text{otherwise,} \end{cases}$$

$$\varphi(\mathbf{x}) = \begin{cases} 1; & \mathbf{x} \in D \\ 0; & \text{otherwise,} \end{cases}$$

and

$$\psi(\mathbf{x}) = \begin{cases} 1; & \mathbf{x} \in C_j \\ 0; & \text{otherwise.} \end{cases}$$

Obviously functions f, g, φ and ψ satisfy the following conditions. f and g are MTP_2 functions with respect to $\sigma = \sigma_1 \times ... \times \sigma_n$. φ and ψ are a pair of monotone functions, by Karlin and Rinott (1980),

$$\left(\int g(\mathbf{x})f(\mathbf{x})d\sigma(\mathbf{x})\right)\left(\int \varphi(\mathbf{x})\psi(\mathbf{x})g(\mathbf{x})f(\mathbf{x})d\sigma(\mathbf{x})\right)$$

$$\geq \left(\int \varphi(\mathbf{x})g(\mathbf{x})f(\mathbf{x})d\sigma(\mathbf{x})\right)\left(\int \psi(\mathbf{x})g(\mathbf{x})f(\mathbf{x})d\sigma(\mathbf{x})\right) \qquad (8.31)$$

Now, notice that

$$\int g(\mathbf{x})f(\mathbf{x})d\sigma(\mathbf{x}) = P(C_{j^*}),$$

$$\int \varphi(\mathbf{x})\psi(\mathbf{x})g(\mathbf{x})f(\mathbf{x})d\sigma(\mathbf{x}) = P(C_1 \cap ... \cap B_j),$$

$$\int \varphi(\mathbf{x})g(\mathbf{x})f(\mathbf{x})d\sigma(\mathbf{x}) = P(D \cap C_{j^*})$$

and

$$\int \psi(\mathbf{x})g(\mathbf{x})f(\mathbf{x})d\sigma(\mathbf{x}) = P(C_j \cap C_{j^*}).$$

Therefore by (8.31)

$$P(C_1 \cap ... \cap C_j)P(C_{j^*}) \geq P(D \cap C_{j^*})P(C_j \cap C_{j^*}).$$

Recalling that $D \cap C_{j^*} = C_1 \cap ... \cap C_{j-1}$, we have

$$P(C_1 \cap ... \cap C_j)P(C_{j^*}) \geq P(C_1 \cap ... \cap C_{j-1})P(C_j \cap C_{j^*}).$$

Thus

$$P(C_j|C_1,...,C_{j-1}) \geq P(C_j|C_{j^*}),$$

which means that $C_1, ..., C_n$ are strong sub-Markovian events.

Similarly (or following the argument of Glaz and Johnson, 1984), for each $t = 1,...,k$, $\{X_{t1} > c_{t1}\}$, ..., $\{X_{tn_t} > c_{tn_t}\}$ are strong sub-Markovian events. Therefore, Theorem 8.2.4 follows. $\qquad \square$

It is related at this point to discuss the optimality property of the hybrid lower bounds. The following theorem shows that the new bound (8.26), although at the cost of computation, is sharper than the optimal Dawson-Sankoff lower bound at certain points.

Theorem 8.2.5 For any set of events $\{A_1, ..., A_n\}$, let S_1 and S_2 be the corresponding first and second degree Bonferroni summations. Denote the integer $k^* = [\frac{2S_2}{S_1}] + 1$, the lower bound (8.26) at $k = k^*$, is sharper than or equal to the Dawson-Sankoff lower bound when the optimal point reaches the edges, $k^* \in \{1, n-1, n\}$.

Proof: We start with the case where $k^* = n - 1$ because under this condition, each set in the partition of the vertex set V contains only one vertex that corresponds to an event in $\{A_1, ..., A_n\}$. It is a natural sense of discrete partition with each element. Notice that when $k^* = n - 1$, $k^* + 1 = n$, and a partition of the vertex set consists of $B_1 = \{A_i\}$, $B_j = \{A_t\}$ with $j = 1, ..., n, j \neq 1$ and $t = 1, ..., n, t \neq i$, for each i, $1 \leq i \leq n$. Now for notational convenience, denoting $B_j = \{A_{t_j}\}$ and $D = \{(u_p, u_q) : u_p, u_q \in B_t \cup B_{t+1}, p < q\}$, the difference between the two lower bounds becomes

$$\frac{2}{k^*+1}[S_1 - \min_i \min_{\tau_i} \frac{1}{2} \sum_{t=1}^{k^*+1} \sum_D P(A_p A_q)] - \frac{2}{k^*+1}S_1 + \frac{2}{(k^*+1)k^*}S_2$$

$$= \frac{2}{k^*+1}[-\min_i \min_{\tau_i} \frac{1}{2} \sum_{t=1}^{k^*+1} \sum_D P(A_p A_q) + S_2/k^*], \tag{8.32}$$

since in the first degree Bonferroni inequality, the Bonferroni summation B_1 is invariant under different permutations of the events $A_1, ..., A_n$.

Now, notice that

$$\min_i \min_{\tau_i} \frac{1}{2} \sum_{t=1}^{k^*+1} \sum_{u_p, u_q \in B_t \cup B_{t+1}, p<q} P(A_p A_q)$$

$$= \min_i \min_{\tau_i} \frac{1}{2} \sum_{j=1}^{n} P(A_{t_j} A_{t_{j+1}}) \quad \text{with } A_{t_{n+1}} = A_{t_1} = A_i$$

$$\leq \frac{1}{2}n \frac{S_2}{\frac{n(n-1)}{2}}$$

(by taking the average over all permutations of $A_1, ..., A_n$)

$$\leq \frac{S_2}{n-1}.$$

Applying the above inequality to (8.32) shows that Bound (8.26) is sharper than or equal to the Dawson-Sankoff bound when the optimal point $k^* = n - 1$.

When the optimal point $k^* = n$, $k^* + 1 = n + 1$, in this case, it is impossible to split n vertexes into $n + 1$ components of the partition of a set of vertexes. Thus we extend the set $\{A_1, ..., A_n\}$ to $\{A_1, ..., A_n, A_{n+1}\}$ with $A_{n+1} = \emptyset$. Denote $B_{j*} = \{A_{n+1}\} = \{\emptyset\}$. For each $j \neq j^*$, the corresponding vertex set in the partition, B_j, contains one vertex that represents an event from $A_1, ..., A_n$.

Now, let $B_j = \{A_{t_j}\}$, for $j \in \{1, ..., n+1\}\setminus\{j^*\}$, $T = \{1, ..., n+1\}\setminus\{j^*, j^*-1\}$ since $B_{j*} = \{\emptyset\}$, by Theorem 8.3.1, we have

$$P(\bigcup_{i=1}^{n} A_i) = P(\bigcup_{i=1}^{n+1} A_i)$$

$$\geq \max_i \max_{\tau_i} \frac{2}{n+1}[S_1 - \frac{1}{2}\sum_{j=1}^{n+1} \sum_{u_p, u_q \in B_j \cup B_{j+1}, p<q} P(A_p A_q)]$$

$$= \max_i \max_{\tau_i} \frac{2}{n+1} [S_1 - \frac{1}{2} \sum_{j \in T} \sum_{u_p, u_q \in B_j \cup B_{j+1}, \, p<q} P(A_p A_q)].$$

In this case, due to the invariance of S_1 and the definition of S_2, the right-hand side of (8.26) becomes

$$\frac{2}{n+1} [S_1 - \min_i \min_{\tau_i} \frac{1}{2} \sum_{j \in T} \sum_{u_p, u_q \in B_j \cup B_{j+1}, \, p<q} P(A_p A_q)]$$

$$= \frac{2}{n+1} [S_1 - \min_{\tau_i} \frac{1}{2} \sum_{j \in T} P(A_{t_j} A_{t_{j+1}})]$$

with $A_{t_{n+1}} = A_{t_1} = A_i$

$$\geq \frac{2}{n+1} [S_1 - \frac{1}{2}(n-1) \frac{S_2}{\frac{n(n-1)}{2}}]$$

(taking the average overall permutations of $A_1, \ldots, A_n$)

$$= \frac{2}{n+1} (S_1 - \frac{S_2}{n}).$$

The last term is actually the Dawson-Sankoff bound when the optimal point $k^* = n$.

Finally, we consider the situation where the optimal point $k^* = \frac{[2S_2]}{S_1} + 1 = 1$. In this case, the partition is formed essentially by two sets of vertexes. The first one contains the event $\{A_i\}$ and the second one contains the rest of the events. $B_1 = \{A_i\}$ and $B_2 = \{A_1, \ldots, A_{i-1}, A_{i+1}, \ldots, A_n\}$, thus we have

$$\min_i \min_{\tau_i} \frac{1}{2} \sum_{t=1}^{k^*+1} \sum_{u_p, u_q \in B_t \cup B_{t+1}, \, p<q} P(A_p A_q)$$

$$= \min_i \min_{\tau_i} \sum_{1 \leq i < j \leq n} P(A_i A_j)$$

$$= S_2.$$

From (8.32) we know that the difference between the two lower bounds is zero, since $k^* = 1$. Thus the new bound (8.26) is the same as the Dawson-Sankoff bound when the optimal point is $k^* = 1$. This completes the proof of Theorem 8.2.5. □

Theorem 8.2.5 shows that with additional information on the intersections of the events, the Hamilton-type lower bound (8.26), when the optimal point $k^* \in \{1, n-1, n\}$, reaches the level at least the same as the Dawson-Sankoff optimal lower bound.

As discussed in Chapter 2, a bound can not be improved with the same bounding information if it is a Fréchet optimal bound. For example, with the same values of S_1 and S_2, the Dawson-Sankoff lower bound can not be sharpened, because it is a Fréchet optimal bound. However, when more specific information on the joint probabilities is used, such as in (8.26), the bound is no longer merely a function of S_1 and S_2. With different bounding information, the Dawson-Sankoff lower bound can be strengthened. The following theorem shows that whenever the Dawson-Sankoff lower bound achieves its optimal value, there is always a set of events such that Bound (8.26) achieves the same value. And Example 8.2.5 confirms that in some situations the improvement of lower bound (8.26) over the Dawson-Sankoff lower bound can be very substantial.

Theorem 8.2.6. Consider a set of events $A_1, \ldots, A_n$, denote the optimal point $k^* = \left[\frac{2S_2}{S_1}\right] + 1$. There exists a set of events $A_1^*, \ldots, A_n^*$ such that

$$\frac{2}{k^*+1}[S_1 - \frac{1}{2}\sum_{j=1}^{k^*+1}\sum_{u_p,u_q \in B_j \cup B_{j+1}, \, p<q} P(A_p^* A_q^*)] = \frac{2}{k^*+1}(S_1 - S_2/k^*), \qquad (8.33)$$

thus for any set of events achieving the Fréchet optimal point in the Dawson-Sankoff lower bound, the bound in (8.26) reaches the same value.

Proof. Notice that for the optimal point of the Dawson-Sankoff bound, denote $k = k^* = \left[\frac{2S_2}{S_1}\right] + 1$,

$$\frac{2S_2}{S_1} \leq k \leq \frac{2S_2}{S_1} + 1.$$

Thus for this integer k, we have

$$S_1 - \frac{2}{k}S_2 \geq 0 \quad \text{and} \quad \frac{2}{k+1}S_2 - \frac{k-1}{k+1}S_1 \geq 0. \qquad (8.34)$$

Therefore, the Dawson-Sankoff lower bound can be decomposed as

$$(S_1 - \frac{2}{k}S_2) + (\frac{2}{k+1}S_2 - \frac{k-1}{k+1}S_1) = \frac{2}{k+1}(S_1 - \frac{S_2}{k}) \leq P(\bigcup_{i=1}^{n} A_i) \leq 1. \qquad (8.35)$$

Based on (8.34) and (8.35), there exist two events A_0 and B_0 satisfying the following conditions:

$$P(A_0 \cap B_0) = 0, \quad P(A_0) = S_1 - \frac{2}{k}S_2 \quad \text{and} \quad P(B_0) = \frac{2}{k+1}S_2 - \frac{k-1}{k+1}S_1. \qquad (8.36)$$

With the two events A_0 and B_0 specified above, we construct a set of events A_j^*, $j = 1, \ldots, n$ as follows. Let $A_j^* = A_0 \cup B_0$ for $j = 1, \ldots, k$; $A_{k+1}^* = B_0$ and $A_j^* = \emptyset$ for $j = k+2, \ldots, n$. For this set of events A_j^*, $j = 1, \ldots, n$, from (8.36), we have

$$S_1(A^*) = kP(A_0) + (k+1)P(B_0) = S_1,$$

$$\begin{aligned} S_2(A^*) &= \frac{k(k-1)}{2}P(A_0) + \frac{k(k+1)}{2}P(B_0) \\ &= S_2, \end{aligned}$$

and

$$P(\bigcup_{i=1}^{n} A_i^*) = P(A_0) + P(B_0) = \frac{2}{k+1}(S_1 - S_2/k).$$

Thus, $\{A_j^*, j = 1,...,n\}$ satisfies the criteria of Fréchet optimality described in Chapter 2. We now verify that the new lower bound (8.26) also reaches the probability of the union of this set of events.

For events $A_1^*, ..., A_n^*$, the bound of (8.26), taking a special Hamilton-type circuit, becomes

$$\max_{i=1,...,n} \max_{\tau_i} \frac{2}{k+1}[S_1 - \frac{1}{2}\sum_{j=1}^{k+1} \sum_{u_p,u_q \in B_j \cup B_{j+1},\, p<q} P(A_p A_q)]$$

$$\geq \frac{2}{k+1}\{S_1 - \frac{1}{2}[\sum_{i=1}^{k} P(A_i^* A_{i+1}^*) + P(A_{k+1} A_1)]\}$$

$$= \frac{2}{k+1}\{S_1 - \frac{1}{2}[(k-1)P(A_0) + (k+1)P(B_0)]\}$$

$$= \frac{2}{k+1}(S_1 - S_2/k),$$

which is the optimal value of the Dawson-Sankoff lower bound. Thus, for the set of events $A_1, ..., A_n$,

$$P(\bigcup_{i=1}^{n} A_i^*) = \max_{i=1,...,n} \max_{\tau_i} \frac{2}{k+1}[S_1 - \frac{1}{2}\sum_{j=1}^{k+1} \sum_{u_p,u_q \in B_j \cup B_{j+1},\, p<q} P(A_p A_q)]. \quad \square$$

Finally, we present a hybrid lower bound that stems from but does not use any of the new concepts (Markovian events or Hamilton-type circuits).

Theorem 8.2.7. For a set of event $A_1, ..., A_n$,

$$P(\bigcup_{i=1}^{n} A_n) \geq \frac{2}{n}[S_1 - \frac{1}{2}\min_{\delta} \sum_{j=1}^{n} P(A_{t_j} A_{t_{j+1}})], \tag{8.37}$$

where $A_{t_{n+1}} = A_{t_1}$ for notational convenience, and $\delta = \{(t_1,...,t_n)\}$ is the set of all permutations of all the elements in the set $\{1,...,n\}$.

Proof: In Theorem 8.2.2, denote $k = n-1$ and let B_i, $i = 1,...,n$, be n distinct vertex sets consisting of only one vertex in each set. In this case, the second term in the bound (8.26) is invariant over different selections of the beginning vertex v_i. But it depends on the permutation of the index set $\{1,...,n\}$. Thus, the optimization over all permutations of elements in $\{1,...,n\}$ in Bound (8.26) yields Bound (8.37). $\square$

In what follows in this section, we use numerical examples to illustrate the tightness of the hybrid lower bounds.

8.2.3   Examples of Hybrid Lower Bounds

We illustrate Theorem 8.2.1 using the following example (see, for example, Block and Chen 2001).

Example 8.2.3 Consider a set of events $E_1, E_2, E_3, E_4, E_5, E_6$ in a probability space

$(\Omega_1, \mathcal{F}, P)$ with the following non-zero probabilities of intersections of elementary conjunctions (which are, in this case, intersections of six events involving $E_1, E_2, ...,$ E_6 and their complements):

$$P(E_1^c E_2^c E_3^c E_4^c E_5^c E_6^c) = 0.2,$$

$$P(E_1 E_2^c E_3^c E_4 E_5^c E_6^c) = 0.1$$

$$P(E_1 E_2^c E_3^c E_4 E_5 E_6^c) = 0.1$$

$$P(E_1 E_2^c E_3^c E_4 E_5 E_6) = 0.1$$

$$P(E_1 E_2 E_3^c E_4 E_5^c E_6^c) = 0.1$$

$$P(E_1 E_2 E_3^c E_4 E_5 E_6^c) = 0.1$$

$$P(E_1 E_2 E_3^c E_4 E_5 E_6) = 0.1$$

$$P(E_1 E_2 E_3 E_4 E_5^c E_6^c) = 0.1$$

$$P(E_1 E_2 E_3 E_4 E_5 E_6^c) = 0.05,$$

$$P(E_1 E_2 E_3 E_4 E_5 E_6) = 0.05.$$

We define three sets of events for the scenario of Theorem 8.2.1. $A_1 = E_1$, $A_2 = E_2$, $A_3 = E_3$, and $B_1 = E_4$, $B_2 = E_5$, $B_3 = E_6$. In this way, we have two sets of events A_i for $i = 1, 2, 3$ and B_j for $j = 1, 2, 3$. We can then consider another set of events C_k for $k = 1, 2, 3$ in $(\Omega_1, \mathcal{F}, P)$ as follows

$$P(C_1^c C_2^c C_3^c) = 0.7$$

$$P(C_1 C_2^c C_3^c) = 0.1$$

$$P(C_1 C_2^c C_3) = 0.1,$$

$$P(C_1 C_2 C_3) = 0.1.$$

With event sets $\{A_i\}$, $\{B_j\}$ and $\{C_k\}$ constructed above, we specify three sets of events in the space of $\Omega_1 \times \Omega_1$:

$$A_{1i} = A_i \times \Omega_1 \qquad i = 1, 2, 3$$

$$A_{2j} = B_j \times \Omega_1 \qquad j = 1, 2, 3$$

$$A_{3k} = \Omega_1 \times C_k \qquad k = 1, 2, 3$$

By the way of construction of the three event sets, we have

$$S_{11}(1,2) = \sum_{i=1}^{3} \sum_{j=1}^{3} P(A_i B_j) = 2.85,$$

$$S_{21}(1,2) = \sum_{1 \le i_1 < i_2 \le 3} \sum_{j=1}^{3} P(A_{i_1} A_{i_2} B_j) = 1.65,$$

$$S_{12}(1,2) = \sum_{1 \le i \le 3} \sum_{1 \le j_1 < j_2 \le 3} P(A_i B_{j_1} B_{j_2}) = 1.80,$$

$$S_{22}(1,2) = \sum_{1 \le i_1 < i_2 \le 3} \sum_{1 \le j_1 < j_2 \le 3} P(A_{i_1} A_{i_2} B_{j_1} B_{j_2}) = 1.$$

Similarly

$$S_{11}(1,3) = 0.9, \quad S_{12}(1,3) = 0.6,$$
$$S_{21}(1,3) = 0.54, \quad S_{22}(1,3) = 0.36$$

$$S_{11}(2,3) = 0.93, \quad S_{12}(2,3) = 0.62$$
$$S_{21}(2,3) = 0.6, \quad S_{22}(2,3) = 0.40.$$

Now, based on the information of S_{ijk} for $i \le 2$, $j \le 2$, $k \le 2$, we want to find the lower bound for

$$x = P(\bigcup_{i=1}^{3} A_{1i} \cap \bigcup_{j=1}^{3} A_{2j} \cap \bigcup_{k=1}^{3} A_{3k}).$$

The set-up of this example guarantees that event sets $B_1 = \cup_i \{A_{1i}\}$, $B_2 = \cup_j \{A_{2j}\}$ and $B_3 = \cup_k \{A_{3k}\}$ are weak sub-Markovian events, the spanning tree T_B in Theorem 8.2.1 may be the tree consisting of $\{(1,2)(2,3)\}$, or $\{(2,1)(1,3)\}$, or $\{(1,3)(3,2)\}$. Thus the lower bound in Theorem 8.2.1 becomes

$$P(\bigcup_i \{A_{1i}\} \cap \bigcup_j \{A_{2j}\} \cap \bigcup_k \{A_{3k}\})$$

$$\ge \quad \max\{L(1,2)L(2,3)/G_2, L(2,1)L(1,3)/G_1, L(1,3)L(3,2)/G_3\} \quad (8.38)$$

where $L(1,2) = 0.6167$ by setting $n = m = 3$ with the algorithm in Section 7.4 where

$$\mathbf{b}' = (1, S_{11}, S_{12}, S_{21}, S_{22}) = (1, 2.85, 1.65, 1.80, 1.00)$$

and the associated $\mathbf{B}$ matrix is

$$\mathbf{B} = \begin{pmatrix} 1 & 1 & 1 & 1 & 1 \\ 0 & 4 & 6 & 6 & 3 \\ 0 & 2 & 6 & 3 & 0 \\ 0 & 2 & 3 & 6 & 3 \\ 0 & 1 & 3 & 3 & 0 \end{pmatrix}$$

and

$$\mathbf{B}^{-1}\mathbf{b} = (0.3833, 0.4000, 0.0833, 0.1167, 0.0167)'.$$

Similarly $L(2,3) = 0.1867$ with $\mathbf{b} = (1, 0.93, 0.62, 0.6, 0.4)'$,

$$\mathbf{B} = \begin{pmatrix} 1 & 1 & 1 & 1 & 1 \\ 0 & 4 & 6 & 6 & 9 \\ 0 & 2 & 6 & 3 & 9 \\ 0 & 2 & 3 & 6 & 9 \\ 0 & 1 & 3 & 3 & 9 \end{pmatrix}$$

and
$$\mathbf{B}^{-1}\mathbf{b} = (0.8133, 0.1100, 0.0367, 0.0300, 0.0100)'.$$

Also, $L(1,3) = 0.1867$ with $\mathbf{b} = (1, 0.9, 0.6, 0.54, 0.36)'$,

$$\mathbf{B} = \begin{pmatrix} 1 & 1 & 1 & 1 & 1 \\ 0 & 4 & 6 & 6 & 9 \\ 0 & 2 & 6 & 3 & 9 \\ 0 & 2 & 3 & 6 & 9 \\ 0 & 1 & 3 & 3 & 9 \end{pmatrix}$$

and
$$\mathbf{B}^{-1}\mathbf{b} = (0.8133, 0.1200, 0.0400, 0.0200, 0.0067)'.$$

By the way of construction, A_{11}^c, A_{12}^c, A_{13}^c are weak sub-Markovian events. The permissible spanning tree can be $(1,2)$ $(2,3)$, thus G_1 in the lower bound described in Theorem 8.2.1, becomes:

$$
\begin{aligned}
G_1 &= 1 - \max \frac{\prod_{(i,j)\in T_1} P(A_{1i}^c \cap A_{1j}^c)}{\prod_{i=1}^{3} P(A_{1i}^c)^{d_1(i)-1}} \\
&= 1 - \frac{P(A_{11}^c \cap A_{12}^c)P(A_{12}^c \cap A_{13}^c)}{P(A_{12}^c)} \\
&= 1 - \frac{0.2 \times 0.5}{0.5}.
\end{aligned}
$$

Therefore
$$G_1 = 0.8. \tag{8.39}$$

Similarly
$$G_2 = 1 - \max\{0.2, 0.143, 0.2\} = 0.8$$

and
$$G_3 = 1 - \max\{0.7, 0.7, 0.622\} = 0.3.$$

Putting the values of $L(1,2)$, $L(2,3)$, $L(1,3)$ as well as G_1, G_2, G_3 into Theorem 8.2.1 yields

$$
\begin{aligned}
P(\bigcup_i \{A_{1i}\} \cap \bigcup_j \{A_{2j}\} \cap \bigcup_k \{A_{3k}\}) \\
\geq \quad \max\{0.144, 0.144, 0.116\} \\
= \quad 0.144. \quad \square
\end{aligned}
$$

Example 8.2.4 We now compare the multivariate bound (Theorem 8.2.4) with inequalities using similar information, because it is not appropriate to compare bounds when the bounding information is different.

Consider bounds eligible for comparisons with the inequality in Theorem 8.2.4

dealing with 3 sets of events with the information of S_{ijk} for $i \leq 2$, $j \leq 2$, $k \leq 2$ only. For the setting in Example 8.2.2, the classical multivariate inequality (Meyer, 1969) reads,

$$
\begin{aligned}
P(\cup_i\{A_{1i}\} \cap \cup_j\{A_{2j}\} \cap \cup_k\{A_{3k}\}) &\geq S_{111} - (S_{112} + S_{121} + S_{211}) \\
&= 1.71 - (1.14 + 1.08 + 0.99) \\
&= -1.5.
\end{aligned}
$$

This is essentially a trivial lower bound because any probability is non-negative.

Alternatively, we may use the independence condition and construct a lower bound as follows. By the independence between event sets $\{A_{1i}, A_{2j}\}$ and $\{A_{3k}\}$ (but $\{A_{1i}\}$ and $\{A_{2j}\}$ are not independent), we have

$$
P(\cup_i\{A_{1i}\} \cap \cup_j\{A_{2j}\} \cap \cup_k\{A_{3k}\}) = P(\cup_i\{A_{1i}\} \cap \cup_j\{A_{2j}\})P(\cup_k\{A_{3k}\})
$$

Since
$$
P(\cup_i\{A_{1i}\} \cap \cup_j\{A_{2j}\}) \geq L(1,2) = 0.6167
$$

and
$$
P(\cup_k\{A_{3k}\}) \geq S_1(C) - S_2(C) = 0.6 - 0.4 = 0.2,
$$

we have a lower bound as

$$
P(\cup_i\{A_{1i}\} \cap \cup_j\{A_{2j}\} \cap \cup_k\{A_{3k}\}) \geq 0.6167 \times 0.2 = 0.1233.
$$

This lower bound is
$$
\frac{0.144 - 0.1233}{0.1233} = 17\%
$$

weaker than the value of the lower bound (0.144) obtained in Example 8.2.2. □

We now use a numerical example to show the improvement of the new bound (8.26) over the Dawson-Sankoff lower bound in Theorem 8.2.2.

Example 8.2.5 Let A_1, ..., A_6 be six events in a probability space. And assume that all non-zero probabilities of elementary conjunctions of intersections involving A_1, ..., A_6 and their complements A_1^c, ..., A_6^C are given as follows:

$$
\begin{aligned}
P(A_1^c A_2^c A_3^c A_4^c A_5^c A_6^c) &= 0.07 & P(A_1 A_2 A_3^c A_4 A_5^c A_6^c) &= 0.01 \\
P(A_1 A_2 A_3 A_4^c A_5^c A_6) &= 0.31 & P(A_1 A_2 A_3^c A_4^c A_5^c A_6^c) &= 0.20 \\
P(A_1^c A_2 A_3 A_4^c A_5^c A_6^c) &= 0.20 & P(A_1^c A_2^c A_3^c A_4 A_5 A_6^c) &= 0.20 \\
P(A_1^c A_2^c A_3^c A_4^c A_5 A_6^c) &= 0.01.
\end{aligned}
$$

The corresponding joint probabilities and individual probabilities are given as follows.

$$
\begin{aligned}
P(A_1) &= 0.01 + 0.31 + 0.20 = 0.52 & P(A_2) &= 0.72 & P(A_3) &= 0.51 \\
P(A_4) &= 0.21 & P(A_5) &= 0.21 & P(A_6) &= 0.31.
\end{aligned}
$$

And

$$P(A_1A_2) = 0.52 \quad P(A_1A_3) = 0.31 \quad P(A_1A_4) = 0.01 \quad P(A_1A_5) = 0.00$$
$$P(A_1A_6) = 0.31 \quad P(A_2A_3) = 0.51 \quad P(A_2A_4) = 0.01 \quad P(A_2A_5) = 0.00$$
$$P(A_2A_6) = 0.31 \quad P(A_3A_4) = 0.00 \quad P(A_3A_5) = 0.00 \quad P(A_3A_6) = 0.31$$
$$P(A_4A_5) = 0.20 \quad P(A_4A_6) = 0.00 \quad P(A_5A_6) = 0.00.$$

Therefore, the first Bonferroni summation is $S_1 = \sum_{i=1}^{6} P(A_i) = 2.48$. Our Theorem 8.2.2 gives a lower bound as

$$\frac{1}{4}[2S_1 - (1.26)] = 0.925,$$

with the Hamilton-type circuit $(\{A_5\}, \{A_2, A_6\}, \{A_4\}, \{A_3, A_1\}, \{A_5\})$.
The Dawson-Sankoff lower bound gives

$$\frac{2S_1}{k^* + 1} - \frac{2S_2}{k^*(k^* + 1)} = 0.825.$$

with

$$S_2 = \sum_{1 \le i < j \le 6} P(A_iA_j) = 2.49$$

and

$$k^* = [\frac{2S_2}{S_1}] + 1 = 3.$$

Thus the hybrid lower bound (0.925) is sharper than the Dawson-Sankoff lower bound in this example. □

8.3 Applications in Outlier Detection

Lower bounds derived beyond the conventional Bonferroni framework have various applications due to the refinement and sharpness over the traditional Bonferroni-type bounds (Galambos and Simonelli (1996)). As an illustrative example, we discuss the detection of a single outlier using refined lower bounds.

According to Barnett and Lewis (1984), an outlier is not just an extreme value, it is also statistically unreasonable even when viewed as an extreme. An outlier occurs when a scratch amplifies the expression level in gene micro-array studies, when an interviewee blindly checks "yes" without reading questions in a survey, or when we deal with contaminated data. It is critical to distinguish the outlier from the extreme value in a data set.

Assume that a random sample $X_1, ..., X_n$ follows a normal population $N(\mu, \sigma^2)$. To detect whether $X_{(n)}$, the largest observation, is an outlier, one of the testing statistics is the discordancy test

$$T = (X_{(n)} - \overline{X})/\sigma$$

where $\overline{X}$ is the sample mean of all the observations. The difficulty is to determine the constant t so that $P(T > t) \leq \alpha$.

As discussed in Barnett and Lewis (1984), denoting $T_i = (X_i - \overline{X})/\sigma$ and letting A_i denote the event $T_i > t$, the ith observation is an outlier. We have

$$P(T > t) = P(\bigcup_i A_i).$$

Thus $P(T > t)$ is approximated by the lower bound:

$$P(T > t) \geq \sum_{i=1}^n P(X_i - \overline{X} > \sigma t) - \sum_{1 \leq i < j \leq n} P(X_i - \overline{X} > \sigma t, X_j - \overline{X} > \sigma t)$$

and upper bound,

$$P(T > t) \leq \sum_{i=1}^n P(X_i - \overline{X} > \sigma t).$$

Under the null hypothesis that there are no outliers, $X_1, ..., X_n$ constitute a random sample, and the probabilities of $P(A_i)$ for $i = 1, ..., n$ are equal, so are the joint probabilities. Thus

$$\begin{aligned}
(T > t) \quad &\geq \quad \sum_{i=1}^n P(X_i - \overline{X} > \sigma t) - \sum_{1 \leq i < j \leq n} P(X_i - \overline{X} > \sigma t, X_j - \overline{X} > \sigma t) \\
&= \quad nP(A_1) - \frac{n(n-1)}{2} P(A_1 \cap A_2).
\end{aligned}$$

And the endpoint of the test statistic is found via

$$nP(A_1) \geq P(T > t) \geq nP(A_1) - \frac{(n-1)n}{2} P(A_1 \cap A_2) \qquad (8.40)$$

as shown in Barnett and Lewis (1984). Note that the upper and lower bounds in (8.40) are limited in the classical Bonferroni inequalities, the upper limit in fact can be improved by the Hunter's bound and the lower limit can be improved by refined lower bounds such as (8.26). Further readings in this regard include Liu (2011), Iyengar (1986), Stefansky (1971, 1972), and Das Gupta (1974), among others.

Chapter 9

Case Studies

In this chapter, we synthesize cases that use univariate and multivariate Bonferroni-type inequalities. As shown in Chapter 1, Bonferroni inequalities have broad applications in different areas. It is impossible (and it is not the purpose of this book) to list all the applications. Instead, we select representative applications that are related to the bounding theory discussed in previous chapters.

9.1 Molecular Cancer Therapy

Consider a cancer of the white blood cells, lymphoma (which occurs when lymphocytes grow abnormally). Lymphoma is a common blood disease of children in the world. Using the data of an experiment reported by Arditti et al. (2005) regarding apoptotic killing effects on two types of lymphoma tumor cells using the therapy of Rituximab-alliinase conjugate, Chen (2008a) applied an inequality combined with a partitioning technique to detect the drug effect that was previously unidentifiable.

Utilizing the mean reduction in the number of tumor cells as a killing effect, Arditti et al. (2005) investigated apoptotic killing effects on B-cell lymphocytic leukemia tumor cells by allicin. The enzyme alliinase catalyzes the transformation of the garlic compound allin into allicin. To ensure that allicin reacts on the surface of tumor cells, Rituximab is used to direct Alliinase to the surface of CD20+ B-cells for apoptotic effects.

The experiment consists of incubating tumor cells (10^6 cells /well) in four different groups: (1) Untreated; (2) Rituximab alone; (3) Alliin alone; and (4) Allicin generate by the conjugate of Rituximab and Alliin. After 48 hours in culture, cells from different treatment wells were counted. Data of the experiment were obtained through fluorescence-activated cell sorting analysis (thereby the FACS readings). The investigators were interested in the killing effects on two different cancer cell lines (SS lymphoma and EBV lymphoma) associated with the exposure to three different treatments: Rx-all, Alliin, and Rtx-all+Alliin, respectively. Chen (2008a) formulated such a bivariate problem in the framework of screening for discernible treatments in terms of two responses across different treatments, and adopted the therapeutic window procedure to solve the bivariate problem. The therapeutic window procedure applies to the simultaneous inference of two endpoints (such as toxicity and efficacy) of a drug, regarding the minimum effective dose and maximum tolerated dose.

Since the sample size is moderate and normality is not assumed in Arditti et

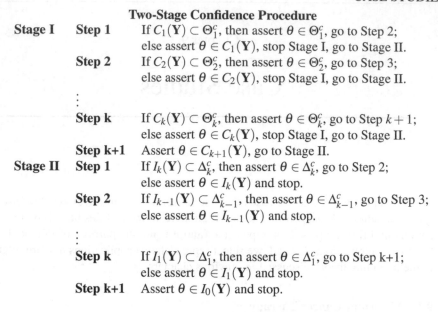

Two-Stage Confidence Procedure

Stage I **Step 1** If $C_1(\mathbf{Y}) \subset \Theta_1^c$, then assert $\theta \in \Theta_1^c$, go to Step 2;
else assert $\theta \in C_1(\mathbf{Y})$, stop Stage I, go to Stage II.

Step 2 If $C_2(\mathbf{Y}) \subset \Theta_2^c$, then assert $\theta \in \Theta_2^c$, go to Step 3;
else assert $\theta \in C_2(\mathbf{Y})$, stop Stage I, go to Stage II.

$\vdots$

Step k If $C_k(\mathbf{Y}) \subset \Theta_k^c$, then assert $\theta \in \Theta_k^c$, go to Step $k+1$;
else assert $\theta \in C_k(\mathbf{Y})$, stop Stage I, go to Stage II.

Step k+1 Assert $\theta \in C_{k+1}(\mathbf{Y})$, go to Stage II.

Stage II **Step 1** If $I_k(\mathbf{Y}) \subset \Delta_k^c$, then assert $\theta \in \Delta_k^c$, go to Step 2;
else assert $\theta \in I_k(\mathbf{Y})$ and stop.

Step 2 If $I_{k-1}(\mathbf{Y}) \subset \Delta_{k-1}^c$, then assert $\theta \in \Delta_{k-1}^c$, go to Step 3;
else assert $\theta \in I_{k-1}(\mathbf{Y})$ and stop.

$\vdots$

Step k If $I_1(\mathbf{Y}) \subset \Delta_1^c$, then assert $\theta \in \Delta_1^c$, go to Step k+1;
else assert $\theta \in I_1(\mathbf{Y})$ and stop.

Step k+1 Assert $\theta \in I_0(\mathbf{Y})$ and stop.

Figure 9.1 *Two-stage procedure for two endpoints*

al. (2005), Chen (2008a) used confidence sets corresponding to the Mann-Whitney statistic to compare the median differences simultaneously in a location shift model.

Let $\theta = (\theta_1, ..., \theta_k, \delta_1, ..., \delta_k)'$ be a parameter vector in Θ. For $i = 1, ..., k$, and $j = 1, ..., k$, let $C_i(\mathbf{Y})$, $D(\mathbf{Y})$ and $I_j(\mathbf{Y})$ be directed confidence sets of θ based on a set of data $\mathbf{Y}$. For notational convenience, let $C_{k+1}(\mathbf{Y}) = C_0(\mathbf{Y}) = I_{k+1}(\mathbf{Y}) = I_0(\mathbf{Y}) = D(\mathbf{Y})$, $\Theta_0 = \Delta_{k+1} = \emptyset$, $\Theta_{k+1} = \Delta_0 = \Theta$. Further, assume that for $i = 0, 1, ..., k, k+1$, and $j = 0, 1, ..., k, k+1$, $C_i(\mathbf{Y})$ and $I_j(\mathbf{Y})$ are constructed according to the following bivariate adjustment

$$P(\theta \in C_i(\mathbf{Y}) \bigcap I_j(\mathbf{Y})) \geq 1 - \alpha. \tag{9.1}$$

Note that the Bonferroni inequality can be applied to make the above bivariate adjustment, and bounds discussed in this book can be consequently applied to refine the adjustment.

In the study of molecular cancer therapies, the two response variables are the extent of apoptosis in two cancers: SS lymphomas and EBV lymphoma; the three different treatments are (1) Rtx-all, (2) Alliin, and (3) Rtz-all combined with Alliin, in which the untreated group serves as the control population. The setting is analogous to the setting for the analysis of the efficacy and toxicity (two responses), where different dose levels are compared with a placebo in dose-response studies in clinical trials. Notice that under this scenario the two readings of interest are the FACS readings corresponding to the reductions of mantle lymphoma tumor cells and the EBV transformed cells.

While the conventional method is unable to draw any conclusion, the procedure listed in Figure (9.1) detects the killing effect differences of the two different cell-lines using simultaneous comparisons of all three treatments with the untreated group. Notice that with the bivariate adjustment in (9.1), the simultaneous confidence interval in this case becomes

$$P(A_1 \cap \cdots \cap A_L \cap B_1 \cap \cdots \cap B_M) \geq 1 - \alpha, \qquad (9.2)$$

where the numbers of events $L(\omega)$ and $M(\omega)$ are two random numbers that depend on the data. The events A_i and B_j take the following forms in this example:

$$A_i = \begin{cases} \Theta_i^c; & \text{if } i < L \\ C_i(\mathbf{Y}); & \text{if } i = L, \end{cases}$$

and

$$B_j = \begin{cases} \Delta_j^c; & \text{if } j < M \\ I_j(\mathbf{Y}); & \text{if } j = M. \end{cases}$$

Now, denoting

$$A_i^* = \begin{cases} A_i; & \text{if } i \leq L \\ \Theta; & \text{if } i > L, \end{cases}$$

and

$$B_j^* = \begin{cases} B_j; & \text{if } j \leq M \\ \Theta; & \text{if } j > M, \end{cases}$$

the condition (9.2) can be written as

$$P(A_1^* \cap \cdots \cap A_k^* \cap B_1^* \cap \cdots \cap B_k^*) \geq 1 - \alpha, \qquad (9.3)$$

which is the same as

$$P(v_{A^c} = 0, v_{B^c} = 0) \geq 1 - \alpha, \qquad (9.4)$$

where v_{A^c} and v_{B^c} are, respectively, the numbers of events of $\theta_i \notin A_i$ for $i = 1, \ldots, k$ and the number of events $\delta_j \notin B_j$ for $j = 1, \ldots, k$. The upper bounds for $P(v_{A^c} = 0, v_{B^c} = 0)$ can be found in Theorems 6.1.1 and 6.2.3 while the corresponding lower bounds can be found in Theorems 7.2.2 and 7.2.3, for instance.

When more than two cancer cell lines are of interest, multivariate bounds discussed in Chapters 5 and 6 (for upper bounds), and Chapters 7 and 8 (for lower bounds) can be applied to facilitate similar analyses.

9.2 Therapeutic Window

Chapter 1 posits a question on the simultaneous inference for the therapeutic window of a drug. In this section, we will use an inequality approach to provide a solution. Recall that the therapeutic window (Chen 2008a) is essentially the dose range in which the treatment effect is larger than a clinically specified threshold δ, and the toxicity is less than a clinically specified threshold γ.

Assume that the ANOVA assumption is plausible for the data. As usual, denote the following for the observations of the experiment.

$$X_{ij} = \eta_i + \psi_{ij} \quad Z_{ij} = \mu_i + \varepsilon_{ij} \quad \text{with} \quad i = 0, 1, ..., k \quad \text{and} \quad j = 1, ..., n_i,$$

where random variables ψ_{ij} $i = 0, 1, ..., k$, $j = 1, ..., n_i$ independently and identically follow a $N(0, \sigma^2)$ distribution, and the errors ε_{ij}, $i = 0, 1, ..., k$, $j = 1, ..., n_i$ independently and identically follow a $N(0, \tau^2)$, with $Corr(\psi_{ij}, \varepsilon_{ij}) = \rho$. Under this setting, we have

$$\frac{\overline{X}_i - \overline{X}_0 - (\mu_i - \mu_0)}{\sigma \sqrt{1/n_i + 1/n_0}} \sim N(0, 1), \quad \text{and} \quad \frac{\overline{Z}_j - \overline{Z}_0 - (\eta_j - \eta_0)}{\tau \sqrt{1/n_j + 1/n_0}} \sim N(0, 1).$$

Denote the t-statistics

$$T_1(v) = \frac{\overline{X}_i - \overline{X}_0 - (\mu_i - \mu_0)}{\hat{\sigma} \sqrt{1/n_i + 1/n_0}} \quad \text{and} \quad T_2(v) = \frac{\overline{Z}_j - \overline{Z}_0 - (\eta_j - \eta_0)}{\hat{\tau} \sqrt{1/n_j + 1/n_0}},$$

where $v = \sum_{i=0}^{k}(n_i - 1)$.

Notice that $corr(X_{ij}, Z_{ij}) = \rho$ and ρ is not necessarily zero, so the joint distribution between $T_1(v)$ and $T_2(v)$ is a bivariate Siddiqui-type t-distribution (not the standard bivariate t-distribution, Siddiqui, 1967), where the correlation structure reads

$$corr(\overline{X}_i - \overline{X}_0, \overline{Z}_j - \overline{Z}_0) = \begin{cases} \rho, & i = j \\ \rho \sqrt{(n_i n_j)/[(n_j + n_0)(n_i + n_0)]}, & i \neq j. \end{cases}$$

Let $d_{\alpha, v, \rho}$ be the critical value satisfying

$$P(T_1(v) < d_{\alpha, v, \rho}, \quad T_2(v) > -d_{\alpha, v, \rho} | \rho) = 1 - \alpha.$$

Chen (2008a) used numerical approximations to compute the value $d_{\alpha, v, \rho}$. When no information is available on the unknown parameter ρ, the Bonferroni inequalities can be applied, and sharper bounds discussed in the previous chapters can be used to refine the result of data analysis.

When the correlation coefficient ρ can be plausibly estimated by its consistent estimator, $\hat{\rho}$, the bivariate t-distribution discussed in the preceding section may be applied to construct bivariate confidence sets. When n is large enough, the bivariate confidence set may be constructed from the bivariate normal distribution with sample correlation coefficient ($\hat{\rho}$). For notational convenience, we denote $d_{\alpha, v, \rho}$ by $d_{\alpha, v}$. The following bivariate set has confidence level $1 - \alpha$.

$$\mu_p - \mu_0 \geq \overline{X}_p - \overline{X}_0 - d_{\alpha, v} \hat{\sigma} \sqrt{\frac{1}{n_p} + \frac{1}{n_0}}; \quad \eta_q - \eta_0 \leq \overline{Z}_q - \overline{Z}_0 + d_{\alpha, v} \hat{\tau} \sqrt{\frac{1}{n_q} + \frac{1}{n_0}},$$

where $d_{\alpha, v}$ is the cut-off value with correlation coefficient $\hat{\rho}$.

With this scenario, we have the following procedure for data satisfying the ANOVA assumption for efficacy-toxicity inference. Define $d = \max(d_1, d_2)$. (If

Therapeutic Window for ANOVA data

Stage I **Step 1** If $\overline{X}_k - \overline{X}_0 - d\hat{\sigma}\sqrt{1/n + 1/n_0} > \delta$
then assert $\mu_k - \mu_0 > \delta$ go to Step 2;
else assert $\mu_k - \mu_0 > \overline{X}_k - \overline{X}_0 - d\hat{\sigma}\sqrt{1/n + 1/n_0}$
stop Stage I, go to Stage II.

Step 2 If $\overline{X}_{k-1} - \overline{X}_0 - d\hat{\sigma}\sqrt{1/n + 1/n_0} > \delta$
then assert $\mu_{k-1} - \mu_0 > \delta$ go to Step 3;
else assert $\mu_{k-1} - \mu_0 > \overline{X}_{k-1} - \overline{X}_0 - d\hat{\sigma}\sqrt{1/n + 1/n_0}$,
stop Stage I, go to Stage II.

$\vdots$

Step k If $\overline{X}_1 - \overline{X}_0 - d\hat{\sigma}\sqrt{1/n + 1/n_0} > \delta$
then assert $\mu_1 - \mu_0 > \delta$ go to Step k+1;
else assert $\mu_1 - \mu_0 > \overline{X}_1 - \overline{X}_0 - d\hat{\sigma}\sqrt{1/n + 1/n_0}$,
stop Stage I, go to Stage II.

Step k+1 Claim $\min_{i=1,\ldots,k} \mu_i - \mu_0 \geq$
$\min_{i=1,\ldots,k} \{\overline{X}_i - \overline{X}_0 - d\hat{\sigma}\sqrt{1/n + 1/n_0}\}$,
go to Stage II.

Stage II **Step 1** If $\overline{Z}_k - \overline{Z}_0 + d\hat{\tau}\sqrt{1/n + 1/n_0} < \gamma$,
then assert $\eta_k < \eta_0 + \gamma$, go to Step 2;
else assert $\eta_k - \eta_0 < \overline{Z}_k - \overline{Z}_0 + d\hat{\tau}\sqrt{1/n + 1/n_0}$ and stop.

Step 2 If $\overline{Z}_{k-1} - \overline{Z}_0 + d\hat{\tau}\sqrt{1/n + 1/n_0} < \gamma$,
then assert $\eta_{k-1} < \eta_0 + \gamma$, go to Step 3;
else assert $\eta_{k-1} - \eta_0 < \overline{Z}_{k-1} - \overline{Z}_0 + d\hat{\tau}\sqrt{1/n + 1/n_0}$, stop.

$\vdots$

Step k If $\overline{Z}_1 - \overline{Z}_0 + d\hat{\tau}\sqrt{1/n + 1/n_0} < \gamma$,
then assert $\eta_1 < \eta_0 + \gamma$, go to Step k+1;
else assert $\eta_1 - \eta_0 < \overline{Z}_1 - \overline{Z}_0 + d\hat{\tau}\sqrt{1/n + 1/n_0}$ and stop.

Step k+1 Assert $\max_{i=1,2,\ldots,k} \eta_i - \eta_0 \leq$
$\max_{i=1,2,\ldots,k} \{\overline{Z}_i - \overline{Z}_0 + d\hat{\tau}\sqrt{1/n + 1/n_0}\}$ and stop.

Figure 9.2 *Efficacy-toxicity procedure*

$\hat{\rho}$ is positive, $d_1 > d_2$, if $\hat{\rho}$ is negative, $d_1 < d_2$.) Also, $C_i(\mathbf{Y}) = (\overline{X}_i - \overline{X}_0 - d\hat{\sigma}\sqrt{1/n + 1/n_0}, \quad \infty)$, and $I_j(\mathbf{Y}) = (-\infty, \quad \overline{Z}_j - \overline{Z}_0 + d\hat{\tau}\sqrt{1/n + 1/n_0})$. The general confidence procedure presented in Figure 9.1 can be further specified in Figure 9.2.

As shown in Figure 9.2, efficacy is scanned from the highest to lowest dose, and toxicity is scanned from the lowest to highest dose. In the procedure in Figure 9.2, assume that the procedure stops at Step P in Stage I and Step Q in Stage II, respectively. If $P = 1$ or $Q = 1$, no dose is effective (or safe), a confidence set for the efficacy of Dose k and toxicity of Dose 1 is given. For example, when $k = 5$, $P = 3$ and $Q = 3$, then $k - P + 2 = 4 > 2 = Q - 1$, the therapeutic window does not exist; but the procedure provides simultaneous estimates on the efficacy and toxicity effects. If $k = 5, P = 4, Q = 5$, then $k - P + 2 = 3 \leq 4 = Q - 1$ (because we claim the efficacy of Doses $k, ..., k - P + 2$ if the procedure stops at Step P with dose level $k - P + 1$), the $(1 - \alpha)100\%$ confidence interval for the therapeutic window is then $[3, 4]$ with confidence estimates of efficacy and toxicity at doses $k - P + 1$ and Q, respectively.

Obviously, when more than two endpoints are of interest (for example, the quality-of-life of the patient, efficacy, and toxicity) the above procedure can not be applied. The new scenario necessitates the use of multivariate inequalities discussed in Chapter 6.

It should be noted that the method discussed in this section is valid under the ANOVA assumption only. Since the ANOVA assumption is valid (especially the equal variance assumption across treatments) only for some experiments, the following section discusses an alternative adjustment when the ANOVA assumption is violated.

9.3   Minimum Effective Dose with Heteroscedasticity

As discussed in Section 1.2.3, Scheffé (1953) pointed out the pitfall of directly applying agriculture-based multiple testing procedures to clinical data or to experiments on human beings. For example, when comparing yields of a crop associated with different plots for different combinations of farming conditions (soil type, fertilizers, type of seeds, moistures) in experiments, the assumption of equal variances across all the plots is plausible, as long as the experiment is well planned. However, the similar assumption is inappropriate (if it is not wrong) when the response variable comes from human beings. For instance, consider clinical measurements on a pain where some patients are calm and stable after being treated with a high dosage of a pain killer, while other patients (treated with lower dosages or in the placebo group) are struggling with the pain or an uncomfortable feeling in the body. It is implausible to assume equal variabilities for these two groups of patients in terms of pain measurements. Under this scenario, multiple testing with unequal variances among treatment groups is needed. An effective and available approach is the use of a probability inequality in conjunction with Chapman's two-stage sampling method (which was introduced in Chapter 6 for the method of successive comparisons with unequal variances), as follows.

In a dose-response study, for dose levels $0, 1, ..., k$ where 0 represents the placebo group. Let the response of the jth patient from the ith treatment at the stage t be

$$Y_{ijt} = \mu_{it} + \varepsilon_{ijt}$$

for $j = 1, ..., n_i$ and $i = 0, 1, ..., k$, where μ_{it} is the unknown treatment effect of dose level i at stage t, $i = 0, 1, ..., k$, $t = 1, ..., T$. The question of interest is to find the simultaneous confidence intervals for $\mu_{it} - \mu_{0t}$ for $i = 1, ..., k$ and $t = 1, ..., T$ when $\varepsilon_{ijt} \sim N(0, \sigma_{it})$ and $\sigma_{it} \neq \sigma_{js}$ when $(i, t) \neq (j, s)$.

One approach to constructing the simultaneous confidence intervals is to find the confidence interval for the individual treatment difference $\mu_i - \mu_0$ and then adjust the multiplicity with Bonferroni inequalities. The formulation of confidence interval for the individual difference $\mu_i - \mu_0$ with unequal variances requires the Behrens-Fisher problem as discussed in Section 6.3. We first consider the two-stage sampling method as in Chapman (1950).

1: Take initial sample size of n_0 observations $Y_{i1}, Y_{i2}, \ldots Y_{in_0}$ and $Y_{01}, Y_{02}, \ldots Y_{0n_0}$ from $N(\mu_i, \sigma_i^2)$ and the placebo $N(\mu_0, \sigma_0^2)$ respectively. Denote $\overline{Y}_i, \overline{Y}_0, S_i, S_0$ the corresponding mean and standard deviations of the initial sample for the two populations, respectively.

2: For any given positive value c, let

$$N_1 = \max(n_0 + 1, [\frac{S_i^2}{c}] + 1), \qquad N_2 = \max(n_0 + 1, [\frac{S_0^2}{c}] + 1), \qquad (9.5)$$

where $[z]$ is the largest integer less than or equal to z. Take additional $N_1 - n_0$ observations $Y_{i,n_0+1}, \ldots, Y_{i,N_1}$ from $N(\mu_1, \sigma_1^2)$ and additional $N_2 - n_0$ observations $Y_{0,n_0+1}, \ldots, Y_{0,N_2}$ from $N(\mu_0, \sigma_0^2)$.

With the above two sets of samples collected in two stages for each population, denote

$$a_1 = \frac{1 - (N_1 - n_0)b_1}{n_0}, \qquad b_1 = \frac{1}{N_1}\left\{1 + \sqrt{\frac{(N_1 c - S_i^2)n_0}{(N_1 - n_0)S_i^2}}\right\}.$$

$$a_2 = \frac{1 - (N_2 - n_0)b_2}{n_0}, \qquad b_2 = \frac{1}{N_2}\left\{1 + \sqrt{\frac{(N_2 c - S_0^2)n_0}{(N_2 - n_0)S_0^2}}\right\}.$$

Let

$$Y_i^* = a_1 \sum_{j=1}^{n_0} Y_{ij} + b_1 \sum_{j=n_0+1}^{N_1} Y_{ij}$$

$$Y_0^* = a_2 \sum_{j=1}^{n_0} Y_{0j} + b_2 \sum_{j=n_0+1}^{N_2} Y_{0j}.$$

Then, by Theorem 6.3.1, we have that

$$T_1 = \frac{Y_i^* - \mu_1}{\sqrt{c}} \quad \text{and} \quad T_2 = \frac{Y_0^* - \mu_0}{\sqrt{c}},$$

are two independent Student's t variables with $n_0 - 1$ degrees of freedom. Thus the confidence interval A_i for the mean difference $\mu_i - \mu_0$ can be constructed.

Notice that a convenient approach for the simultaneous confidence intervals of $\mu_{it} - \mu_{0t}$ for $i = 1, \ldots, k$ and $t = 1, \ldots, T$ is the Bonferroni approach. Let A_{it} be the $1 - \alpha^*$ confidence intervals for the differences $\mu_{it} - \mu_{0t}$, respectively. With the setting of (9.3) and by the multivariate Bonferroni inequality in Chapter 8, for example, the lower bound,

$$P(v_1 = r_1 \ldots v_d = r_d) \geq \sum_{t=\Sigma r_j}^{\Sigma r_j + 2k + 1} \sum_{\Sigma i_j = t} (-1)^{t - \Sigma r_j} \prod_{j=1}^{d} \binom{i_j}{r_j} S_{i_1, \ldots, i_J}$$

and the upper bound in Chapter 6 such as (6.1),

$$P(v_1 = m_1 \ldots v_J = m_J) \leq \sum_{t=\Sigma m_j}^{\Sigma m_j + 2k} \sum_{\Sigma i_j = t} (-1)^{t-\Sigma m_j} \prod_{j=1}^{J} \binom{i_j}{m_j} S_{i_1, \ldots, i_J}, \qquad (9.6)$$

in which $m_i = 0$, for $i = 1, \ldots, J$ and $J = T$ for this example.

For illustration, assume that there are three stages of a drug development ($T = 3$), then the value of α^* is adjusted according to

$$P(v_1 = 0, v_2 = 0, v_3 = 0)$$

$$\geq \sum_{t=0}^{1} \sum_{\Sigma_j i_j = t} (-1)^t \prod_{j=1}^{3} \binom{i_j}{0} S_{i_1 i_2 i_3}$$

$$= 1 - (S_{001} + S_{010} + S_{100}).$$

To satisfy the condition of $P(v_1 = 0, v_2 = 0, v_3 = 0) \geq 1 - \alpha$, we can then correspondingly adjust the value $\alpha^* = \alpha/(3k)$ to guarantee that the overall coverage probability is at least $1 - \alpha$. On the other hand, the simultaneous coverage probability can be bounded by the upper bound (as in Chapter 6):

$$P(v_1 = 0, v_2 = 0, v_3 = 0)$$

$$\leq \sum_{t=0}^{2} \sum_{\Sigma_j i_j = t} (-1)^t \prod_{j=1}^{3} \binom{i_j}{0} S_{i_1 i_2 i_3}$$

$$= 1 - (S_{001} + S_{010} + S_{100}) + S_{110} + S_{101} + S_{011} + S_{200} + S_{020} + S_{002},$$

where the degree-two joint probabilities can be computed using R, *Mathlab*, or *Mathematica*.

Certainly, the above derivation can be improved by a higher degree inequality when the associated information is available. The application of a sharper inequality leads to a more powerful inference procedure while maintaining the overall coverage of the simultaneous confidence intervals.

9.4   Simultaneous Inference with Binary Data

Another type of data that violates the ANOVA assumption is the binary data. Chen (2008b) used inequalities to propose a stepwise procedure for detecting the minimum effective dose of a drug when the response takes the form of binary data.

The model for binary data can be formulated as follows. Denote $i = 0$ the placebo group and $i = 1, \ldots, k$ the treatment of the ith dosage. Let X_{ij} be the dichotomous response of the jth subject recruited in the ith group in a double-blinded experiment, where $X_{ij} = 0$ (or 1) represents the status of a disease (or no disease correspondingly) with $P(X_{ij} = 1) = p_i$. For a pre-specified threshold $\delta > 0$, since p_0 represents the proportion of non-disease subjects in the placebo group, let $\theta_s = p_s - p_0$, for $s = 1, \ldots, k$, and $\Theta_s^c = \{\theta : \theta_s > \delta\}$. In this setting, for each value $s(s = 1, \ldots, k)$, θ_s is the

Figure: Stepwise Procedure for Dichotomous Data

1: Compute the lower confidence bounds $L(i)$ for $i = 1, ..., k$, where k is the total number of doses tested.

$$L(i) = [t_i - c_i - z_\alpha \sqrt{(n_i + n_0 + z_\alpha^2)/(4n_i n_0) - (t_i - c_i)^2/(n_i + n_0)}]/(1 + z_\alpha^2/(n_i + n_0)),$$

where $t_i = x_i/n_i - x_0/n_0$, z_α is the upper $1 - \alpha$ percentile of the standard normal model, $c_i = c_i(n_i, x_i, n_0, x_0)$ is a continuity correction equal to half the absolute difference between the sample value t_i and the next smaller possible value of the random variable T_i, where $T_i = X_i/n_i - X_0/n_0$.

2: Starting at k to search for the largest index M (if it exists) such that $L(M) > \delta$ and $L(M - 1) < \delta$. For the two extreme cases, if $L(k) < \delta$ then no dose is effective (and no MED can be identified); if $L(1) > \delta$ ($M = 1$), then all doses are effective.

3: Dose M is the MED and $L(M - 1)$ is the lower confidence bound for the effect of Dose $M - 1$.

4: Once M has been determined, the bounds corresponding to Dose $M - 2$ and below are irrelevant and do not even have to be computed.

Figure 9.3 *Binary response procedure*

efficacy of the dose level s. $\theta = (\theta_1, ..., \theta_k)$ is the vector consisting of risk differences between the treatments and the placebo.

Denote $X_i = \sum_{j=1}^{n_i} X_{ij}$ for $i = 0, 1, ..., k$, where X_i is the number of responses to dose level i. Assume that a drug is effective at dose i if $p_i - p_0 > \delta$ holds, which implies that the drug reduces the risk of developing a disease by a clinically pre-specified proportion, δ. Since the purpose of the inference of the minimum effective dose of a drug is to estimate the smallest effective dose, making inferences on MED is equivalent to estimating M such that $p_i - p_0 > \delta$ for all $i \geq M$ and $p_{M-1} - p_0 \not> \delta$. This formulation originates from the fact that the minimum effective dose is the smallest dose M from which any equal or higher dose reduces the risk of developing the disease by, at least, δ. Under this setting the following procedure selects the minimum effective dose of a drug when the response is binary.

In a case study, Chen (2008b) applied the inequality-based procedure in Figure-9.3 to an experiment on acetaminophen, originally posited by Tallarida (2000). In the experiment, acetaminophen, a widely applied analgesic, was injected into the spinal cord of male and female mice with the abdominal irritant test. The four different dosage levels applied in the experiment were: the baseline dose, $91 \mu g$, $151 \mu g$ and $227 \mu g$. Assessment of dose effects was based on the protection from an irritant injection during an observation period following its administration. The outcome of the experiment is summarized in Table-9.1.

As shown in Table-9.1, we have the individual 95% one-sided confidence bound as $p_1 - p_0 \geq 0.0275$ with $c_1 = 0.005$, $p_2 - p_0 \geq 0.2915$ with $c_2 = 0.0011$, and $p_3 - p_0 \geq 0.2085$ with $c_3 = 0.0100$.

Table 9.1 *Acetaminophen effect at spinal cord*

Treatment	No. protected	No. tested	Risk difference
Baseline	5	50	
$91\mu g$	7	20	$p_1 - p_0$
$151\mu g$	11	18	$p_2 - p_0$
$227\mu g$	6	10	$p_3 - p_0$

When the pre-specified risk improvement is set to $\delta = 0.20$, we assume that a dose level is effective if it increases the chance of protection by 20%, which is normally predetermined by medical experts. With the 95% confidence lower bound, we use the procedure in Figure 9.3 and conclude in the second step of the stepwise procedure (for the highest dosage) that dose level $227\mu g$ is effective in the sense that it increases the rate of protection by 20%. In Step 2 of the procedure, for the next dose level we similarly conclude that dose level $151 \mu g$ is effective. For the lowest dose in Step 2 we cannot conclude that dose level $91\mu g$ is effective in increasing the rate of protection by 20%, but we can conclude that it is more than 2.75%. Thus, according to the procedure, the 95% simultaneous confidence set is: $p_1 - p_0 \geq 0.0275$; $p_2 - p_0 \geq 0.20$ and $p_3 - p_0 \geq 0.20$. In this case, the minimum effective dose is estimated to be $151 \mu g$ and the dose effect at $91\mu g$ level increases the rate of protection by about 2.75%.

The above case contains only one clinical outcome. When two or more endpoints are of interest in an experiment, hybrid inequalities discussed in the previous chapters can be applied to improve the above analysis. For example, considering (6.22) with $k = 0$ and $t = 1$,

$$
\begin{aligned}
P(\bigcup_{i=1}^{n} A_i) &= P(v_1 \geq 1, v_2 \geq 0) \\
&\leq S_{10} - \max_{\pi_1,\pi_2} \sum_{\substack{1 \leq i_1 < \ldots < i_t \leq n \\ 1 \leq j_1 < \ldots < j_k \leq m}} \max_{\substack{k_1 \geq 1 \\ k_2 \geq 1}} P(A_{k_1} D^* \bigcup B_{k_2} D^*) \\
&= S_1 - \max_k \sum_i P(A_k \bigcap A_i).
\end{aligned}
$$

And hybrid lower bounds such as Theorems 8.2.2 and 8.2.7 can be applied to evaluate the joint probability for the overall error rates.

When the data is binary and the experimenter is interested in estimating the therapeutic window (toxicity and efficacy simultaneously), the above method can be combined with the pairwise confidence intervals (for example, Agresti and Coull, 1998) and multivariate Bonferroni inequalities in the previous chapters, to construct the simultaneous confidence sets.

9.5   Post-thrombotic Syndrome and Rang Regression

Chen and Comerota (2012) proposed a new regression method, range regression, to separate the source of data variation. The method effectively analyzes the association between percentage of thrombolysis and post-thrombotic syndrome. A discernible feature of range regression is the detection of the association between residual thrombus and post-thrombotic classification of venous diseases.

The study focuses on iliofemoral deep venous thrombosis (IFDVT, which refers to complete or partial thrombosis of the iliac vein or the common femoral vein). Current research shows that IFDVT predicts a higher risk of poor clinical outcomes (including venous claudication, physiological abnormalities, venous ulcers, and impaired quality of life). Treatments for IFDVT include anticoagulation or catheter-directed thrombolysis (CDT), referring to the infusion of a thrombolytic agent directly into the venous thrombus via multiple-side-hole catheters, with the use of imaging guidance for original placement (for example, Comerota et al. (2012), Kahn et al. (2008)). Due to the concern that catheter-based methods may lead to embolic and bleeding complications, it is necessary to weigh the risks versus benefits of CTD. The project was partly motivated by the interest on whether the quantity of clot lysed in IFDVT patients correlates with clinical outcomes on the quality of life after the operation. The task was correspondingly transferred to an evaluation on the association between the CEAP score (a measurement of clinical outcomes) with residual thrombus (quantitative thrombolysis) in IFDVT patients.

Due to sample fluctuation and unexpected variabilities in the data, the conventional approach of linear regression fails to catch the association between the two variables. As shown in the plot in Figure 9.4, the female CEAP scores vary greatly across different levels of residual thrombus. The change of the outcome variable (CEAP score) is strongly affected by additional noises, and is not adequately explained by residual thrombus. Denote Y the CEAP score, X the residual thrombus, so the model governing the data can be expressed as

$$Y = a + bX + \varepsilon + \eta,$$

where ε and η are two different sources of random fluctuations. ε denotes the normal fluctuation of the patients away from the mean CEAP scores and η denotes all the random sources associated with each level of residual thrombus. As shown in Figure 9.4, at the same level of residual thrombus, some patients have higher CEAP scores while some have lower CEAP scores. The trend between the two variables (CEAP scores and residual thrombus) is blurred by the existence of other sources of random effects.

To resolve the problem related to more than one component of variation and focus on the subject variability at similar amounts of clot lysed, Chen and Comerota (2012) provided an approach named range regression. The main idea of range regression is to stratify patients with similar amount of clot lysed into one group and then identify a measure that bundles subject variability within each stratum.

For the IFCVI patients in the investigation, Chen and Comerota (2012) took a range of 10% clot lysed as a criterion to stratify patients. This partly takes care of

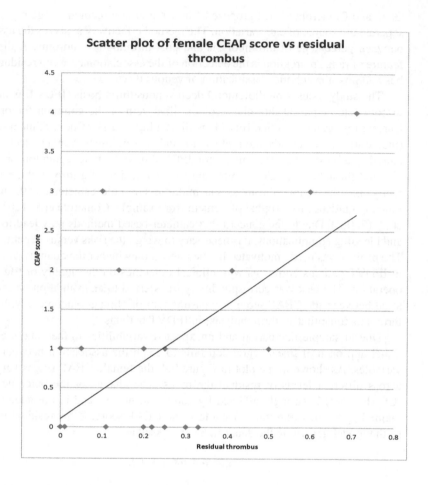

Figure 9.4 *Post-thrombotic syndrome and linear regression*

the impact of data variability due to confounding factors associated with each range of clot lysed, because patients in the same stratum have similar quantities (10%) of thrombus removal. The individual variability is then bundled with other patients in the same percentage range to seek the association between clot lysed and CEAP score.

After stratifying, the sample mean of all CEAP scores for patients at the same stratum (which contains patients having similar amounts of clot lysed) is computed as an indicator for the clinical outcome. For example, denote $y =$ Mean female CEAP

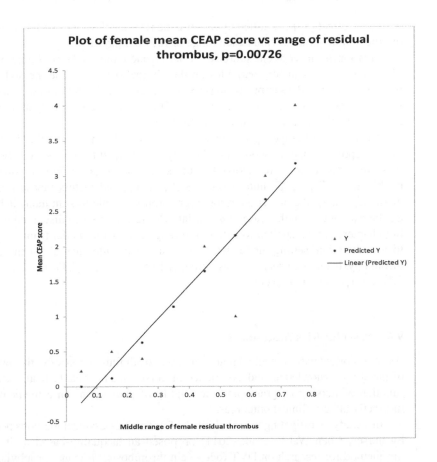

Figure 9.5 *Post-thrombotic syndrome and range regression*

score, x = Middle range of residual thrombus, ε = random effect, we have a regression line

$$Y = -0.646 + 5.09X + \varepsilon,$$

for the female patients in the data set. Since the sample mean is an asymptotically unbiased estimator of the population mean, range regression essentially models the conditional expected value of CEAP scores for each fixed range of residual thrombus as a linear trend of the residual thrombus.

$$E(CEAP|residual\ thrombus) = a + b(range\ of\ residual\ thrombus) + \varepsilon,$$

where the effect of η is implied by ranging the residual thrombus and then averaging out the output variable within each range of residual thrombus.

This approach keeps one source of variation and removes the confounding effects of intervening explanatory variables on the clinical outcome. The method of range regression models the sample mean (group average) of a discrete variable (the CEAP score) across different ranges of clot lysed. The asymptotic normality assumption of linear regression follows by the central limit theorem.

The method of range regression successfully unveils the association between the mean response of CEAP scores and the range of residual thrombus. However, when more than one factors are involved in the study, for instance, indexes of patients' health status, X_1, ..., X_n, nutrition status Y_1, ..., Y_m, and factors measuring quality of life Z_1, ..., Z_t, the univariate range regression is unable to simultaneously model all the parameters. In this case, multivariate Bonferroni bounds discussed in preceding chapters can be applied to extend the range regression model from a univariate to a multivariate setting, under which we can make inference about the effect of thrombolysis on three sets of measurements $\{X_i, i = 1, ..., n\}$; $\{Y_j, j = 1, ..., m\}$; and $\{Z_k, k = 1, ..., t\}$ simultaneously.

9.6 Vascular Risk Assessment

As an important part of medical management and administration, effective evaluation of the ability, knowledge, and experience of nurses in a hospital critically affects the function of nursing programs and the quality of patient care in a medical unit. It indirectly affects clinical outcomes.

In a study investigating the reliability of DVT risk assessment methods performed by nurses, Chen Walsh, et al. (2011) proposed an inequality-based multiple testing method for research on DVT (deep vein thrombosis) risks and prophylaxis. The method can be extended to analyze multiple risk factors with the use of multivariate inequalities.

The study consists of two sources of data. The first source is the risk assessment score obtained by nurses according to their knowledge and experience. The second source of data consists of the master score, which is the review and risk assessment performed by an experienced nurse (the master nurse) independently. We want to test the hypothesis that the nurse risk assessment score is lower than the master score on average for two categories of vascular risks.

Without assuming the specific distribution of the data, the Wilcoxon signed rank statistic is used to compare the median score difference for matched observations between the nurse and the master scores; the p-value is 0.036 for the median score difference in risk category 1, and 0.042 for the median score difference in risk category 2. For this problem, Holm's procedure (Section 5.4) fails to make any significant conclusion and the Hochberg's step-up procedure (Hochberg (1988), Simes (1986)) is invalid due to the lack of the normality assumption.

Normally, venous thrombosis occurs when red blood cells and others (such as fibrin, platelets and leukocytes) form a mass within an intact vein. Pulmonary embolism occurs when a piece of thrombus detaches from a vein wall, travels to the lungs, and lodges within the pulmonary arteries. Clinically, the two categories of risk factors are different knowledge, which distinguishes an experienced nurse in vascular practices.

Using a degree-one inequality, Chen, Walsh, et al. (2011) proposed a new confidence procedure that strongly controls the familywise error rate.

Table 9.2 *Two-step confidence procedure*

Step 1	If $\hat{P}_{(2)} < \alpha$,	reject $H_{(2)}$, go to Step 2
	else	claim $\eta_{(2)} > L_{(2)}(\mathbf{Y})$, stop.
Step 2	If $\hat{P}_{(1)} < \alpha/2$,	reject $H_{(1)}$, stop
	else	claim $\eta_{(1)} > L_{(1)}(\mathbf{Y})$, stop.

In the above procedure, $\eta_{(1)}$ and $\eta_{(2)}$ are ordered medians corresponding to the ordered p-values $R_{(1)}$ and $R_{(2)}$, while $L_{(1)}$ and $L_{(2)}$ are the upper confidence limits associated with individual 95% confidence level for each risk category.

The new method concludes that the median nurse score is significantly lower than the median master score for risk category 2, and that the median nurse risk assessment score is lower than two units of the master score for risk category 1. The conclusion is coherent with the individual inference that the median nurse risk score is lower than that of the master score in risk category 1, and the same conclusion for the risk category 2. While Holm's testing method is unable to make any conclusion, the new method is able to reject the hypothesis that is associated with the largest p-value.

When more risk factors or risk categories are included in a data set, methods of multivariate bounds discussed in the previous chapters can be applied similar to Sections 9.3 and 9.4. The new multiple testing procedures will improve the power of the test and enhance the coherence between individual testing and multiple testing.

9.7 Big-data Analysis

Big-data refers to periodic occurrences of huge amounts of information that cause difficulties in the storage and analysis of the complete data. Such a phenomenon occurs, for example, when analyzing customer transaction data collected daily in all Wal-Mart stores nationwide. Big-data analysis consists of a process examining huge volume of data covering large varieties of features of interest with the content of the data accumulating at a large velocity. Such a data mine contains information for hidden patterns and unknown correlations that leads to effective marketing strategies and revenues for companies. However, mining such huge data requires new thinking frameworks in statistics, because the conventional statistical methods become useless in front of big-data. For a simple analogy, a random sample in traditional statistics is just like a drop of water in the ocean of big-data. Big-data includes information captured from a variety of data sources such as sensors, web-related data, and cell phone records. In front of a tsunami of data, although most conventional statistical methodologies become meaningless, an appropriate application of stratification in conjunction with multivariate Bonferorni-type inequalities is able to take the *big* out of big-data. With appropriate probability inequalities, big-data (including unstructured data) is no longer big, but statistically manageable.

The idea of handling big-data with inequalities partly stems from David (1981), David and Newell (1965), Galambos and Simonelli (1996) on the detection of seasonal effects on the occurrences of events over disjoint discrete time intervals. Related references in the setting of time series analysis can be found, for example, in Móri (1990), Worsley (1979), and Ku and Seneta (1994, 1998).

Assume that we can divide the time interval of interest into $2n$ small periods such that the volume, variety, and velocity of data are manageable and the number of occurrences of events of interest in each period is observable. This can always be achieved because we may reduce the time interval from days to hours, to minutes, or even to seconds.

For k features of interest (such as special sales volumes, price ranges, customer preferences, etc.), let $\mathbf{m}_i$ be the vector denoting the number of occurrences of the k features over n periods starting from period i, $i = 1, ..., n$. Notice that the vector $\mathbf{m}$ contains k elements $m(j)$, $j = 1, ..., k$ reflecting the k features of interest. In this case, the statistic of interest is a $k \times 1$ vector $\mathbf{z}$ with elements $Z(j)$, $j = 1, ..., k$, and

$$Z(j) = \max_{1 \le i \le n} \left| \frac{m(j)_i - (m^*(j) - m(j)_i)}{\sqrt{m^*(j)}} \right|,$$

where $m^*(j)$ is the total number of occurrences through the time intervals of interest. Under the assumption that the variables $m(j)_1, ..., m(j)_n$ are binomial random variables with parameters $m^*(j)$ and $p = 1/2$, the $Z(j)_i = (2m(j)_i - m^*(j))/\sqrt{m^*(j)}$ are asymptotically standard normal distributed with correlation coefficients between $Z(j)_i$ and $Z(j)_{i+h}$ as $\rho_{i,i+h} = 1 - (2h)/n$.

Different from the previous discussions where the Bonferroni inequalities refer to bounds on the probability of a union of finite events, in the case of big-data with the above partitioning of the longitudinal space, an upper bound that we may use is the following (Casella and Berger, 2002),

$$P\left(\bigcap_{i=1}^{\infty} A_i\right) \le \sum_{i=1}^{\infty} P(A_i), \tag{9.7}$$

when the series in the right-hand side of (9.7) converges.

As discussed in Casella and Berger (2002), inequality (9.7) can be proved as follows.

Denote $B_1 = A_1$, and $B_i = A_i \backslash (\bigcup_{j=1}^{i-1} A_j)$ for $i \ge 2$. We have

$$P\left(\bigcup_{i=1}^{\infty} A_i\right) = P\left(\bigcup_{i=1}^{\infty} B_i\right), \tag{9.8}$$

Since $B_i \cap B_j = \emptyset$ when $i \neq j$, the probability of the union of a sequence of mutually exclusive events equals the summation of the probabilities of individual events.

$$P(\bigcup_{i=1}^{\infty} B_i) = \sum_{i=1}^{\infty} P(B_i). \tag{9.9}$$

Now putting (9.9) into (9.8) gets

$$P(\bigcup_{i=1}^{\infty} A_i) = \sum_{i=1}^{\infty} P(B_i). \tag{9.10}$$

Recalling $B_i \subset A_i$ (thus $P(B_i) \leq P(A_i)$), in conjunction with (9.10), gets (9.7).

Inequality (9.7) bridges conventional statistical approaches with the analysis of big-data by partitioning the time space into manageable subintervals. For example, letting $A_{ji} = \{|Z(j)_i| > c\}$ for $j = 1, ...k$ and $i = 1, ..., \infty$, transfers the problem from patterns and trends in dynamically changing data into the inference of trends and patterns in k sets of events where the conventional statistics are valid.

Example 9.7.1 Consider the analysis of sensing imagery data from remote sensors in the detection of rare earth elements. Assume that the chance of the detection of a rare element can be characterized by a Poisson model for certain positive value λ in a small enough time interval,

$$P(A_i) = e^{-\lambda_i} \frac{\lambda_i^j}{j!},$$

where j is an integer depending on A_i and

$$A_i = \max_{t \in (s_i, s_{i+1})} (|\mathbf{z}| > c).$$

Then across the time span in which the screening process takes place, the chance of the detection of a rare element over the whole time period $(0, T)$ is

$$
\begin{aligned}
P(\sup_{t \in (0, T)} |\mathbf{z}_t| > c) &= P(\bigcup_{i=1}^{\infty} (\max_{t \in (s_i, s_{i+1})} (|\mathbf{z}| > c))) \\
&= P(\bigcup_{i=1}^{\infty} A_i) \\
&\leq \sum_{i=1}^{\infty} P(A_i) \\
&= \sum_{i=1}^{\infty} e^{-\lambda_i} \frac{\lambda_i^j}{j!}.
\end{aligned}
$$

Notice that the sub time intervals $(s_1, s_2), ..., (s_i, s_{i+1}), ...$ constitute a partition of the time span $(0, T)$ such that in each sub time interval the rate of the occurrence can be estimated for all the features included in the vector **z**.

The above example shows the way the Bonferroni inequality bridges the probability associated with unmanageable big-data and the probabilities of the detection of a rare element in manageable sub time intervals.

Example 9.7.2 Consider the analysis of a huge volume of customer transaction data for the identification of a hidden trend connecting customer responses with a sales promotion of a substitute item over a week. Assuming that we can partition the time period such that the chance of detecting the hidden trend in each sub time interval $P(A_i)$ can be bounded by a value such as $(\frac{1}{10})^i$, then we have

$$
\begin{aligned}
P(\text{hidden trend over}(0,\ T)) &= P\left(\bigcup_{i=1}^{\infty}(\text{hidden trend over}(s_i, s_{i+1}))\right) \\
&= P\left(\bigcup_{i=1}^{\infty} A_i\right) \\
&\leq \sum_{i=1}^{\infty} P(A_i) \\
&= \sum_{i=1}^{\infty}\left(\frac{1}{10}\right)^i \\
&= \frac{1}{9}.
\end{aligned}
$$

This implies that the chance of having the overall hidden trend across the period $(0, T)$ in which big-data occurs, is $\frac{1}{9}$.

Following the idea of the above two examples, the multivariate inequalities can be applied in the analysis of big data as follows. Denote v_j the number of events that occur in group j. This could be the number of times when a conjectured trend is significant at the ith time interval. Then, for any integers m_j, $j = 1, ..., k$, with a discussion similar to Chapter 6, when the information on the joint probabilities are available, we have for any integer k_n, $n = 1, 2, ...$

$$
\begin{aligned}
&P(v_1^{(t)} = m_1^{(t)}, \ldots, v_k^{(t)} = m_k^{(t)} \quad t \in (0, T)) \\
&= P\left(\bigcup_{n=1}^{\infty}\{\mathbf{v}_n = \mathbf{m}_n\}\right) \\
&\leq \sum_{n=1}^{\infty} \sum_{t=\sum m_j^n}^{\sum m_j^n + 2k_n} \sum_{\sum i_j = t} (-1)^{t - \sum m_j} \prod_{j=1}^{k}\binom{i_j}{m_j} S_{i_1, \ldots, i_k}.
\end{aligned}
$$

The upper bound above leads to an estimate of $P(\mathbf{v}_n = \mathbf{m}_n)$. For big-data, n tends to ∞, so the limit of the upper bound (if it exists) provides an estimate of the probability that a particular trend occurs in the whole data set for the k features.

Bibliography

[1] A. Agresti and B. A. Coull. Approximate is better than "exact" for interval estimation of binomial proportions. *The American Statisticians*, 52:119–126, 1998.

[2] F. Arditti, A. Rabinkov, and T. Miron et al. Apoptotic killing of B-chronic lymphocytic leukemia tumor cells by allicin generated in situ. *Molecular Cancer Therapeutics*, 4:325–332, 2005.

[3] B. Bailey. Tables of the Bonferroni t statistic. *Journal of American Statistics*, 72:469–478, 1977.

[4] K. Balasubramanian and R.B. Bapat. Identities for order statistics and a theorem of Rényi. *Statistics and Probability Letters*, 12:141–143, 1991.

[5] V. Barnett and T. Lewis. *Outliers in Statistical Data*. Wiley, New York, 1984.

[6] H. Block and J. T. Chen. Multivariate product-type lower bounds. *Journal of Applied Probability*, 38:407–420, 2001.

[7] H. W. Block, T. Costigan, and A. Sampson. Product-type probability bounds of higher order. *Probability in the Engineering and Information Sciences*, (6):349–370, 1992.

[8] H. W. Block, T. Costigan, and A. Sampson. Optimal product-type bounds. *Journal of Applied Probability*, (30):675–691, 1993.

[9] H. W. Block, T. Costigan, and A. Sampson. Bonferroni-type inequalities and the methods of indicators and polynomials. *Advances in the Theory and Practice of Statistics: A Volume in Honor of Samuel Kotz, eds, N.L. Johnson and N. Balakrishnan, John Wiley and Sons*, pages 535–550, 1997.

[10] C. E. Bonferroni. Teoria statistica delle classi e calcolo delle probabilit/'a. *Pubblicazioni del R. Istituto Superiore di Scienze Economiche e Commerciali di Firenze*, 8:1–62, 1936.

[11] G. Boole. *Laws of Thought*. American reprint of 1854 ed., New York, 1984.

[12] E. Boros and A. Prékopa. Closed form two-sided bounds for probabilities that at least r and exactly r out of n events occur. *Math. Operations Res.*, 14:317–342, 1989.

[13] D. Brent, M. Baugher, J. Bridge, J. T. Chen, and L. Chiappetta. Age and sex-related risk factors for adolescent suicide. *Journal of the American Academy of Child and Adolescent Psychiatry*, 38(12):1497–1505, 1999.

[14] J. Bukszár. Upper bounds for the probability of a union by multitrees. *Ad-*

vances in Applied Probability, 33:437–452, 2001.

[15] J. Bukszár. Hypermultitrees and sharp Bonferroni inequalities. *Mathematical Inequalities and Applications*, 4:727–743, 2003.

[16] J. Bukszár, G. Mádi-Nagy, and T. Szántai. Computing bounds for the probability of the union of events by different methods. *Annals of Operations Research*, 201:63–81, 2012.

[17] G. Casella and R. L. Berger. *Statistical Inference*. Duxbury, Pacific Grove, CA, 2002.

[18] E. Castillo. *Extreme Value Theory in Engineering*. Academic Press, New York, 1988.

[19] F. Y. Chan, L. K. Chan, and G. D. Lin. On consecutive k-out-n: F systems. *European Journal of Operations Research*, 36:207–216, 1988.

[20] D. G. Chapman. Some two sample tests. *Ann. Math. Statist.*, 21:601– 606, 1950.

[21] J. Chen and J. Glaz. Approximations and bounds for moving sums of discrete random variables. *Applied Sequential Methodologies*, 173:105–121, 2004.

[22] J. Chen and J. Glaz. Approximations for multiple scan statistics. *Recent Advances in Applied Probability*, pages 97–114, 2005.

[23] J. T. Chen. Direct product Bonferroni-type optimality. *Bulletin of the International Statistical Institute*, Book 1:189–190, 1995.

[24] J. T. Chen. Multivariate Bonferroni-type inequalities and optimality. *Bulletin of the Australian Mathematical Society*, 58:167–168, 1998.

[25] J. T. Chen. Optimal lower bounds for bivariate probabilities. *Advances in Applied Probability*, 30:476 – 492, 1998.

[26] J. T. Chen. Letter to the Editor: On the estimation of attributable risk in case-control studies. *Statistics in Medicine*, 20:979–982, 2001.

[27] J. T. Chen. A lower bound using Hamilton-type circuit and its applications. *Journal of Applied Probability*, 40:1121–1132, 2003.

[28] J. T. Chen. A two-stage estimation procedure. *Biometrics*, 64:406–412, 2008a.

[29] J. T. Chen. Inference on the Minimum Effective Dose Using Binary Data. *Communication in Statistics*, 38:2124–2135, 2008b.

[30] J. T. Chen. A New Skew-normal Model for the Application-oriented Skew-t Model. *European Journal of Pure and Applied Mathematics, special issue on Granger Economics and Statistical Modeling*, 3:531–540, 2010.

[31] J. T. Chen and A. J. Comerota. Detecting the Association between Residual Thrombus and Post-thrombotic Classification of Chronic Venous Disease with Range Regression. *Reviews on Recent Clinical Trials*, 7:329–334, 2012.

[32] J. T. Chen, A. K. Gupta, and C. Troskie. Distribution of stock returns when the market is up (down). *Communications in Statistics-Theory and Methods*, 32:1541–1558, 2003.

[33] J. T. Chen and F. M. Hoppe. Simultaneous confidence intervals. *The Encyclopedia of Biostatistics*, 5:4114–4116, 1998.

[34] J. T. Chen and F. M. Hoppe. A connection between successive comparisons and ranking procedures. *Statistics and Probability Letters*, 67:19–25, 2004.

[35] J. T. Chen, F. M. Hoppe, S. Iyengar, and D. Brent. A hybrid logistic regression model for case-control studies. *Methodology and Computing in Applied Probability*, 5:419–426, 2003.

[36] J. T. Chen, S. Iyengar, and D. Brent. Constraint Estimation for the Population Attributable Risk. *Journal of Applied Probability and Statistics*, 2:251–265, 2007.

[37] J. T. Chen and E. Seneta. A note on bivariate Dawson-Sankoff-type bounds. *Statistics and Probability Letters*, 24:99–104, 1995a.

[38] J. T. Chen and E. Seneta. Multivariate identities, and permutation and Bonferroni upper bounds. *Combinatorics, Probability and Computing*, 4:331–342, 1995b.

[39] J. T. Chen and E. Seneta. Multivariate Bonferroni-type lower bounds. *Journal of Applied Probability*, 33:729–740, 1996.

[40] J. T. Chen and E. Seneta. Lower bounds for the probability of intersection of several unions of events. *Combinatorics, Probability and Computing*, 7:353–364, 1998.

[41] J. T. Chen and E. Seneta. A Stepwise rejective test procedure with strong control of familywise error rate. *Bulletin of the International Statistical Institute*, 52nd session Helsinki:241–242, 1999.

[42] J. T. Chen and E. Seneta. A refinement of multivariate Bonferroni-type inequalities. *Journal of Applied Probability*, 37:276–282, 2000.

[43] J. T. Chen and E. Seneta. A Fréchet-optimal strengthening of the Dawson-Sankoff lower bound. *Methodology and Computing in Applied Probability*, 8(2):255–264, 2006.

[44] J. T. Chen, E. Walsh, and A. Comerota et al. A new multiple test approach for nursing care administration of deep vein thrombosis patients. *Istanbul University Journal of the School of Business Administration*, 1:22–34, 2011.

[45] K. L. Chung. On the probability of the occurrence of at least m events among n arbitrary events. *Annals of Mathematics and Statistics*, 12:328–338, 1941.

[46] K. L. Chung. On fundamental systems of probabilities of a finite number of events. *Annals of Mathematics and Statistics*, 14:123–133, 1943a.

[47] K. L. Chung. Further results on probabilites of a finite number of events. *Annals of Mathematics and Statistics*, 14:234–237, 1943b.

[48] K. L. Chung and P. Erdos. On the application of the Borel-Cantelli lemma. *Trans. Amer. Math. Soc.*, 72:179–186, 1952.

[49] W. G. Cochran. The distribution of the largest of a set of estimated variances as a fraction of their total. *Ann. Eugen., Lond.*, 11:47–52, 1941.

[50] A. J. Comerota, N. Grewal, J. T. Martinez, and J. T. Chen et al. Post-thrombotic morbidity correlates with residual thrombus following catheter-directed thrombolysis for iliofemoral deep vein thrombosis. *Journal of Vascular Surgery*, 55:768–773–426, 2012.

[51] M. Csorgo and J. Gastwirth R. Zitikis. Asymptotic confidence bands for the Lorenz and Bonferroni curves based on the empirical Lorenz curve. *Journal of Statistical Planning and Inference*, 74(1):65–91, 1998.

[52] H. A. David. On the application to statistics of an elementary theorem in probability. *Biometrika*, 43:85–91, 1956.

[53] H. A. David. *Order Statistics 2nd edition*. Wiley, New York, 1981.

[54] H. A. David. On recurrence relation for order statistics. *Statistics and Probability Letters*, 23, 1996.

[55] H. A. David and D. J. Newell. The identification of annual peak periods for a disease. *Biometrics*, 21:645–650, 1965.

[56] D. A. Dawson and D. Sankoff. An inequality for probabilities. *Proc. Am. Math. Soc.*, 33:504–507, 1967.

[57] A. Dmitrienko, A. Tamhane, and F. Bretz. *Multiple Testing Problems in Pharmaceutical Statistics*. Chapman and Hall/CRC, 2009.

[58] K. Dohmen. *Improved Bonferroni inequalities via abstract tubes*. Springer-Verlag, Berlin, 2003.

[59] E. J. Dudewicz and S. U. Ahmed. New exact and asymptotically optimal solution to the Behrens-Fisher Problem. *J. Math. Management Sci*, 18:359–426, 1998.

[60] E. J. Dudewicz, Y. Ma, S. E. Mai, and H. Su. Exact solutions to the Behrens-Fisher problem: Asymptotically optimal and finite sample efficient choice among. *Journal of Statistical Planning and Inference*, 137:1584–1605, 2007.

[61] O. J. Dunn. Estimation of the means of dependent random variables. *Annals of Mathematics and Statistics*, 29:1095–1111, 1958.

[62] O. J. Dunn. Confidence intervals for the means of dependent, normally distributed intervals. *Journal of American Statistics Association*, 54:613–621, 1959.

[63] O. J. Dunn. Multiple comparisons among means. *Journal of American Statistics Association*, 56:52–64, 1961.

[64] C. Dunnett. A multiple comparison procedure for comparing several treatments with a control. *Journal of American Statistical Association*, 50:1096–1121, 1955.

[65] M. L. Eaton. A review of selected topics in multivariate probability inequalities. *Annals of Statistics*, 10(1):11–43, 1982.

[66] C. H. Edwards. *Advanced Calculus of Several Variables*. Academic Press, New York, 1973.

[67] P. Erdos. Some remarks about additive and multiplicative fractions. *Bulletin*

of American Mathematics Society, 52:527–537, 1946.

[68] P. Erdos and M. Kac. The Gaussian law of errors in the theory of additive number theoretic functions. *American Journal of Mathematics*, 62:738–742, 1940.

[69] P. Erdos, J. Neveu, and A. Rényi. An elementary inequality between the probabilities of events. *Mathematica Scandinavica*, 13:99–104, 1963.

[70] P. Erdos and A. Rényi. On the evolution of random graphs. *Magyar Tud. Akad. Math. Kut. Int. Kozl.*, 5:17–61, 1960.

[71] R. W. Falk. Hommel's Bonferroni-type inequality for unequally spaced levels. *Biometrika*, 76(1):189–191, 1989.

[72] W. Feller. *An Introduction to Probability Theory and Its Applications. vol. 1, 2nd edition*. Wiley, New York, 1957.

[73] M. Fréchet. *Les probabilités associées l'a un systéme d' événéments compatibles et dépendants. Ie Partie*. Hermann, Paris, 1940.

[74] M. Fréchet. *Les probabilités associées l'a un systéme d' événéments compatibles et dépendants. IIe Partie*. Hermann, Paris, 1943.

[75] J. Galambos. On the sieve methods in probability theory. *Studia Sci. Math. Hungar*, 1:39–50, 1966.

[76] J. Galambos. Quadratic inequalities among probabilities. *Ann. Univ. Sci. Budapest, Sectio Math.*, 12:11–16, 1969.

[77] J. Galambos. On the distribution of the maximum of random variables. *Annals of Mathematics and Statistics*, 43:516–521, 1972.

[78] J. Galambos. A limit theorem with applications in order statistics. *Journal of Applied Probabilities*, 11:219–222, 1974.

[79] J. Galambos. Methods for proving Bonferroni type inequalities. *Journal of London Mathematical Society*, 2:561–564, 1975a.

[80] J. Galambos. Order statistics of samples from multivariate distributions. *Journal of American Statistics Association*, 70:674–680, 1975b.

[81] J. Galambos. Bonferroni inequalities. *Annals of Statistics*, 5(4):577–581, 1977.

[82] J. Galambos. *The asymptotic Theory of Extreme Order Statistics. Wiley Series in Probability and Mathematical Statistics*. John Wiley & Sons, New York-Chichester-Brisbane, 1978.

[83] J. Galambos. The role of exchangability in the theory of order statistics. *Exchangability in probability theory*, pages 75–86, 1982.

[84] J. Galambos. Order Statistics. *Handbook of Statistics*, pages 359–382, 1984.

[85] J. Galambos. A new bound on multivariate extreme value distributions. *Ann. Univ. Sci. Budapest, Sectio Math.*, 27:37–40, 1985.

[86] J. Galambos. *The asymptotic theory of extreme order statistics. 2nd ed.* Krieger, Malabar, FL, 1987.

[87] J. Galambos. Variant of the graph dependent model in extreme value theory. *Communications in Statistics, Theory and Methods*, 17:2211–2221, 1988.

[88] J. Galambos. Bonferroni-type inequalities in statistics: a survey. *Journal of Applied Statistics and Science*, 1:195–209, 1994.

[89] J. Galambos. *Advanced Probability Theory. 2nd ed.* Dekker, New York, 1995.

[90] J. Galambos. Univariate extreme value theory and applications. Order statistics: theory and methods. *Handbook of Statistics*, 16:315–333, 1998.

[91] J. Galambos and M. Lee. Extensions of some univariate Bonferroni-type inequalities to multivariate setting. Probability theory and applications. *Kluwer Acad. Publ., Dordrecht*, 80:143–154, 1992.

[92] J. Galambos and M. Lee. Further studies of bivariate Bonferroni-type inequalities. Studies in applied probability. *Journal of Applied Probability*, 31A:63–69, 1994.

[93] J. Galambos and R. Mucci. Inequalities for linear combinations of binomial moments. *Publ. Math. Debrecen*, 27:263–269, 1980.

[94] J. Galambos and A. Rényi. On quadratic inequalities in the theory of probability. *Studia Scient. Math. Hungar.*, 3:351–358, 1968.

[95] J. Galambos and I. Simonelli. *Bonferroni-type inequalities with applications. Probability and its Applications.* Springer-Verlag, New York, 1996.

[96] J. Galambos and I. Simonelli. Characterizations of probability distributions by properties of products of random variables. *Journal of Applied Statistical Science*, 13:1–10, 2004.

[97] J. Galambos and Y. Xu. A new method for generating Bonferroni-type inequalities by iteration. *Math. Proc. Camb. Phil. Soc*, 107:601–607, 1990a.

[98] J. Galambos and Y. Xu. Regular varying expected residual life and domains of attraction of extreme value distributions. *Ann. Univ. Sci. Budapest, Sectio Math.*, 33:105–108, 1990b.

[99] J. Galambos and Y. Xu. Some optimal Bonferroni-type bounds. *Proceedings of the American Mathematical Society*, 117:523–528, 1993.

[100] J. Galambos and Y. Xu. Bivariate extension of the method of polynomials for Bonferroni-type inequalities. *Journal of Multivariate Analysis*, 52(1):131–139, 1995.

[101] J. Galambos and Y. Xu. Two sets of multivariate Bonferroni-type inequalities. *Statistical theory and applications, Springer, New York*, pages 29–36, 1996.

[102] J. Glaz. A comparison of improved Bonferroni-type and product-type inequalities in presence of dependence. *Topics in statistical dependence*, 16:223–235, 1990.

[103] J. Glaz, M. Guerriero, and R. Sen. Approximations for a three dimensional scan statistic. *Methodol. Comput. Appl. Probab.*, 12:731–747, 2010.

[104] J. Glaz and B. Johnson. Probability inequalities for multivariate distributions with dependence structures. *Journal of American Statistical Association*,

79:436–440, 1984.

[105] J. Glaz and V. Pozdnyakov. A repeated significance test for distributions with heavy tails. *Sequential Analysis*, 24:77–98, 2005.

[106] J. Glaz, V. Pozdnyakov, and S. Wallenstein. *Scan Statistics. Methods and applications*. Birkhauser Boston Inc, Boston MA, 2009.

[107] J. Glaz and N. Ravishanker. Simultaneous prediction intervals for multiple forecasts based on Bonferroni and product-type inequalities. *Statistics and Probability Letters*, 12:57–63, 1991.

[108] J. Glaz and Z. Zhang. maximum scan score-type statistics. *Statistics and Probability Letters*, 76:1316–1322, 2006.

[109] M. Guerriero, V. Pozdnyakov, J. Glaz, and P. Willett. A repeated significance test with applications to sequential detection in sensor networks. *IEEE Trans. Signal Process*, 58:3426–3435, 2010.

[110] E. J. Gumbel. *Statistics of extremes*. Columbia University Press, New York, 1958.

[111] W. Guo. A note on adaptive Bonferroni and Holm procedures under dependence. *Biometrika*, 96(4):1012–1018, 1995.

[112] A. Gupta and J. T. Chen. Goodness of fit test for the skew-normal distribution. *Communication in Statistics*, 30(4):907–930, 2001.

[113] A. Gupta and J. T. Chen. On the sample characterization criterion for normal distributions. *Journal of Statistical Computation and Simulation*, 73:155–163, 2003.

[114] A. Gupta and J. T. Chen. A class of multivariate skew-normal models. *The Annals of the Institute of Statistical Mathematics*, 56(2):305–315, 2004.

[115] S. Das Gupta. Probability inequalities and errors in classification. *Ann. Statist.*, 2:751–762, 1974.

[116] G. Hadley. *Linear Programming*. Addison-Wesley Publishing Co., Inc., Reading, Mass.-London, 1962.

[117] T. Hailperin. Best possible inequalities for the probability of a logical function of events. *American Mathematical Monthly*, 72:343–359, 1965.

[118] Y. Hochberg. A sharper Bonferroni procedure for multiple tests of significance. *Biometrika*, 75(4):800–802, 1988.

[119] Y. Hochberg and A. C. Tamhane. *Multiple comparison procedures*. Wiley, New York, 1987.

[120] B. Holland and M. Copenhaver. An improved sequentially rejective Bonferroni test procedure. *Biometrika*, 43(2):417–423, 1987.

[121] S. Holm. A simple sequentially rejective multiple test procedure. *Scand. J. Statist.*, 6(2):65–70, 1979.

[122] G. Hommel. A comparison of two modified Bonferroni procedures. *Biometrika*, 76(3):624–625, 1989.

[123] G. Hommel and G. Bernhard. Bonferroni procedures for logically related hypotheses. Multiple comparisons (Tel Aviv, 1996). *Journal of Statistical Planning and Inference*, 82(1-2):119–128, 1996.

[124] F. M. Hoppe. Iterating Bonferroni bounds. *Statistics and Probability Letters*, 3(3):121–125, 1985.

[125] F. M. Hoppe. *Multiple Comparisons, Selection Procedures and Applications in Biometry*. Dekker, New York, 1993.

[126] F. M. Hoppe. Improving probability bounds by optimization over subsets. *Discrete Math.*, 306(5):526–530, 2006.

[127] F. M. Hoppe. The effect of redundancy on probability bounds. *Discrete Math.*, 309(1):123–127, 2009.

[128] F. M. Hoppe and M. Nediak. Fréchet optimal bounds on the probability of a union with supplementary information. *Statistics and Probability Letters*, 78(3):311–319, 2008.

[129] F. M. Hoppe and E. Seneta. Bonferroni-type inequalities and the methods of indicators and polynomials. *Advances in Applied Probability*, 22(1):341–246, 1990.

[130] F. M. Hoppe and E. Seneta. Gumbel's identity, binomial moments, and Bonferroni sums. *International Statistical Review*, 80:269–292, 2012.

[131] J. C. Hsu. *Multiple Comparisons: Theory and Methods*. London, Chapman and Hall/CRC, 1996.

[132] J. C. Hsu and R. Berger. Stepwise confidence intervals without multiplicity adjustment for dose-response and toxicity studies. *Journal of American Statistical Association*, 94:468 – 482, 1999.

[133] D. Hunter. An upper bound for probability of a union. *Journal of Applied Probability*, 13:597–603, 1976.

[134] S. Iyengar. On a lower bound for the multivariate normal Mill's ratio. *Annals of Probability*, 4:1399–1403, 1986.

[135] S. Iyengar. Evaluation of normal probabilities of symmetric regions. *SIAM Journal on Scientific and Statistical Computing*, 3:418–423, 1988.

[136] D.R. Jensen and M.Q. Jones. Simultaneous confidence intervals for variances. *Journal of American Statistics Association*, 64:324–332, 1969.

[137] K. Jogdeo. Bonferroni-type inequalities and the methods of indicators and polynomials. *Annals of Statistics*, 5(3):495–504, 1977.

[138] B. D. Kaehler and R. A. Maller. A generalized skewness statistic for stationary ergodic martingale differences. *Mathematical Methods of Statistics*, 19:267–282, 2010.

[139] S. R. Kahn, I. Shrier, J. A. Julian, and T. Ducruet et al. Determinants and time course of the postthrombotic syndrome after acute deep venous thrombosis. *Ann Intern Med*, 149:698–707, 2008.

[140] S. Karlin and Y. Rinott. Classes of orderings of measures and related cor-

relation inequalities. I. Multivariate totally positive distributions. *Journal of Multivariate Analysis*, 10:467–498, 1980.

[141] E. G. Kounias. Bounds for the probability of a union of events, with applications. *Annals of Mathematics and Statistics*, 39:2154–2158, 1969.

[142] S. Kounias and J. Marin. Best Linear Bonferroni bounds. *SIAM Journal of Applied Mathematics*, 30:307–323, 1976.

[143] S. Kounias and K. Sotirakoglou. Bonferroni bounds revisited. *Journal of Applied Probability*, 26:231–241, 1989.

[144] S. Ku and E. Seneta. The number of peaks in a stationary sample and orthant probabilities. *Journal of Time Series Analysis*, 15:385– 403, 1994.

[145] S. Ku and E. Seneta. Practical estimation from the sum of AR(1) processes. *Communications in Statistics — Simulation and Computation*, 27:981– 998, 1998.

[146] S. M. Kwerel. Most stringent bounds on aggregates probabilities of partially specified dependent probability systems. *Journal of American Mathematical Association*, 70:472–479, 1975a.

[147] S. M. Kwerel. Bounds on the probability of the union and intersection of m events. *Advanced Applied Probability*, 7:431–438, 1975b.

[148] S. M. Kwerel. Most stringent bounds on the probability of the union and intersection of m events for systems partially specified by $S_1, S_2,..., S_k, 2 \leq k \leq m$. *Journal of Applied Probability*, 12:612–619, 1975c.

[149] M. Y. Lee. Bivariate Bonferroni inequalities. *Aequationes Math.*, 44:220–225, 1992.

[150] M. Y. Lee. Improved bivariate Bonferroni-type inequalities. *Statistics and Probability Letters*, 4:359–364, 1997.

[151] R. Lee and J. Spurrier. Successive comparisons between ordered treatments. *Journal of Statistical Planning and Inference*, 43(3):323–330, 1995.

[152] W. Liu. *Simultaneous Inference in Regression*. Chapman and Hall CRC, Boca Raton, 2011.

[153] E. Margaritescu. A note on Bonferroni's inequalities. *Biometrical Journal*, 28:937–943, 1986.

[154] E. Margaritescu. On some Bonferroni Inequalities. *Stud. cerc. Mat. Tom*, 39:246–251, 1987.

[155] E. Margaritescu. Improved Bonferroni inequalities. *Rev. Roum. Math. Pures Appl.,*, 33:509–515, 1988.

[156] E. Margaritescu. Some best linear Bonferroni inequalities. *Sutd. Cerc. Mat. Tom*, 41:33–39, 1989.

[157] E. Margaritescu. Optimal bounds of degree two for some inequalities of Bonferroni type. *Rev. Roumaine Math. Pures Appl.*, 35:631–638, 1990a.

[158] E. Margaritescu. Optimal properties for some Bonferroni type inequalities.

Rev. Roumaine Math. Pures Appl., 35:541–548, 1990b.

[159] B. H. Margolin and W. Maurer. Tests of the Kolmogorov-Smirnov type for exponential data with unknown scale, and related problems. *Biometrika*, 63:149–160, 1976.

[160] R. M. Meyer. Note on a 'multivariate' form of Bonferroni's inequalities. *Annals of Mathematical Statistics*, 40(2):692–693, 1969.

[161] J. Mi and A. Sampson. A comparison of the Bonferroni and Scheffe bounds. *Journal Statistical Planning and Inference.*, 36(1):101–105, 1993.

[162] T. Mori. Limit laws for maxima and second maxima for strong mixing processes. *Annals of Probability*, 4:122–126, 1976.

[163] T. F. Móri. More on the waiting time till each of some given patterns occurs as a run. *Canadian Journal of Mathematics*, 5:915–932, 1990.

[164] T. F. Móri. On the waiting time till each of some given patterns occurs as a run. *Probability Theory and Related Fields*, 83:313–323, 1991.

[165] T. F. Móri. Bonferroni-type inequalities and deviations of discrete distributions. *Journal of Applied Probability*, 1:115–121, 1996.

[166] T. F. Móri and G. J. Székely. On the Erdos-Rényi generalization of the Borel-Cantelli lemma. *Studia Sci. Math. Hungar*, 18:173–182, 1983.

[167] T. F. Móri and G. J. Székely. A note on the background of several Bonferroni-Galambos-type inequality. *Journal of Applied Probability*, 22:836–843, 1985.

[168] D. Q. Naiman and H. Wynn. Inclusion-exclusion-Bonferroni identities and inequalities for discrete tube-like problems via Euler characteristics. *Annals of Statistics*, 20(1):43–76, 1992.

[169] T. Nakamura and H. Douke. Multivariate multiple comparison procedure by using the improved Bonferroni inequality. *Proceedings of the School of Science of Tokai University*, 38:59–69, 2003.

[170] T. T. Nguyen, J. T. Chen, A. K. Gupta, and K. T. Dinh. A proof of the conjecture on positive skewness of generalized inverse Gaussian distributions. *Biometrika*, 90:245–250, 2003.

[171] I. Olkin and M. Sobel. Integral expressions for tail probabilities of the multinomial and negative multinomial distributions. *Biometrika*, 52:167–179, 1965.

[172] V. Pozdnyakov and J. Glaz. A nonparametric repeated significance test with adaptive target sample size. *Journal of Statistical Planning and Inference*, 137:869–878, 2007.

[173] V. Pozdnyakov, J. Glaz, and M. Kulldorff. A martingale approach to scan statistics. *Annals of the Institute of Statistical Mathematics*, 57:21–37, 2005.

[174] A. Prékopa. Boole-Bonferroni inequalities and linear programming. *Operations Research*, 36:145–162, 1988.

[175] A. Prékopa. The discrete moment problem and linear programming. *Discrete Applications of Mathematics*, 27:235–254, 1990a.

[176] A. Prékopa. Sharp bounds on probabilities using linear programming. *Operations Research*, 38:227–239, 1990b.

[177] E. Recsei and E. Seneta. Bonferroni-type inequalities. *Advanced Applied Probability*, 19:508–511, 1987.

[178] R. D. Reiss and M. Thomas. *Statistical Analysis of Extreme Values with Applications*. Birkhuauser Verlag, Berlin, 1997.

[179] A. Rényi. On mixing sequences of sets. *Acta Math. Acad. Sci. Hungar*, 9:472–479, 1958.

[180] A. Rényi. Remarks on the Poisson process. *Studia Sci. Math. Hungar*, 2:119–123, 1967.

[181] D. Rom. A sequentially rejective test procedure based on a modified Bonferroni inequality. *Biometrika*, 77(3):663–665, 1990.

[182] K. Rosen. *Discrete Mathematics and Its Applications*. McGraw-Hill Publishing Company, 2011.

[183] S. Ross. *Introduction to Probability Models*. Academic Press, Boston, MA, 1993.

[184] E. Samuel-Cahn. Is the Simes improved Bonferroni procedure conservative? *Biometrika*, 83(4):928–933, 1996.

[185] S. K. Sarkar. Some probability inequalities for ordered MTP_2MTP2 random variables: a proof of the Simes conjecture. *Annals of Statistics*, 26(2):494–504, 1998.

[186] S. Sathe, M. Pradhan, and S.P. Shah. Inequalities for the occurrence of at least m out of n events. *Journal of Applied Probability*, 17:1127–1132, 1980.

[187] H. Scheffé. A method for judging all contrasts in the Analysis of Variance. *Biometrika*, 40:87– 104, 1953.

[188] E. Seneta. On a genetic inequality. *Biometrics*, 29:810–814, 1973.

[189] E. Seneta. An inequality from genetics. *Advances in Applied Probability*, 18(3):860–861, 1986.

[190] E. Seneta. Degree, iteration and permutation in improving Bonferroni-type bounds. *Australian Journal of Statistics*, 30A:27–38, 1988.

[191] E. Seneta. On the history of the strong law of large numbers and Boole's inequality. *Historia Mathematica*, 19(1):24–39, 1992.

[192] E. Seneta. Probability inequalities and Dunnett's test. *Multiple Comparisons, Selection, and Applications in Biometry, Edited by F. M. Hoppe, Dekker, New York*, 134:29–35, 1993.

[193] E. Seneta and J. T. Chen. Frechét optimality of upper bivariate Bonferroni-type bounds. *Theory of Probability and Mathematical Statistics*, 52:147–152, 1996.

[194] E. Seneta and J. T. Chen. A sequentially rejective test procedure. *Theory of Stochastic Processes*, 3(19)(3-4):393–402, 1997.

[195] E. Seneta and J. T. Chen. Multivariate Sobel-Uppuluri-Galambos type bounds. *Ukrainian Mathematical Journal*, 52(9):1283–1293, 2000.

[196] E. Seneta and J. T. Chen. On explicit and Fréchet optimal lower bounds. *Journal of Applied Probability*, 39:81–90, 2002.

[197] E. Seneta and J. T. Chen. A generator for explicit univariate lower bounds. *Statistics and Probability Letters.*, 75:256–266, 2005.

[198] E. Seneta and J. T. Chen. Simple stepwise tests of hypotheses and multiple comparisons. *International Statistical Review*, 73(1):21–34, 2005.

[199] E. Seneta and N. C. Weber. Attainable bounds for expectations. *Journal of the Australian Mathematical Society*, 33:411 – 420, 1982.

[200] R. J. Serfling. Probability inequalities for the sum in sampling without replacement. *Annals of Statistics*, 2:39–48, 1974.

[201] R. J. Simes. An improved Bonferroni procedure for multiple tests of significance. *Biometrika.*, 73(3):751–754, 1986.

[202] C. Sison and J. Glaz. Simultaneous confidence intervals and sample size determination for multinomial proportions. *Journal of American Statistics Association*, 90(429):366–369, 1995.

[203] M. Sobel and V.R.R. Uppuluri. On Bonferroni-type inequalities of the same degree for the probability of unions and intersections. *Annals of Mathematical Statistics*, 43:1549–1558, 1972.

[204] W. Stefansky. Rejecting outliers by maximum normed residual. *Annals of Mathematics and Statistics*, 42:35–45, 1971.

[205] W. Stefansky. Rejecting outliers in factorial designs. *Technometrics*, 14:469–479, 1972.

[206] R. J. Tallarida. *Drug Synergism and Dose-effect Data Analysis*. New York, Chapman and Hall/CRC, 2000.

[207] A. Tamhane and C. W. Dunnett. Stepwise multiple test procedures with biometric applications. *Journal of Statistical Planning and Inference*, 82:55 – 68, 1999.

[208] X. Tan and Y. Xu. Some inequalities of Bonferroni-Galambos type. *Statistics and Probability Letters*, 8:17–20, 1989.

[209] J. Tao, N. Shi, J. Guo, and W. Gao. Stepwise procedures for the identification of minimum effective dose with unknown variances. *Statistics & Probability Letters*, 57:121–131, 2002.

[210] J. Tydeman and R. Mitchell. A note on the Kounias and Marin method of best linear Bonferroni bounds. *SIAM Journal of Applied Mathematics*, 39:173–177, 1980.

[211] S. Wallenstein, J. Naus, and J. Glaz. Power of the scan statistic in detecting a changed segment in a Bernoulli sequence. *Biometrika*, 81(3):595–601, 1994.

[212] N. C. Weber. U-statistics. *Internal Encyclopedia of statistical science*, 21:1634–1635, 2011.

[213] H. Wegner. Stirling numbers of the second kind and Bonferroni's inequalities. *Elemente der Mathematik*, 3:124–129, 2005.

[214] P. Westfall, W. Johnson, and J. Utts. A Bayesian perspective on the Bonferroni adjustment. *Biometrika*, 84(2):419–427, 1997.

[215] W. Wong and X. Shen. Probability inequalities for likelihood ratios and convergence rates of sieve MLEs. *Annals of Statistics*, 23(2):330–362, 1995.

[216] K. J. Worsley. On the likelihood ratio test for a shift in location of normal populations. *Journal of American Statistics Association*, 74:365–368, 1979.

[217] K. J. Worsley. An improved bonferroni inequality and applications. *Biometrika*, 69(2):297–302, 1982.

[218] K. J. Worsley. Bonferroni (improved) wins again. *American Statistician*, 3:235, 1985.

[219] Y. Xu. Bonferroni-type inequalities via interpolating polynomials. *Proceedings of the American Mathematical Society*, 107:825–831, 1989.

[220] A. M. Zubkov. Inequalities for the distribution of sums of functions of independent random variables. *Rossiskaya Akademiya Nauk. Matematicheskie Zametki*, 22:745–758, 1977.

Index

Milton Keynes UK
Ingram Content Group UK Ltd.
UKHW040448071024
449327UK00020B/1073